INTERMEDIATE ALGEBRA

INTERMEDIATE ALGEBRA

Fourth Edition

PAUL K. REES

Louisiana State University

FRED W. SPARKS

Texas Tech University

McGraw-Hill Book Company

New York   St. Louis   San Francisco   Düsseldorf   Johannesburg
Kuala Lumpur   London   Mexico   Montreal   New Delhi   Panama
Rio de Janeiro   Singapore   Sydney   Toronto

Library of Congress Catalog Card Number 71-155788

07-051673-1

1 2 3 4 5 6 7 8 9 0 H D M M 7 9 8 7 6 5 4 3 2 1

This book was set in Caledonia by Textbook Services,
Inc., printed on permanent paper by Halliday Li-
thograph Corporation, and bound by The Maple Press
Company. The designer was Merrill Haber; the draw-
ings were done by John Cordes, J. & R. Technical Services, Inc. The editors were Roland S. Woolson, Joan
A. DeMattia, and James W. Bradley. Robert R. Laffler
supervised production.

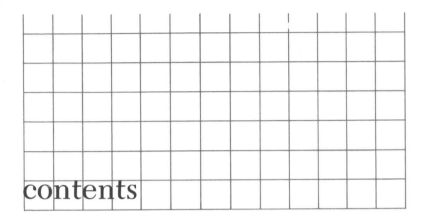

contents

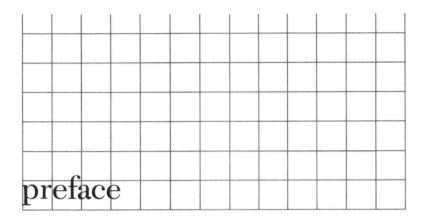

preface

The fourth edition of Intermediate Algebra was prepared for students who have had one year of high school algebra, and it contains the basic material necessary for more advanced courses in mathematics and for the required courses in the sciences and in statistics. The text was planned for a course of the usual three semester hours but it contains sufficient material for longer courses.

The subject matter is essentially the same as that in the former editions. However, since most freshmen today have had some experience with the reformed precollege mathematics, the authors have incorporated in this edition more of the modern spirit, terminology, and methods. Thus several chapters have been completely rewritten, others have been carefully revised, the subject matter has been somewhat reorganized, and several new topics have been added.

A brief but adequate discussion of the theory of sets is presented in the first chapter, and the terminology of sets is used throughout the book. The axioms of a field are introduced as needed and are consistently used to define and develop the properties of the real number system and to prove the theorems necessary for all operations and procedures. For the convenience of both student and teacher, a list of axioms, theorems, and important formulas is printed on the endpapers of the book.

The concepts of an open sentence and the truth set are discussed and used to clarify the meaning of an equation and of an inequality and of the solution set of each. Since the methods for solving equations and inequalities are similar, the discussion of inequalities is integrated with the material on equations. Furthermore, the concept of the union and intersection of sets is used in expressing the solution set of an inequality.

The chapter on functions, graphs, and the inverse of a function was rewritten. A function is defined as a set of ordered pairs, and the relation between a function and its inverse is carefully explained. The graph is used consistently in the discussion of equations and in the method for solving inequalities. Two topics not appearing in previous editions are a graphical treatment of systems of inequalities in two variables and the method for obtaining the maximum and minimum values of a linear binomial in two variables over a convex region of the plane bounded by straight lines.

Determinants of orders 2 and 3 and Cramer's rule are discussed in Chapter 8. The expansion of a determinant of order 3 is defined in terms of the cofactors of the elements of a row or column rather than by the "crisscross" method.

In Chapter 13 the treatment of logarithms is preceded by discussions of approximations, significant figures, the method for rounding off a number that is an approximation, and scientific notation. Scientific notation is used in deriving the rule for obtaining the characteristic of the common logarithm of a number.

The following features of the earlier editions have been retained:

1 The introduction of new concepts is postponed until they are needed. Definitions are stated concisely and are illustrated.
2 Every process is illustrated by carefully explained examples.
3 The authors have attempted to help the student over the difficulty of using equations to solve word problems by providing detailed directions for setting up the equations and by including numerous fully explained illustrative examples. The exercises contain a good variety of problems from geometry, physics, engineering, and so forth.
4 The chapter on simultaneous quadratic equations contains methods for solving most of the problems that the student will encounter in later courses. The method of eliminating a variable by substitution is emphasized.
5 One difficulty that the inexperienced teacher encounters is the matter of judging the length of an assignment. The authors have attempted to help the teacher over this difficulty by organizing the book so that combining the text material between two exercises with every fourth problem from the exercise that follows it constitutes a good assignment.
6 In a course in intermediate algebra, considerable drill work is necessary. For this reason, the authors have given special attention to the preparation of the exercises and believe that the following salient points are worthy of consideration.

a Approximately 2800 carefully graded problems are provided.

b In the selection of problems every effort has been made to avoid triviality and, at the same time, provide problems in which a mass of tedious computation does not obscure the principle involved.

c The problems in each exercise are arranged in groups of four that are on about the same level of difficulty, with the difficulty increasing gradually through the exercise. A good coverage of each exercise may be obtained by assigning every fourth problem. This makes it possible for the book to be used in several sections simultaneously or for several consecutive semesters without repeating assignments.

d The more difficult problems are placed at the end of each exercise to serve as a challenge for the more advanced students.

e Answers are provided for all problems except those whose numbers are divisible by 4.

The authors are indebted to numerous teachers who have used previous editions of this text. Their constructive criticism and suggestions have been very helpful in the preparation of this edition.

Paul K. Rees
Fred W. Sparks

INTERMEDIATE ALGEBRA

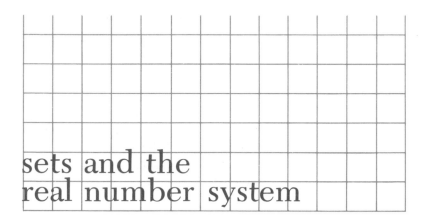

sets and the real number system

If one opens a technical journal or turns to the science section of a popular magazine he will probably find a statement, a formula, or an equation expressed in numerals, letters, and other mathematical symbols. Some knowledge of algebra is necessary for understanding such statements. In this text we present topics of algebra necessary for further progress in mathematics and in fields that use mathematics.

The basis for our discussion consists of (1) a few undefined terms, (2) definitions of other terms and operations, and (3) a set of axioms. The axioms are the rules that we agree to follow. From these we develop the properties of numbers and operations and the rules of procedure.

The real number system is used in the first few chapters of this text, and the concept and terminology of sets is used throughout the book. We discuss these in this chapter.

1.1
SETS

One of the basic and useful concepts of mathematics is denoted by the word *set*. This word is used every day in such phrases as "a set of dishes," "a croquet set," "a set of drawing instruments," and in other expressions referring to a *collection* of objects. We assume that the reader is familiar with the word *collection* and define a set as follows:

Set Defined

■ A *set* is a collection of well-defined objects called *elements*.

By "well defined" we mean that there exists a criterion that enables us to make one of the following decisions about an object, or an element, that we shall designate by a:

1 a belongs to the set.
2 a does not belong to the set.

We now list three sets of well-defined objects. We designate each set by the letter S and state the criterion that defines an element of S.

S is the football squad of Trinity College. The criterion that determines the membership of the squad is the list of names selected by the coach.

S is the herd of sheep in the south pasture of the Bar X ranch. If an animal is a sheep and is in the specified pasture, it is an element of the set. A sheep not in the specified pasture is not an element of the set, and an animal that is not a sheep is not in the set.

S is the set of counting numbers less than 7 that are divisible by 2. The description of the set establishes the criterion. The numbers 2, 4, and 6 are elements of S since each is divisible by 2 and is less than 7. Numbers not divisible by 2 and those equal to, or greater than, 7 are not elements of S.

Designation of Sets

As implied above, we frequently use a capital letter to stand for a set. A set is described in two ways. In one, we list the elements of the set, such as letters, numerals, or the names of objects, and enclose the list in braces { }. In the other, we enclose a descriptive phrase in braces and understand that the elements of the set are those, and only those, that satisfy the description. For example, if W is the set of the names of the days of a week, then we designate W as follows:

$W = \{$Sunday, Monday, Tuesday, Wednesday, Thursday, Friday, Saturday$\}$

or

$W = \{$the names of the days of a week$\}$

or

$W = \{x \mid x$ is the name of a day of a week$\}$

The last line is the method most often used in the descriptive method. The vertical line | is read "such that."

Equality of Sets

■ Two sets are equal if each element of each is an element of the other.

The relation of equality does not require that the elements of the sets be arranged in the same order. For example,

$$\{s, t, a, r\} = \{r, a, t, s\} = \{t, a, r, s\}$$

In the sets $A = \{a, b, c, d, e\}$ and $B = \{a, c, e\}$, each element of B is an element of A. This situation illustrates the following definition:

Subset and Proper Subset

■ If each element of a set B is an element of a set A, then B is a *subset* of A. Furthermore, if each element of B is an element of A but there are elements in A that are not elements of B, then B is a *proper subset* of A.

We use the notation $B \subseteq A$ to indicate that B is a subset of A, and $B \subset A$ to denote that B is a proper subset of A.

EXAMPLE 1
If $A = \{1, 2, 3, 4, 5\}$, $B = \{1, 3, 5\}$, and $C = \{3, 5, 1\}$, then $B \subset A$ and $C \subseteq B$.

EXAMPLE 2
If $T = \{x \mid x$ is a member of a football squad$\}$ and $S = \{x \mid x$ is a member of the squad who plays end$\}$, then $S \subset T$.

If a is an element of the set S, we say that a *belongs to* S and express the statement by the notation $a \in S$. The notation $a \notin S$ means that a does not belong to S.

Now if $B \subset A$ and $A \subset B$, it follows that each element of B belongs to A and each element of A belongs to B. Hence, $A = B$. Therefore the definition of equality of A and B can be stated in this way:

■ If $B \subset A$ and $A \subset B$, then $A = B$.

It may happen that a subset of A is also a subset of B. For example, if $A = \{1, 2, 3, 4, 5, 6\}$, and $B = \{2, 4, 6, 8, 10\}$, then the set $\{2, 4, 6\}$ is a subset of A and of B. This set is called the *intersection of A and B* and illustrates the following definition.

Intersection of Two Sets

■ The *intersection* of the sets A and B is designated by $A \cap B$ and consists of all elements of A that also belong to B. The notation $A \cap B$ is read "A cap B," or the intersection of A and B.

This definition may also be stated as follows:

■ $A \cap B = \{x \mid x \in A \text{ and } x \in B\}$

EXAMPLE 3

If $A = \{a, c, e, g\}$ and $B = \{c, a, f, e\}$, then $A \cap B = \{a, c, e\}$.

EXAMPLE 4

If $A = \{x \mid x \text{ is an alderman of Alton}\}$ and $B = \{x \mid x \text{ is a member of the Alton Kiwanis Club}\}$, then $A \cap B = \{x \mid x \text{ is an alderman and a Kiwanian}\}$.

Empty or Null Set

If in Example 4, no alderman is a Kiwanian, then $A \cap B$ contains no elements and illustrates the following definition:

■ The *empty* or *null* set is designated by $\varnothing$, and is the set that contains no elements.

Other examples of the null set are

1 $\{x \mid x \text{ is a woman who has been president of the United States}\}$
2 $\{x \mid x \text{ is a two-digit number less than } 10\}$
3 $\{x \mid x \text{ is a former governor of California}\} \cap \{x \mid x \text{ is a former governor of Texas}\}$

Disjoint Sets

■ If $S \cap T = \varnothing$, then the sets S and T are *disjoint* sets.

Another concept associated with the theory of sets is the complement of one set with respect to another. As an example, if $A = \{x \mid x \text{ is a student of a given college}\}$ and $B = \{x \mid x \text{ is on the football squad of that college}\}$, then the complement of B with respect to A is $C = \{x \mid x \text{ is a student of the college who is not on the football squad}\}$. This illustrates the following definition:

Complement of a Set

■ The *complement* of the set B with respect to A is designated by $A - B$, and $A - B = \{x \mid x \in A \text{ and } x \notin B\}$.

As a second example, we consider the sets $T = \{x \mid x \text{ is a co-ed in college } C\}$ and $S = \{x \mid x \text{ is a member of the senior class of } C\}$, then $T - S = \{x \mid x \text{ is a co-ed not classified as a senior}\}$. Note that in this case, S is not a subset of T.

The totality of elements that are involved in any specific situation is called the *universal* set, and is designated by the capital letter U. For example, the states in the United States are frequently classified into sets, such as the New England states, the Midwestern states, the Southern states, and in several other ways. Each of these sets is a subset of the universal set, which, in this example, is composed of all the states in the United States.

Venn Diagrams

A method for picturing sets and certain relations between them was devised by the Englishman John Venn, 1834 to 1923, who used a simple plane figure to represent a set. We shall illustrate the method by the use of circles and shall define the universal set U as all points within and on the circumference of a circle C. The various subsets of U will be represented by circles wholly within the circle C. In Fig. 1.1 we show the Venn diagrams for $A \cap B$, $A - B$, and the situation in which $A \cap B = \emptyset$.

If $S = \{1, 2, 3, 4, 5, 6\}$ and $T = \{2, 4, 6, 8, 10\}$, the elements 1, 3, and 5 belong to S but not to T; the elements 8 and 10 belong to T but not to S; and the elements 2, 4, and 6 belong to both S and T. Hence the elements of $V = \{1, 2, 3, 4, 5, 6, 8, 10\}$ are in S or in T or are in both S and T. The set V is called the *union of the sets S and T* and illustrates the following definition:

Union of Two Sets

■ The *union* of the sets S and T is designated by $S \cup T$ and is the set of all elements x such that $x \in S$ or $x \in T$ or $x \in$ both S and T. The notation $S \cup T$ is read "S cup T."

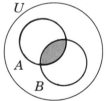

(a) The shaded area is $A \cap B$

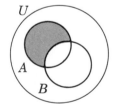

(b) The shaded area is $A - B$

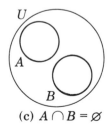

(c) $A \cap B = \emptyset$

FIGURE 1.1

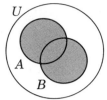

The shaded area is $A \cup B$

FIGURE 1.2

Figure 1.2 shows the Venn diagram for the union of two sets. The following examples are illustrations of the union and the intersection of two sets and of the complement of one set with respect to another.

EXAMPLE 5
If $A = \{m, r, t\}$ and $B = \{r, t, s\}$, then

$A \cup B = \{m, r, t, s\}$

$A \cap B = \{r, t\}$

$A - B = \{m\}$

$B - A = \{s\}$

EXAMPLE 6
If $A = \{1, 3, 5\}$, $B = \{2, 4, 6, 7\}$, and $C = \{3, 4, 5\}$, then

$A \cup B = \{1, 2, 3, 4, 5, 6, 7\}$

$A \cap B = \varnothing$

$A - B = A$

$A \cup C = \{1, 3, 4, 5\}$

$A - C = 1$

EXAMPLE 7 Using a Venn diagram, prove that

$T \cup (S - T) = S \cup T$

SOLUTION We represent S and T by the circles in Fig 1.3. Now $S - T$ is the set of all points in the shaded region bounded by the two circles. Furthermore, all points in the shaded region together with all points in T constitute the set of all points in the region bounded by the two circles. Therefore, $T \cup (S - T) = S \cup T$.

FIGURE 1.3

EXERCISE 1.1

Operations on Sets

1   If $A = \{2, 3, 4, 7, 8\}$ and $B = \{2, 4, 8\}$, find $A \cup B$, $A \cap B$, and $A - B$.

2   If $A = \emptyset$ and $B = \{1, 3, 5, 7, 9\}$, find $A \cup B$, $A \cap B$, and $A - B$.

3   If $A = \{x \mid x$ is a member of the marching band$\}$ and $B = \{x \mid x$ plays a trumpet in the marching band$\}$, find $A \cup B$, $A \cap B$, and $A - B$.

4   If $W = \{a, c, e, g, h\}$ and $P = \{b, c, d, e, f\}$, find $W \cup P$, $W \cap P$, and $W - P$.

5   If $C = \{x \mid x$ is in a French class$\}$ and $D = \{x \mid x$ is in an algebra class$\}$, find $C \cup D$, $C \cap D$, and $C - D$.

6   If $M = \{x \mid x$ has black hair$\}$ and $N = \{x \mid x$ likes strawberry ice cream$\}$, find $M \cup N$, $M \cap N$, and $M - N$.

7   If $A = \{x \mid x$ is a college student who has red hair$\}$ and $B = \{x \mid x$ is a co-ed who has red hair$\}$, find $A \cup B$, $A \cap B$, and $A - B$.

8   If $A = \{x \mid x$ is a senator who weighs 170 lb or more$\}$ and $\{B = x \mid x$ is a senator who weighs less than 170 lb$\}$, find $A \cup B$, $A \cap B$, and $A - B$.

The following sets are used in Probs. 9 to 12:

$A = \{x \mid x$ is a counting number less than, or equal to, 11$\}$
$B = \{x \mid x \in A$ and x is not divisible by 2$\}$
$C = \{x \mid x \in A$ and x is divisible by 3$\}$ $3, 6, 9$
$D = \{x \mid x \in A$ and x is not divisible by 3$\}$
$E = \{3, 5, 7, 9\}$

9   Find $A \cup C \cup D \cup E$, $A \cap B \cap D$, $A \cap D \cap E$.

10   Find $(A - C) \cup (D - E)$, $(A \cap C) - (B \cup D)$.

11   Find $(A \cup B) - (B \cap C)$.

12   Prove that $C \cup (D \cap E) = (C \cup D) \cap (C \cup E)$.

Using a Venn diagram, show that each of the following statements is true.

13   $(S - T) \subset S$, $T \neq \emptyset$

14   $T \cup (T - S) = T$

15   $T \cap (T - S) = T - S$

16   $T \cap (T \cup S) = T$

1.2
CONSTANTS AND VARIABLES

In the remainder of this text we shall use letters and other symbols to stand for numbers. Symbols used in this way are called *variables* and are defined more precisely below.

Variable and Replacement Set

■   A letter or a symbol that stands for a number that is an element of a given set is a *variable*, and the given set is the *replacement set*.

Constant

■   If the replacement set for a given letter or symbol contains only one element, then that letter or symbol is a *constant*.

For example, the Greek letter π stands for the ratio of the circumference of a circle to the diameter, and is approximately equal to 3.1416. Hence, π is a constant since there is only one number in the replacement set. Also the symbol for each real number such as 2, -3, $\frac{3}{4}$, or $\sqrt{3}$ is a constant. Furthermore, if v stands for the total number of votes cast in the 1968 United States presidential election, then v is a constant. However, if v stands for the total number of votes cast in a United States presidential election, then v is a variable.

1.3
THE SET OF NATURAL NUMBERS

The numbers used in counting are the natural numbers. The concepts of a natural number and the process of counting are based on the notion of one-to-one correspondence between the elements of two sets. We illustrate this notion by considering the sets $A = \{a, b, c, d, e\}$ and $B = \{$Tom, Dick, Harry, Joe, and Jim$\}$. These sets are not equal since the elements of A are letters, and the elements of B are names. We can, however, match each element of either set with one, and only one, element of the other. This matching can be set up in many ways.

One-to-One Correspondence

One method is to match the elements in the order from left to right in which they appear. This method is diagrammed as follows:

$$
\begin{array}{cccccc}
A\colon & a & b & c & d & e \\
 & | & | & | & | & | \\
B\colon & \text{Tom} & \text{Dick} & \text{Harry} & \text{Joe} & \text{Jim}
\end{array}
$$

We therefore say that these two sets have the same number of elements. The Hindu-Arabic symbol for this number is 5, and the English name for it is *five*.

We also say that the sets A and B are equivalent, and they illustrate the following definition:

■ Two sets are *equivalent* if, and only if, there exists a one-to-one correspondence between the elements of the sets.

We next use the notion of equivalence of sets to illustrate the meaning of a natural number. We assume that the Hindu-Arabic number symbols and their order of succession are known, and we shall use them as elements of sets.

Now we say that every set that is equivalent to $\{1\}$ contains *one* element and every set that is equivalent to $\{1, 2\}$ contains *two* elements. In general, if n stands for a natural number expressed in Hindu-Arabic notation, then every set that is equivalent to $\{1, 2, 3, \ldots, n\}$ contains n elements. (The dots indicate that the sequence of number symbols is continued from 3 to n.) Consequently,

■ The *natural number* n is defined as the number associated with any set that is equivalent to $\{1, 2, 3, \ldots, n\}$.

1.4
GRAPHICAL INTERPRETATION OF THE REAL
NUMBER SYSTEM

A graphical interpretation of the real number system assists in understanding the nature of numbers, the meaning of a signed number, and the meaning of the relations "greater than" and "less than" between two numbers. To obtain this interpretation we start with the set of natural numbers and associate each number with one, and only one, point on a line. As we proceed, we define the other subsets of the set of real numbers and interpret them graphically.

We start with a straight line L of unlimited length, a portion of which is shown in Fig. 1.4, and a unit length u. We assume that the unit u can be laid off on L any desired number of times and that a straight line segment can be separated into any desired number of equal parts.

We choose the reference point P on L, and then, starting at P, we lay off successive intervals of length u to the right of P. Next we assign 1 to the right end of the first interval to the right of P, 2 to the right end of the second interval, and so on for 3, 4, 5, $\ldots$, where dots mean that the sequence of

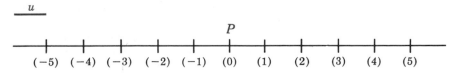

FIGURE 1.4

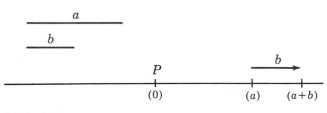

FIGURE 1.5

numbers can be counted indefinitely. In this way, we associate a point on the line L with each of the natural numbers. We indicate the point that is associated with one of these numbers by writing the numeral enclosed in parentheses. Thus, by the notation (5) we mean the point on L that is associated with 5.

Binary Operation

Now we discuss a graphical interpretation of the binary operation of addition in the set of natural numbers.

A *binary operation* in a set S is a rule that assigns to every pair of elements of S, taken in a prescribed order, another element of S.

For example, the operation of addition assigns the number 5 to the sum of 2 and 3 since $2 + 3 = 5$.

The symbol $+$ indicates addition. Thus, $a + b$ means that b is added to a. Furthermore, $a + b$ is called the *sum* of a and b.

We obtain the point $(2 + 3)$ associated with $2 + 3$ by starting at (2) and moving over three intervals to the right of (2), and thus arriving at the point (5). In general, the point $(a + b)$ is the point on L that is b units to the right of (a), as indicated in Fig. 1.5.

Since the point $(a + b)$ is obtained by starting at (a) and counting b units to the right, then $a + b$ is obtained by counting and is therefore a natural number. Furthermore, since there is no end to the natural numbers, the number $a + b$ exists regardless of the magnitude of b. Hence, the set of natural numbers is said to be *closed* under the operation of addition.

Zero

We now define the number zero and interpret it graphically.

■ The number *zero*, designated by 0, is the number such that $0 + a = a$.

Obviously, if we start at the point P on L and move a units to the right, we arrive at (a). Hence it is logical to assign the number 0 to P.

We use the symbol $>$ to represent "greater than" and say that $a > b$ if (a) is to the right of (b) on L. Similarly, $c < d$ means that c is less than d and that (c) is to the left of (d).

If $a > b$, we interpret $a + (-b)$ to mean that we start at (a) and move b units to the left of (a) and thus arrive at $(a + (-b))$, as indicated in Fig. 1.6. Consequently $5 + (-3)$ is associated with the point (2), and $9 + (-5)$ with the point (4).

10

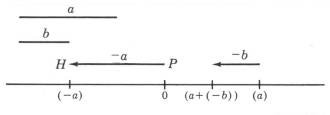

FIGURE 1.6

Negative of a Number

■ The *negative of a number a* is the number $-a$ such that $a + (-a) = 0$.

Reflection of a Point

We now consider (-8). We get the point $7 + (-8)$ by moving eight intervals to the left of (7), and we thus arrive at a point that is 1 unit to the left of (0). So far, we have associated no numbers with the points on L at the left of (0), and this is our next task. We first obtain the point H in Fig. 1.6 that is at a distance of a units to the left of (0). We call this point the *reflection* of the point (a) with respect to (0). We now assign the number $-a$ to the reflection of the point (a). The points (-1), (-2), (-3), and so on, in Fig. 1.4 are the reflections of the points (1), (2), (3), etc., with respect to (0).

Positive and Negative Integers

Hereafter we shall refer to the set of natural numbers as the *set of positive integers*, and the set of negatives of the natural numbers as the *set of negative integers*. Thus in the terminology of sets, {natural numbers} = {positive integers}, and {negatives of the natural numbers} = {negative integers}. We define the set of integers as the set whose elements are the set of positive integers, zero, and the set of negative integers. Hence,

{Integers} = {positive integers} $\cup$ {0} $\cup$ {negative integers}

The symbol $+$ is the positive sign, and the symbol $-$ is the negative sign. The sign that precedes a numeral identifies the sense of the number. For example $+3$ is positive and -3 is negative. Likewise, if a stands for a positive number, $+a$ is positive and $-a$ is negative. If no sign appears before a numeral or letter, it is understood that the sign is $+$. For example, we usually write $+3$ and $+a$ as 3 and a, respectively. If, however, n stands for an integer, we do not know the sense of either n or $-n$.

The graphical interpretation of positive and negative numbers assigns a direction to each of them, and when we place the negative sign $-$ before a numeral, we reverse the direction. Frequently, we encounter numbers such as $-(-3)$. The negative sign before the parentheses means that we are to reverse the direction of the number -3 that is inside the parentheses. Therefore $-(-3) = 3$ and, in general, $-(-a) = a$. The number $-(-a)$ is often written $--a$.

11

Frequently, it is desirable to disregard the directional aspect of a number and consider only the *absolute value of* the number. Interpreted graphically, the absolute value of n is the number that expresses the measure of the length of the line segment of L measured from (0) to (n) if n is positive, and from (n) to (0) if n is negative. More precisely, we define the absolute value of n, designated by $|n|$ as follows:

Absolute Value

■ If n is a positive number or 0, $|n| = n$, and if n is a negative number, $|n| = -n$.

For example, $|5| = 5$, $|-5| = -1--5 = 5$, and $|0| = 0$.

As stated previously, if $a > b$, then the point (a) lies to the right of (b) on L; and if $c < d$, then (c) is at the left of (d). The notation $c < b < d$ means that b is between c and d. If n is a positive number, the point (n) is at the right of (0). Hence, $n > 0$. Similarly $-n < 0$. Examples of this notation are

$$10 > 7, 3 < 5, -10 < -3, -1 > -9, 3 < 5 < 6, \text{ and } -6 < -2 < -1.$$

Multiplication

The notation 3×5 stands for "three times 5," and means $5 + 5 + 5 = 15$. In general, if a is a positive integer,

$$a \times b = \underbrace{b + b + b + \cdots + b}_{a \text{ terms}}$$

Product

The binary operation indicated by $a \times b$ is called *multiplication*. The result is called the *product*. Usually $a \times b$ is written as ab.

Frequently, a raised dot or parentheses are used to indicate multiplication. For example, $a \cdot b$ and $a(b)$ mean $a \times b$, and $a(b + c)$ means that the sum of b and c is to be multiplied by a.

Multiple, Factor

If $n = ab$, where a and b are nonzero integers, then n is a *multiple* of a and of b, and a and b are *factors* of n.

We shall discuss multiplication more fully in the next chapter.

Quotient

The quotient of a and b is indicated by $a \div b$, $\dfrac{a}{b}$, or a/b and is defined as follows:

■ If $b \neq 0$, the quotient a/b is the number x such that $bx = a$. The symbol $\neq$ means "is not equal to."

For example, $\frac{12}{3} = 4$ since $3 \times 4 = 12$. The fractions in arithmetic such as $\frac{1}{2}$, $\frac{2}{3}$, and $\frac{10}{7}$ are quotients since each is in the form $\dfrac{a}{b}$.

We have used the number line L, the set of natural numbers, and the operation of addition to define the set of integers. We now use the set of integers and the definition of a quotient to define an additional set of numbers.

Rational Number

■ A *rational number* is a number that can be expressed as the quotient of two integers.

For example, $\frac{3}{5}, \frac{7}{2}, \frac{10}{5}$, and in general, a/b are rational numbers.

Now if a is a multiple of b, then a/b is equal to an integer. For example, $\frac{10}{2} = 5$, and we shall show later that $a/1 = a$. Hence, any integer can be expressed as the quotient of two other integers. Therefore, the set of integers is a subset of the set of rational numbers, or in the terminology of sets, {integers} ⊂ {rational numbers}.

To illustrate the method for associating a point on the line L with a rational number, we shall find the point $(\frac{3}{4})$. For this purpose we use Fig. 1.7 which shows the portion of L from (-2) to (2) with the scale enlarged so that the details can be seen more clearly. First, we subdivide the interval from (0) to (1) into four equal parts; then the point at the right end of the third subinterval is the point $(\frac{3}{4})$. We call the length of each of these subintervals $\frac{1}{4}$ unit. To get the point associated with $\frac{7}{4}$ we lay off seven intervals of length $\frac{1}{4}$ starting at (0), then the right end of the seventh interval is the point $(\frac{7}{4})$. Similarly, theoretically at least, we can associate any given rational number with a point on L. The points associated with the negative rational numbers are the reflections of the points associated with the positive rational numbers. The points $(-\frac{3}{4})$ and $(-\frac{7}{4})$ are shown in Fig. 1.7.

It is proved in arithmetic that a rational number that is not an integer can be expressed as a terminating or a nonterminating repeating decimal fraction. For example, $\frac{3}{4} = .75$ and $\frac{4}{11} = .363636\ldots$, where the sequence of digits 36 is repeated indefinitely. Conversely, both terminating and nonterminating repeating decimal fractions can be expressed as the quotient of two integers.

Irrational Number

We now consider $\sqrt{18}$, which cannot be expressed as the quotient of two integers.† By means of the square root process, however, we can obtain a decimal approximation to $\sqrt{18}$, although the decimal never terminates and never starts repeating. We therefore say that $\sqrt{18}$ is $4.24264\ldots$, where the decimal is obtained by the square root process and the dots mean that it is continued indefinitely.

†For proof of this statement see P. K. Rees and F. W. Sparks, College Algebra, 5th ed., p. 12, McGraw-Hill Book Company, New York, 1967.

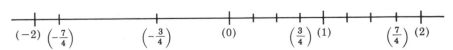

FIGURE 1.7

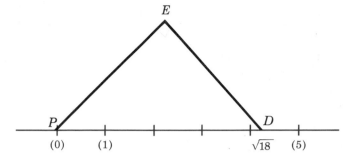

FIGURE 1.8

The number $\sqrt{18}$ is an example of an irrational number. Other examples are $\sqrt{5}$, $\sqrt{3}$, $\sqrt{7}$, and $\pi = 3.1416. \ldots$ Such numbers are defined as follows:

■ An *irrational number* is a number that can be expressed as a nonterminating, nonrepeating decimal.

We now explain a geometrical method for locating the point $(\sqrt{18})$ on L. At the point P, or (0), in Fig. 1.8 we construct the segment PE which is 3 units in length and makes an angle of $45°$ with L. Then at E, we construct a line perpendicular to PE that intersects L at D. By elementary geometry, the segment ED is also 3 units in length. Furthermore, by the Pythagorean theorem,

$$(PD)^2 = (PE)^2 + (ED)^2$$

$$= 9 + 9 = 18$$

Hence,

$$PD = \sqrt{18}$$

Therefore, we associate the point D with the irrational number $\sqrt{18}$.

Real Numbers

We now define the set of real numbers as follows:

■ The set of *real numbers* is the union of the set of rational numbers and the set of irrational numbers.

In set terminology,

■ {Real numbers} = {rational numbers} ∪ {irrational numbers}

We use another geometrical construction to locate the point $(\sqrt{48})$ on L. In Fig. 1.9 we construct the segment PA that is perpendicular to L and is 4 units in length. Then at A, we construct the line PC with the angle PAC equal to $60°$. It is proved in geometry that if one acute angle of a right trian-

14

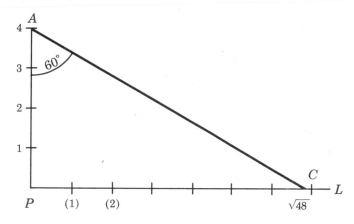

FIGURE 1.9

gle is 60°, then the hypotenuse is twice the side adjacent to 60°. Therefore, $AC = 8$ units. Consequently, by the Pythagorean theorem we have

$$PC = \sqrt{(AC)^2 - (PA)^2} = \sqrt{8^2 - 4^2} = \sqrt{64 - 16} = \sqrt{48}$$

Hence, C is the point ($\sqrt{48}$).

It is proved in advanced mathematics that each point on the number line can be associated with one, and only one, real number, and that each real number can be associated with only one point on the number line.

This completes the definition and graphical interpretation of the set of real numbers. A diagrammatic summary is given in Fig. 1.10.

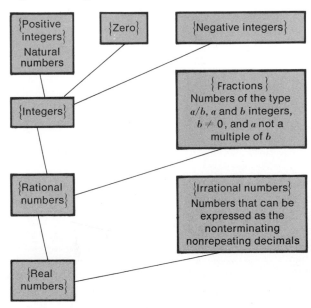

FIGURE 1.10

15

EXERCISE 1.2

What name selected from the diagram in Fig. 1.10 best describes the set of numbers in each of Probs. 1 to 8?

1 $1, 5, 7, 12$ **2** $0\ 3, 8, 10$

3 $-3, -2, 1, 8$ **4** $\frac{1}{2}, \frac{2}{3}, \frac{4}{5}$

5 $-\frac{3}{4}, 1, 0, .333\ldots$ **6** $-\frac{2}{3}, -\frac{6}{7}, -\frac{1}{4}, -\frac{3}{7}$

7 $\sqrt{3}, \sqrt{5}, \sqrt{7}$ **8** $-2, \frac{3}{5}, \sqrt{2}, 2.151515\ldots$

Copy each pair of numbers in Probs. 9 to 12, and place the symbol $>$ or $<$ between the numbers in each pair so that the resulting statement will be true.

9 $25\quad 16,\quad 2\quad 7,\quad 11\quad 6$

10 $-1\quad 4,\quad 5\quad -8,\quad -10\quad -1$

11 $\frac{1}{5}\quad \frac{1}{3},\quad \frac{2}{3}\quad -\frac{5}{3},\quad -\frac{5}{2}\quad -\frac{7}{2}$

12 $-.512\quad -.524,\quad -3\quad \frac{1}{2},\quad .001\quad -.1$

Find the sum of the numbers in each of Probs. 13 to 20. Then draw the number line, and locate the point that is associated with each number.

13 $7 + (-2)$ **14** $(-5) + 8$

15 $(-6) + 1$ **16** $9 + |-1|$

17 $|-4| + |-2|$ **18** $|2 + (-3)|$

19 $|-4 - 3|$ **20** $|6| + |-2|$

Using constructions similar to those in Figs. 1.8 and 1.9, locate a point on the number line that is associated with the number in each of Probs. 21 to 28.

21 $\sqrt{8}$ **22** $\sqrt{50}$ **23** $\sqrt{32}$ **24** $\sqrt{98}$

25 $\sqrt{12}$ **26** $\sqrt{75}$ **27** $\sqrt{27}$ **28** $\sqrt{108}$

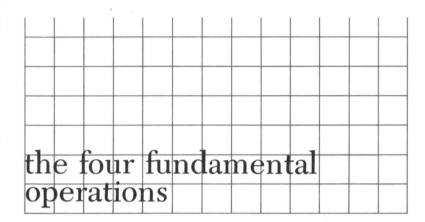

the four fundamental operations

The four fundamental operations of algebra are addition, subtraction, multiplication, and division. We discussed a graphical interpretation of these operations in Chap. 1. In this chapter, we shall discuss them more fully and shall state the axioms and prove the theorems on which the procedures are based.

2.1
DEFINITIONS

In Chap. 1 we stated that the sum of a and b is written $a + b$, the product as ab, and the quotient as $\dfrac{a}{b}$ or a/b. In the product ab, the numbers a and b are called *factors*.

Difference of a and b

In the discussion of addition, we shall have occasion to refer to the difference of two numbers. By the *difference of a and b* we mean the number x such that $b + x = a$, and we shall show later that this difference is equal to $a - b$.

17

The nth Power

The product of two or more equal numbers is a power of that number. We illustrate the notation for such products as follows:

$a \cdot a = a^2$ and is read "*a* square"

$a \cdot a \cdot a = a^3$ and is read "*a* cube"

In general,

$a \cdot a \cdot a \cdots$ to *n* factors $= a^n$ and is read "*a* to the *n*th" or the "*n*th power of *a*"

We summarize the foregoing statements as follows:

■ The number a^n is called the *nth power of a*. The number *n* is an *exponent* and the number *a* is the *base*.

If two or more numbers are combined by one or more of the fundamental operations, the result is called an *expression*.

If more than one operation appears in an expression, then parentheses ( ), brackets [ ], or braces { } are frequently used to indicate the order in which the operations are to be performed. For example, $2 + (3 + 6) = 2 + 9 = 11$ and $4 + 2(8 - 7) = 4 + 2(1) = 6$.

EXAMPLE 1   Examples of expressions are

$$2c + 3d, \qquad (2x + y)(4x - 3y), \qquad \text{and} \qquad \frac{4a - 2b}{3a + 7b}$$

An expression that does not involve addition or subtraction is a *monomial*.

EXAMPLE 2   Examples of monimials are

$$a, \qquad 2ab, \qquad \text{and} \qquad \frac{3x^2 y}{2ab}$$

The sum or difference of two or more monomials is a *multinomial*.

Each monomial in a multinomial together with the sign that precedes it is a *term* of the multinomial.

A multinomial consisting of exactly two terms is a *binomial*. A multinomial containing exactly three terms is a *trinomial*. If each term of a multinomial is a positive integral power of a number symbol or is the product of integral powers of two or more number symbols, the multinomial is called a *polynomial*.

EXAMPLE 3   Examples of polynomials are

$$2a^4 + a^3 - 4a^2 + 5a + 3 \qquad \text{and} \qquad 2x^4 y + x^3 y - 4x^2 y^2 + 3xy^2 - 5y^4$$

If a monomial is expressed as the product of two or more symbols, each of the symbols is called the *coefficient* of the product of the others.

EXAMPLE 4

In $3ab$, 3 is the coefficient of ab, a is the coefficient of $3b$, and b is the coefficient of $3a$. In $3ab$, 3 is the numerical coefficient. Usually when we refer to the coefficient in a monomial, we mean the *numerical coefficient*.

Two monomials or two terms are *similar* if they differ only in their numerical coefficients.

EXAMPLE 5

The monomials $3a^2b$ and $-2a^2b$ are similar, and the terms are similar in

$$4\left(\frac{3a}{5b}\right) + 2\left(\frac{3a}{5b}\right)$$

2.2
RELATION OF EQUALITY AND ORDER RELATIONS

Probably the relation most frequently used in mathematics is the relation of equality denoted by the symbol $=$. We shall attempt no definition of this relation but shall state several axioms that define its essential properties.

Reflexive Axiom

■ $a = a$ (2.1)

Symmetric Axiom

■ If $a = b$, then $b = a$ (2.2)

Transitivity Axiom

■ If $a = b$ and $b = c$, then $a = c$ (2.3)

Additivity Axiom

■ If $a = b$, then $a + c = b + c$ (2.4)

Multiplicativity Axiom

■ If $a = b$, then $ac = bc$ (2.5)

Replacement Axiom

■ If $a = b$, then a can be replaced by b in any statement involving mathematical expressions without affecting the truth or falsity of the statement (2.6)

EXAMPLE 1

If $a = b$, $b = c$, and $c = e$, prove that $a = e$.

PROOF   Since $a = b$ and $b = c$, then $a = c$ by the transitivity axiom (2.3); then since $c = e$, it follows by (2.3) that $a = e$.

EXAMPLE 2

If $a = b$, and $c = d$, prove that $a + c = b + d$.

PROOF

$$a = b \qquad \text{given}$$

$$a + c = b + c \qquad \text{by additivity axiom (2.4)}$$

$$= b + d \qquad \text{by (2.6), since } c = d$$

The order relations are indicated by $>$ for "greater than" and $<$ for "less than." They relate the relative magnitudes of the numbers. We shall, however, assume nothing about them that is not stated in the axioms that follow.

■ $b < a$ if and only if $a > b$ $\hspace{6cm}$ (2.7)

Trichotomy Axiom

■   Only one of the statements $a = b$, $a > b$, $a < b$ is true $\hspace{2.5cm}$ (2.8)

Transitivity Axiom

■ $a > b$ and $b > c$, then $a > c$ $\hspace{5cm}$ (2.9)

Additivity Axiom

■   If $a > b$, then $a + c > b + c$ $\hspace{5cm}$ (2.10)

We illustrate the use of these axioms in proving the following theorem:

■   If $a < b$, then $a + c < b + c$ $\hspace{5cm}$ (2.11)

PROOF   We are given that $a < b$. Hence, by (2.7), $b > a$, and it follows from the additivity axiom (2.10) that $b + c > a + c$. Consequently, by using (2.7) again, we have $a + c < b + c$.

EXAMPLE 3

If $a = b$, and $c > d$, prove that $c + a > d + b$.

PROOF

$$c > d \qquad \text{given}$$

$$c + a > d + a \qquad \text{by additivity axiom (2.10)}$$

$$c + a > d + b \qquad \text{by (2.6), since } a = b$$

2.3
ADDITION

In this section we shall present the axioms of addition and in later sections explain the methods for combining similar monomials and for obtaining the sum of two or more polynomials.

We first deal with the property of *closure*. If the result obtained by performing an operation on any two numbers in a set B is a number in B, then the set B is said to be closed under the operation. In Chap. 1 we stated that the addition of two integers depends basically on counting, and since there is no end to the counting numbers, the set of integers is closed under addition. We assume that this property holds in the set of real numbers and state the assumption in the following axiom:

Closure Axiom

■ If a and b are numbers in the set of real numbers, there exists a unique real number c such that $a + b = c$ (2.12)

We find the sum $2 + 3 + 5$ by combining two of the numbers and adding this sum to the third, and it is readily verified that the way in which the first two are chosen and the order in which they are combined is immaterial. For example,

$$2 + 3 + 5 = (2 + 3) + 5 = (2 + 5) + 3 = (3 + 5) + 2 = 10$$

We assume that this property is true for all real numbers and state the assumption in the following two axioms.

The Commutative Axiom

■ $a + b = b + a$ (2.13)

The Associative Axiom

■ $(a + b) + c = a + (b + c)$ (2.14)

An application of these axioms is illustrated in Example 1.

EXAMPLE 1

$$
\begin{aligned}
(a + b) + c &= c + (a + b) && \text{by commutative axiom} \\
&= (c + a) + b && \text{by associative axiom} \\
&= b + (c + a) && \text{by commutative axiom} \\
&= (b + c) + a && \text{by associative axiom}
\end{aligned}
$$

By continuing the argument in Example 1, we can combine any two of the numbers in $a + b + c$ and then add the third, and the result will be equal to $(a + b) + c$. Therefore, $a + b + c$ is a unique number.

The following axioms enable us to combine two or more similar monomials into a single monomial.

The Right-hand Distributive Axiom

◼ $(a + b)c = ac + bc$ (2.15)

The Left-hand Distributive Axiom

◼ $c(a + b) = ca + cb$ (2.16)

The distributive axioms can be extended to cover situations in which there are more than two terms in the parentheses. For example,

$$(a + b + d)c = [(a + b) + d]c$$
$$= (a + b)c + dc \qquad \text{by right-hand distributive axiom}$$
$$= ac + bc + dc \qquad \text{by right-hand distributive axiom}$$

EXAMPLE 2
Combine terms in $4ab + 6ab - 3ab$ into a single term.

SOLUTION

$$4ab + 6ab - 3ab = (4 + 6 - 3)ab \quad \text{by right-hand distributive axiom}$$
$$= 7ab \qquad \text{since } 4 + 6 - 3 = 7$$

Zero in Addition

In Sec. 1.4 we defined 0 as the number such that $0 + a = a$, and the negative of a as the number $-a$ such that $a + (-a) = 0$. We now interpret $a + (-a)$ as $a - a$; then by the commutative axiom we have

◼ $0 + a = a + 0 = a$ (2.17)

Sum of a and −a

◼ $a + (-a) = a - a = -a + a = 0$ (2.18)

The Number 1 in Multiplication

We now define the number 1 as the number such that

◼ $a \cdot 1 = 1 \cdot a = a$ (2.19)

Since $a + 0 = a$ and $a \cdot 1 = a$, 0 and 1 are called the *identity elements* in addition and multiplication, respectively.

We are now in a position to prove the following very useful theorem.

■   If $a + b = a + c$, then $b = c$ $\qquad$ (2.20)

PROOF

$\quad a + b = a + c \qquad$ given

$a + b - a = a + c - a \qquad$ by additivity axiom **(2.4)**

$a - a + b = a - a + c \qquad$ by commutative axiom

$\quad\quad 0 + b = 0 + c \qquad$ since $a - a = b - b = 0$

$\quad\quad\quad b = c \qquad$ by **(2.17)**

The next corollary follows at once from (2.20).

Corollary of the Cancellation Theorem

■   If $a + b = d$ and $a + c = d$, then $b = c$ $\qquad$ (2.21)

We now use (2.21) to prove that

■ $--a = a$ $\qquad$ (2.22)

and

■ $-(a + b) = -a - b$ $\qquad$ (2.23)

PROOF OF (2.22)   Since $--a = -(-a)$, then $--a$ is the negative of $-a$. Therefore,

$-a + (--a) = 0 \qquad$ by **(2.18)**

Also,

$-a + a = 0 \qquad$ by **(2.18)**

Therefore,

$--a = a \qquad$ by **(2.21)**

PROOF OF (2.23)   The term $-(a + b)$ is the negative of $a + b$. Hence,

$a + b + [-(a + b)] = 0$

Furthermore,

$a + b + (-a - b) = (a - a) + (b - b) \qquad$ by commutative and associative axioms

$\quad\quad\quad\quad = 0 \qquad$ since $a - a = b - b = 0$

Therefore, we have

$$-(a + b) = -a - b \qquad \text{by (2.21)}$$

By a similar argument it can be proved that Theorem (2.23) is true when there are more than two terms in the parentheses. This theorem is useful for removing parentheses from an expression.

We shall use two theorems in deriving the *law of signs for addition*. They are stated and proved as follows:

■ $a > b$ if and only if $a - b > 0$ (2.24)

PROOF We first assume that $a > b$ and proceed as follows:

$$a > b \qquad \text{assumed}$$
$$a + (-b) > b + (-b) \qquad \text{by additivity axiom (2.10)}$$
$$a - b > 0 \qquad \text{since } a + (-b) = a - b \text{ and } b + (-b) = 0$$

Next, we assume that $a - b > 0$, and prove that $a > b$.

$$a - b > 0 \qquad \text{assumed}$$
$$a - b + b > 0 + b \qquad \text{by additivity}$$
$$a > b \qquad \text{since } -b + b = 0, a + 0 = a, \text{ and } 0 + b = b$$

By a similar argument we can prove that

■ $a < b$ if, and only if, $a - b < 0$ (2.25)

Our next theorem deals with the role of zero in multiplication, and it is needed in many addition problems.

Zero in Multiplication

■ $0 \cdot a = a \cdot 0 = 0$ (2.26)

PROOF

$$a + 0 \cdot a = (1 + 0)a \qquad \text{by right-hand distributive axiom}$$
$$= a \qquad \text{since } 1 + 0 = 1 \text{ and } 1 \cdot a = a$$

Also,

$$a + 0 = a \qquad \text{by (2.17)}$$

Hence,

$$0 \cdot a = 0 \qquad \text{by (2.21)}$$

By starting with the left-hand distributive axiom and using a similar argument, we can prove that $a \cdot 0 = 0$.

Law of Signs for Addition

We conclude this section by stating and proving the following law of signs for addition:

■ The absolute value of the sum of two positive numbers or of two negative numbers is the sum of their absolute values. The sum is positive or negative according as the two addends are both positive or both negative. The absolute value of the sum of a positive and a negative number is the difference of their absolute values. The sum is positive or negative according as the number with the greater absolute value is positive or negative.

PROOF In the proof, we shall use the two positive numbers a and b. We first consider $a + b$ and proceed as follows:

$a > 0$	since a is positive
$a + b > 0 + b$	by additivity axiom (2.10)
$a + b > 0$	since $0 + b = b$ and $b > 0$
$\lvert a + b \rvert = a + b$	since $a + b$ is positive
$\quad = \lvert a \rvert + \lvert b \rvert$	since $\lvert a \rvert = a$ and $\lvert b \rvert = b$

We use Theorem (2.23) for dealing with $-a - b$, and have

$-a - b = -(a + b)$	by (2.23)
$-a - b < 0$	since $a + b$ is positive, it follows that $-(a + b)$ is negative

Furthermore,

$\lvert -a - b \rvert = -(-a - b)$	since $-a - b$ is negative
$\quad = --a - (-b) = a + b$	by (2.23) and (2.22)
$\quad = \lvert a \rvert + \lvert b \rvert$	since $\lvert a \rvert = a$ and $\lvert b \rvert = b$

This completes the proof of the statements in the first two sentences of the law of signs.

We now consider $a + (-b) = a - b$ with $a > b$. Since

$a > b$	given

then,

$a - b > 0$	by (2.24)

Hence,

$$a + (-b) > 0$$

Furthermore,

$$|a + (-b)| = a + (-b) \qquad \text{since } a + (-b) > 0$$
$$= a - b$$
$$= |a| - |b| \qquad \text{since } a = |a| \text{ and } b = |b|$$

If

$$a < b$$

then,

$$a - b < 0 \qquad \text{by (2.25)}$$

Hence,

$$a + (-b) < 0$$

Also,

$$|a + (-b)| = -[a + (-b)] \qquad \text{since } a + (-b) < 0$$
$$= -a - (-b) \qquad \text{by (2.23)}$$
$$= -a + b \qquad \text{by (2.22)}$$
$$= -|a| + |b|$$
$$= |b| - |a|$$

This completes the proof of the law of signs.

EXAMPLE 3

$$4 + 5 = 9$$
$$-3 - 6 = -(3 + 6) = -9$$
$$7 + (-2) = 7 - 2 = 5$$
$$3 + (-8) = 3 - 8 = -5$$

2.4
ADDITION OF MONOMIALS

In Example 2 of Sec. 2.3, we illustrated the use of the distributive axiom for obtaining the sum of three similar monomials. We further illustrate the method in Example 1 of this section.

EXAMPLE 1

Find the sum of $3a^2b$, $6a^2b$, $-8a^2b$, and $-5a^2b$.

SOLUTION   We express this sum and proceed as follows:

$3a^2b + 6a^2b + (-8a^2b) + (-5a^2b)$

$\quad = [(3+6) + (-8-5)]a^2b \qquad$ by associative and distributive axioms

$\quad = [9 + (-13)]a^2b \qquad$ by law of signs

$\quad = -4a^2b \qquad$ by law of signs

If a polynomial contains two or more sets of similar terms, we use the commutative and distributive axioms to combine the terms in each set, as illustrated in Example 2.

EXAMPLE 2

$7y^2 + 3xy + 4x^2 + 5x^3 - 2xy + 2x^2 - 6x^3 - xy - x^2 - 2y^2$

$\quad = 7y^2 - 2y^2 + 3xy - 2xy - xy + 4x^2 + 2x^2 - x^2 + 5x^3 - 6x^3$

$\qquad\qquad\qquad\qquad\qquad$ by commutative axiom

$\quad = (7-2)y^2 + (3-2-1)xy + (4+2-1)x^2 + (5-6)x^3$

$\qquad\qquad\qquad\qquad\qquad$ by distributive axiom and the fact
$\qquad\qquad\qquad\qquad\qquad$ that $-xy = -1 \cdot xy$ and $-x^2 = -1 \cdot x^2$

$\quad = 5y^2 + 0 \cdot xy + 5x^2 - x^3 \qquad$ by law of signs

$\quad = 5y^2 + 5x^2 - x^3 \qquad$ since $0 \cdot xy = 0$ by (2.26)

2.5
SYMBOLS OF GROUPING

We stated earlier that symbols of grouping, ( ), [ ], { }, are used to make the meaning of certain expressions clear and to indicate the order in which operations are performed.

Frequently, it is desirable to remove the symbols of grouping from an expression, and we use Theorem (2.23), the right-hand distributive axiom (2.15), and the left-hand distributive axiom (2.16) for this purpose. We illustrate the procedure with two examples.

EXAMPLE 1   Remove the parentheses from

$3x - (2x + 3y) + 2(3x - 4y)$

and combine similar terms.

SOLUTION

$-(2x + 3y) = -2x - 3y \qquad$ by (2.23)

and

$$2(3x - 4y) = 6x - 8y \qquad \text{by left-hand distributive axiom}$$

Consequently,

$$3x - (2x + 3y) + 2(3x - 4y) = 3x - 2x - 3y + 6x - 8y$$
$$= 3x - 2x + 6x - 3y - 8y \qquad \text{by commutative axiom}$$
$$= 7x - 11y$$

This example illustrates the fact that if a pair of grouping symbols preceded by a minus sign is removed, the sign of every term enclosed by the symbols must be changed. However, a pair of grouping symbols preceded by a plus sign can be removed without affecting the signs of the enclosed terms.

Frequently, one or more pairs of grouping symbols are enclosed in another pair. In such cases it is advisable to remove the innermost symbols first.

EXAMPLE 2 Remove the symbols of grouping from

$$3x^2 - \{3x^2 - xy - [5(x^2 - xy) - 3(x^2 - y^2)] + 4xy\} - 3y^2$$

SOLUTION We first apply the left-hand distributive axiom to the expression in the brackets and get

$$3x^2 - \{3x^2 - xy - [5x^2 - 5xy - (3x^2 - 3y^2)] + 4xy\} - 3y^2$$

$$= 3x^2 - \{3x^2 - xy - [5x^2 - 5xy - 3x^2 + 3y^2] + 4xy\} - 3y^2 \qquad \text{removing parentheses}$$

$$= 3x^2 - \{3x^2 - xy - [2x^2 - 5xy + 3y^2] + 4xy\} - 3y^2 \qquad \text{combining terms in brackets}$$

$$= 3x^2 - \{3x^2 - xy - 2x^2 + 5xy - 3y^2 + 4xy\} - 3y^2 \qquad \text{removing brackets}$$

$$= 3x^2 - \{x^2 + 8xy - 3y^2\} - 3y^2 \qquad \text{combining terms in braces}$$

$$= 3x^2 - x^2 - 8xy + 3y^2 - 3y^2 \qquad \text{removing braces}$$

$$= 2x^2 - 8xy \qquad \text{combining similar terms}$$

EXERCISE 2.1

Using the axioms and theorems of this chapter, prove that the statements in Probs. 1 to 4 are true.

  1 $a + (b + c) = b + (c + a)$

  2 $a + (b + c) = (c + a) + b$

3 If $a + b = a$, then $b = 0$.

4 If $a + b = 0$, then $b = -a$.

Find the sum of the numbers in each of Probs. 5 to 12.

5	36	**6**	413	**7**	817	**8**	462
	47		214		214		871

9	37	**10**	427	**11**	−48	**12**	−613
	−24		−342		38		213

Perform the operations indicated in Probs. 13 to 20.

13 $12 + 13 + 21 + 35$

14 $24 + 36 + 75 + 87$

15 $17 - 25 + 60 - 40$

16 $-38 + 42 - 62 + 58$

17 $24a - 60 + 32a + 40$

18 $42b + 67 - 52 - 20b$

19 $76 - 29x + 40 - 21x$

20 $37y - 53 + 19 - 70y$

Combine similar terms in Probs. 21 to 40.

21 $4ab - 2ac + 3ab - 2ac$

22 $3x^2 + 4y^2 - 5x^2 - 6y^2$

23 $4x^2y - 2xy^2 + 4xy^2 - 6x^2y$

24 $7xy - 4xz - 8xy - 2xz$

25 $8a - 3c + 4c - 5a - 7c + 6a$

26 $10a^2b + 4ab^2 - 2a^2b - 6ab^2 + 8a^2b - 5ab^2$

27 $10ab - 8bc + 2bc - 7ab + 6bc - 3ab$

28 $15x^2 - 4y^2 - 7x^2 + 9y^2 + 3x^2 - 5y^2$

29 $3a + 3b - 2c + 4b - 2a + 2c$

30 $2a + 4b - 3c - 3a + 5c$

31 $3x - 4y - 2z + 2y - 6x + 4z$

32 $3w - 5x + 7w + 3z + y - 3x - 2y - 4z$

33 $2ab + a^2b + 3ab^2 - 4ab + a^2b + 2a^2b^2$

34 $7x^2y - 4xy^2 + 7xy + 6xy^2 - 9x^2y - 7xy$

35 $5a^2 - 3c^2 + 3b^2 - 3c^2 + 4a^2 + 7b^2 - 6c^2$

36 $3x^2y + 3xy^2 - 2xy + 4xy - 2x^2y + 2xy - 2xy^2 + 4x^2y - xy^2$

37 $10a^2bc - 12ab^2c - 4a^2bc - 16abc^2 - 8ab^2c - 11abc^2$

38 $11x^2yz - 10xy^2z - 7x^2yz - 12xyz^2 - 6xy^2z - 9xyz^2$

39 $3ab + 5a^2b + 3ab^2 - 5a^2b - ab^2 - ab + 2ab - 3ab^2 - 5a^2b$

40 $ac + 2ac - 3bc + 2ab + 5bc - 3bc + 2ac + 6bc - 2ab$

Remove the symbols of grouping, and combine similar terms in Probs. 41 to 52.

41 $(x - y + z) - (x + 2y - z)$

42 $(2a - 4b) - (3a + 2b) + (4a - 5b)$

43 $2(3a^2 - 2b^2) - (4a^2 + 3b^2)$

44 $3(2x - 2y) + (-2x + y)$

45 $2a + [2a - (b + a)] + 2b$

46 $3c - [c - (2d - c) + 4c]$

47 $x - [3y - 4z - (2x - 3y + 4z) + 2x] - (y + z)$

48 $2a - (3b - 4c) - [a + 2b - (c - 2a + b) - c] - (a - b) + c$

49 $3r - \{2s - t + [6r - (4s - t)] - 2r\} + s$

50 $2c - \{3d - [4e - (2d - 4e) + 2c] - 5e\}$

51 $4x - \{3x - 2[3x - 2(3x - 2) + 3x] + 2\}$

52 $3a - 2\{2a - 3[2(2a - 3) - (4a - 3) + a - 2] - 10\} + 5$

2.6
ADDITION OF POLYNOMIALS

The process of adding two or more polynomials makes use of the commutative axiom of addition, (2.13), and the left-hand distributive axiom (2.16). The commutative axiom enables us to rearrange the terms in the sum so that similar terms are together, and the distributive axiom enables us to combine the similar terms. The process is illustrated in Example 1.

EXAMPLE 1
Find the sum of $3x^2 - 2xy + y^2$, $2xy - 3y^2 - 2x^2$, and $4y^2 - 5x^2 + 4xy$.

SOLUTION   We first write the sum of the three binomials and then rearrange and combine terms as follows:

$3x^2 - 2xy + y^2 + 2xy - 3y^2 - 2x^2 + 4y^2 - 5x^2 + 4xy$

$= 3x^2 - 2x^2 - 5x^2 - 2xy + 2xy + 4xy + y^2 - 3y^2 + 4y^2$

$= x^2(3 - 2 - 5) + xy(-2 + 2 + 4) + y^2(1 - 3 + 4)$

$= x^2(-4) + xy(4) + y^2(2)$

$= -4x^2 + 4xy + 2y^2$

   This process justifies the following shorter procedure, which is the one ordinarily used. Rewrite the expressions so that each one after the first is below the preceding, and at the same time rearrange terms so that those containing the same literal numbers and powers thereof form vertical columns. Finally, draw a horizontal line below the last expression; then combine like terms and write the result below the line. When this is done, we have

$$3x^2 - 2xy + y^2$$
$$-2x^2 + 2xy - 3y^2$$
$$-5x^2 + 4xy + 4y^2$$
$$\overline{-4x^2 + 4xy + 2y^2}$$

EXAMPLE 2

Find the sum of $3a^2 - 2a - 2b^2$, $2ab - 3b - 2a^2$, and $3b^2 - 4a + 4a^2 - 2b$.

SOLUTION We proceed in this problem as we did in Example 1. However, since the first expression contains neither a b term nor an ab term and the second contains neither an a term nor a b^2 term, when we write the second below the first, we leave the spaces under $-2a$ and $-2b^2$ blank and write $-3b$ and $2ab$ at the right. Thus, we have

$$3a^2 - 2a - 2b^2$$
$$-2a^2 \qquad\qquad - 3b + 2ab$$
$$\underline{4a^2 - 4a + 3b^2 - 2b}$$
$$5a^2 - 6a + \;\; b^2 - 5b + 2ab$$

2.7
SUBTRACTION

In elementary arithmetic we learned that if we subtract 3 from 5, we obtain 2. This result can be interpreted in two ways. It is the number of elements in the set that remain if a subset of three is removed from a set of five elements. It is also the number that must be added to 3 in order to obtain 5. We use the second interpretation in the definition of algebraic subtraction stated below.

Definition of Subtraction

■ The operation of subtracting the number b from the number a is the process of determining x such that

$$b + x = a$$

The number a is called the *minuend*, b is the *subtrahend*, and x is the *difference*.

Rule for Subtraction

■ In order to obtain the value of x such that $b + x = a$, we add $-b$ to each member of the equation and get

$$b + x + (-b) = a + (-b)$$

or

$$b + (-b) + x = a + (-b) \qquad \text{by (2.13)}$$
$$x = a + (-b) \qquad \text{since } b + (-b) = 0$$

If $a = m + n - p$ and $b = r - s + t$, then $-b = -1(b) = -1(r - s + t) = -r + s - t$. Therefore, $x = m + n - p + (-r + s - t)$. Consequently, we have the following rule for algebraic subtraction:

■ In order to subtract one number (or one polynomial) from another, we change the sign (or the signs) of the subtrahend and then proceed as in addition.

EXAMPLE 1

Subtract $8a^2$ from $6a^2$.

SOLUTION   In accordance with the above rule, the solution is

$$6a^2 + (-8a^2) = 6a^2 - 8a^2 = -2a^2$$

EXAMPLE 2

Subtract $-4xy$ from $-8xy$.

SOLUTION   If we change the sign of the subtrahend and add, we have

$$-8xy + (+4xy) = -8xy + 4xy = -4xy$$

EXAMPLE 3

Subtract $3a^2 - 2a + 4ab + 3b^2$ from $4a^2 - 2ab - b^2 + 2b$.

SOLUTION   We first write the minuend and then place the subtrahend below it so that like terms in the two expressions are together; thus,

$$
\begin{array}{ll}
4a^2 - 2ab - \ \ b^2 + 2b & \text{minuend} \\
3a^2 + 4ab + 3b^2 \qquad - 2a & \text{subtrahend} \\
\hline
& \text{difference}
\end{array}
$$

Mentally, we now change the sign of each term in the subtrahend, add it to the like term in the minuend, and write the result. The completed problem thus appears:

$$
\begin{array}{l}
4a^2 - 2ab - \ \ b^2 + 2b \\
3a^2 + 4ab + 3b^2 \qquad - 2a \\
\hline
a^2 - 6ab - 4b^2 + 2b + 2a
\end{array}
$$

EXERCISE 2.2
Addition and Subtraction of Polynomials

Add the expressions in Probs. 1 to 12.

1 $\quad 3a - 2b + 7c$	**2** $\quad 4x^2 - 5y^2 + 3z^2$
$\quad\ \ -4a + 5b - 5c$	$\quad\ \ -5x^2 - 2y^2 + 6z^2$
$\quad\ \ \ \ 5a - 3b + \ \ c$	$\quad\ \ \ \ 2x^2 + 6y^2 - 4z^2$

3 $5a - 3b + 8c$
$-2a + 4b - 6c$
$6a - 5b + \ c$

4 $-3ab + 4cd - 6ac$
$4ab - 7cd + 2ac$
$-2ab + 3cd + 4ac$

5 $11xy - 6xz + 4yz$
$3xy - 3xz - 5yz$
$-6xy + 9xz - 2yz$

6 $10xy - 7xz + 3yz$
$4xy - 2xz - 6yz$
$-5xy + 9xz - 8yz$

7 $6a^2 - 4ab + 2c^2$
$4a^2 + 3ab - 7c^2$
$8a^2 - \ ab + \ c^2$

8 $2x^2 + 3xy - \ y^2$
$-3x^2 + 3xy + 2y^2$
$x^2 - 4xy - 4y^2$

9 $-6c^2d^2 \qquad - 4c^2d$
$3c^2d^2 + 2cd^2 + 2c^2d$
$\qquad - 5cd^2 + 2c^2d$

10 $5x^3 - 2x^2 + \ 4x$
$4x^2 + \ 6x$
$-7x^3 \qquad - 10x$

11 $5a^2 - 6b^2$
$3b^2 - 4c^2$
$-2a^2 \qquad + 3c^2$

12 $7hk + 4ik$
$- 6ik + 8jk$
$-10hk - 2ik - 3jk$

In Probs. 13 to 24, write the expressions so that similar terms form columns, and then find the sum.

13 $3x - 5y + 5z,\ 2x - 5z + 2y,\ 3z - 4x - 5y$

14 $4r + 2s - rs,\ 2s + 2rs + t,\ 3r - 3rs + 2t$

15 $5a + 2b - 3ab,\ 3a + ab - c,\ 2a - 5ab + 3c$

16 $2x - 3xy - 2y,\ -7x - 3xy + 2xz,\ 5xy - 3xz$

17 $3x - 2xy + y,\ -4x - 2xy + xz,\ 3xy - 2xz$

18 $6x^3 - 4x^2 - 9x,\ -2x - 5x^3 + 3x^2,\ -10x^2 + 12x + 8x^3$

19 $6a^2b^3 + 7ab^4 - 8a^3b^2,\ 5a^3b^2 + 3a^2b^3 - 8ab^4,\ -4ab^4 + 7a^3b^2 - 9a^2b^3$

20 $ab - a + c,\ -3ab + a + 2ac,\ 4a - 5c - 4ac$

21 $2ab - a + 2c,\ -5ab + 2a - 3ac,\ 5a + 6c - 3ac$

22 $3x - 4y,\ 5y - 2z,\ 6x + 2z,\ 5x - 3y + 5z$

23 $2a - 3b,\ 4b - 3c,\ 5a + 2c,\ 3a - 5b + 7c$

24 $2c^3 - c^2d + d^3,\ -c^2d - cd^2 - d^3,\ c^3 + cd^2 - 3d^3,\ 2c^2d - 3cd^2 - d^3$

In each of Probs. 25 to 28, subtract the second number from the first.

25 461
-324

26 -537
248

27 -632
-431

28 -236
-537

In each of Probs. 29 to 40, subtract the second expression from the first.

29 $-2a + b - 2c,\ -3a + 2b + c$

30 $-3x + 2y - 3z,\ -4x + 3y - z$

31 $7x^2 - 3xy + 5y^2,\ 8x^2 + 2xy - 2y^2$

32 $4x^2 + z^2 - w^2,\ 6x^2 - 5z^2 - 3w^2$

33 $5a^2 + c^2 - d^2, 6a^2 - 4b^2 + c^2$

34 $4a^2 - 2ab + 3b^2, -3a^2 + 2b^2$

35 $3a^2 - ab + 2b^2, -2a^2 + 3b^2$

36 $6x^2 - 4y^2 + z^2, 4x^2 + z^2 - w^2$

37 $3a - 2b, 4a + 3b - 3c$

38 $8c^4 + 4c^3d - 8c^2d^2 + 3d^3, 7c^4 - 12c^3d + 4c^2d^2$

39 $9x^3 + 2x^2y - 3xy^2 + 2y^3, 6x^3 - 3x^2y + 2xy^2 - y^3$

40 $3a^2 + 4ab - 5d^2, 5a^2 + 6ab - 2ad - 3d^2$

2.8
AXIOMS AND THEOREMS OF MULTIPLICATION

We stated previously that the product of a and b is written $a \times b$, $a \cdot b$, $a(b)$, $(a \times b)$, and most often simply as ab. Furthermore, we stated that a and b are *factors* of ab. We also stated that the identity element in multiplication is the number 1 and that

■ $a \cdot 1 = 1 \cdot a = a$ (2.19)

Moreover, we proved that

■ $a \cdot 0 = 0 \cdot a = 0$ (2.26)

The right- and left-hand distributive axioms are essential for multiplication, and we rewrite them here for ready reference.

■ $(a + b)c = ac + bc$ (2.15)

■ $c(a + b) = ca + cb$ (2.16)

Closure Axiom

We now assume the closure axiom for multiplication stated as follows:

■ If a and b are real numbers, there exists a real number c
such that $ab = c$ (2.27)

It is readily verified that $3 \cdot 4 = 4 \cdot 3 = 12$ and that $(3 \cdot 4)2 = 3(4 \cdot 2) = 24$. We assume that these properties hold for all real numbers and thus have the axioms now stated.

Commutative Axiom

■ $a \cdot b = b \cdot a$ (2.28)

Associative Axiom

■ $(a \cdot b)c = a(b \cdot c)$ (2.29)

$$(a \cdot b)c = b(c \cdot a)$$

PROOF

$$
\begin{aligned}
(a \cdot b)c &= (b \cdot a)c & \text{by (2.28)} \\
&= b(a \cdot c) & \text{by (2.29)} \\
&= b(c \cdot a) & \text{by (2.28)}
\end{aligned}
$$

By using arguments similar to that of Example 1, we can prove that the product of three numbers can be obtained by choosing two of them and then multiplying the product of these two by the third. Furthermore, the product is the same regardless of the choice of the first two or the order in which the multiplication is performed. For example, $a \cdot b \cdot c = (ab)c = a(bc) = (ac)b = b(ca)$ and so on. Therefore, $a \cdot b \cdot c$ is a unique number.

The associative axiom can be extended so that it applies to four or more factors. We illustrate the extension in Example 2.

EXAMPLE 2

Prove that $a \cdot b \cdot c \cdot d = (ab)(cd)$.

PROOF   We stated above that $a \cdot b \cdot c$ is a unique number and is equal to $(ab)c$. Hence,

$$a \cdot b \cdot c \cdot d = [(ab)c]d$$

We now replace ab by p and have

$$
\begin{aligned}
a \cdot b \cdot c \cdot d &= [pc]d \\
&= p(cd) & \text{by (2.29)} \\
&= (ab)(cd) & \text{replacing } p \text{ by } ab
\end{aligned}
$$

Law of Exponents for Multiplication

We defined a^n, where n is a positive integer, as follows:

$$a^n = a \cdot a \cdot a \cdots \text{ to } n \text{ factors}$$

It follows that

$$a^m a^n = \underbrace{(a \cdot a \cdot a \cdots a)}_{m \text{ factors}} \underbrace{(a \cdot a \cdot a \cdots a)}_{n \text{ factors}}$$

Hence, since the expression on the right is the product of $m + n$ factors each of which is a, we have

$$a^m a^n = a^{m+n} \qquad \text{m and n are positive integers} \tag{2.30}$$

EXAMPLE 3

Find the product of $4a^3$ and $6a^5$.

SOLUTION

$$4a^3 \cdot 6a^5 = 4 \cdot 6 \cdot a^3 \cdot a^5 \qquad \text{by commutative axiom (2.28)}$$

$$= 24a^{3+5} \qquad \text{by law of exponents (2.30)}$$

$$= 24a^8$$

Power of a Power

We use (2.30) for finding the power of a power. For example,

$$(x^2)^3 = x^2 \cdot x^2 \cdot x^2 = x^{2+2+2} = x^6$$

In general,

$$(x^m)^n = \underbrace{x^m \cdot x^m \cdot x^m \cdots x^m}_{n \text{ factors}}$$

$$= x^{m+m+m+\cdots \text{ to } n \text{ terms}}$$

$$= x^{nm}$$

Consequently, we have

■ $\quad (x^m)^n = x^{nm} \qquad$ x real, m and n positive integers, $\qquad\qquad$ (2.31)

Power of a Product

Furthermore, since

$$(ab)^n = \underbrace{ab \cdot ab \cdot ab \cdots ab}_{n \text{ factors}}$$

$$= \underbrace{(a \cdot a \cdot a \cdots a)}_{n \text{ factors}}\underbrace{(b \cdot b \cdot b \cdots b)}_{n \text{ factors}} \qquad \text{by associative and commutative axioms}$$

$$= a^n b^n$$

We have the following law for the power of a product:

■ $\quad (ab)^n = a^n b^n \qquad$ a and b real, n is a positive integer $\qquad\qquad$ (2.32)

Multiplicativity Axiom for the Order Relation

If $a > c$ and $b > 0$, it can be verified that $ab > cb$ if a, b, and c are replaced with any three positive integers with the replacement for a greater than the replacement for c. Hence, it is reasonable to assume the following axiom:

■ If $a > c$ and $b > 0$, then $ab > bc$. (2.33)

Law of Signs for Multiplication

Our next task is to derive the law of signs for multiplication. We consider the positive numbers a and b, and first prove that ab is positive.

PROOF

$a > 0$	since a is positive
$a \cdot b > 0 \cdot b$	by (2.33)
> 0	since $0 \cdot b = 0$

Hence, $a \cdot b$ is positive. Furthermore, since

$$a > 0, \quad b > 0, \quad \text{and} \quad ab > 0$$

then

$$|a| = a, \quad |b| = b, \quad \text{and} \quad |ab| = ab$$

Hence,

$$|ab| = ab = |a| \cdot |b|$$

We now consider the product $a(-b)$, $b > 0$, and use the distributive axiom to show that $a(-b) = -ab$.

PROOF

$ab + a(-b) = a(b - b)$	by left-hand distributive axiom
$= a(0)$	since $b - b = 0$ by (2.18)
$= 0$	since $a(0) = 0$ by (2.26)

Furthermore, since $-ab$ is the negative of ab, then,

$$ab + (-ab) = 0 \quad \text{by (2.18)}$$

Hence,

$$a(-b) = -ab \quad \text{by (2.21)}$$

Therefore, since ab is positive, $-ab$ and $a(-b)$ are negative. Also,

$|-ab| = ab$	by definition of absolute value
$= |a| \cdot |b|$	since $|a| = a$ and $|b| = b$

We also use the distributive axiom to prove that $(-a)(-b) = ab$. We start with

$$-a(b) + (-a)(-b) = -a[b + (-b)] \qquad \text{by left-hand distributive axiom}$$
$$= -(a)(0) \qquad \text{since } b + (-b) = 0$$
$$= 0$$

Moreover,

$$-(ab) + ab = 0 \qquad \text{by (2.18)}$$

Hence,

$$(-a)(-b) = ab \qquad \text{by (2.21)}$$

Therefore, $(-a)(-b)$ is positive since ab is positive. Furthermore,

$$|(-a)(-b)| = |ab|$$
$$= |a| \, |b|$$

Consequently, we have the following law of signs for multiplication:

■ The absolute value of the product of two numbers is equal to the product of their absolute values. The product is positive if both of the numbers are positive or both are negative. The product is negative if one number is positive and the other is negative.

EXAMPLE 4

$$(3x^2)(4x^3) = (3)(4)(x^2)(x^3) \qquad \text{by associative and commutative axioms}$$
$$= 12x^5 \qquad \text{since } x^2 \cdot x^3 = x^5$$

EXAMPLE 5

$$(2a)(-3b)(4c) = (2)(-3)(4)a \cdot b \cdot c$$
$$= (-6)(4)a \cdot b \cdot c \qquad \text{by associative axiom and law of signs}$$
$$= -24abc$$

EXAMPLE 6

$$(-2a^2)(-8a^3) = (-2)(-8)a^2 \cdot a^3$$
$$= 16a^5 \qquad \text{since } (-2)(-8) = 16 \text{ and } a^3 \cdot a^2 = a^5$$

EXAMPLE 7

$$(-3x^2y)(-4xy^3)(-5x^4y^2) = (-3)(-4)(-5)(x^2x^1x^4)(y^1y^3y^2)$$

$$= -60x^{2+1+4}y^{1+3+2}$$

$$= -60x^7y^6$$

EXAMPLE 8

$$(3a^2b^4c^3)^5 = 3^5(a^2)^5(b^4)^5(c^3)^5 \qquad \text{by (2.32)}$$

$$= 243a^{10}b^{20}c^{15} \qquad \text{by (2.31)}$$

2.9
PRODUCTS OF MONOMIALS AND POLYNOMIALS

Examples 4 to 7 of Sec. 2.8 illustrate the procedure for obtaining the product of two or more monomials.

We use either the left-hand or right-hand distributive axioms to obtain the product of a monomial and a polynomial, as illustrated in Examples 1 and 2.

EXAMPLE 1

Find the product of $2x^3y - 5x^2y^2 + 6xy^3$ and $3x^2y^3$.

SOLUTION We write the product, use the left-hand distributive axiom, and proceed as follows:

$3x^2y^3 (2x^3y - 5x^2y^2 + 6xy^3)$

$\quad = 3(2)x^2x^3y^3y^1 + 3(-5)x^2x^2y^3y^2 + 3(6)x^2x^1y^3y^3$ by left-hand distributive
axiom and commutative axiom

$\quad = 6x^5y^4 - 15x^4y^5 + 18x^3x^6$ by law of signs and law of
exponents

EXAMPLE 2 Perform the multiplication in

$$(2a^3b^2 - 4a^2b^3 + 7ab^4)3a^2b^3 - 2ab^4(4a^4b - 6a^3b^2 + a^2b^3) \qquad (1)$$

SOLUTION By use of the right-hand distributive axiom, the associative and commutative axioms, and the law of signs, we have

$$(2a^3b^2 - 4a^2b^3 + 7ab^4)3a^2b^3 = 6a^5b^5 - 12a^4b^6 + 21a^3b^7$$

Similarly,

$$-2ab^4(4a^4b - 6a^3b^2 + a^2b^3) = -8a^5b^5 + 12a^4b^6 - 2a^3b^7$$

Consequently, the expression (1) is equal to

$$6a^5b^5 - 12a^4b^6 + 21a^3b^7 - 8a^5b^5 + 12a^4b^6 - 2a^3b^7$$

$$= 6a^5b^5 - 8a^5b^5 - 12a^4b^6 + 12a^4b^6 + 21a^3b^7 - 2a^3b^7 \qquad \text{by associative and commutative axioms}$$

$$= -2a^5b^5 + 19a^3b^7 \qquad \text{combining similar terms}$$

The following example illustrates the use of the distributive axioms for removing symbols of grouping:

EXAMPLE 3   Remove the symbols of grouping from

$$3x^3 - \{2x^3 - x^2y - 3x[x(x-y) - y(2x-y)] + 4xy^2\} - 3xy^2 \qquad (2)$$

and combine similar terms.

SOLUTION   As stated earlier, it is advisable to remove the innermost symbols first. We therefore proceed as follows:
The expression (2) is equal to

$$3x^3 - \{2x^3 - x^2y - 3x[x^2 - xy - 2xy + y^2] + 4xy^2\} - 3xy^2 \quad \text{removing parentheses by distributive axiom}$$

$$= 3x^3 - \{2x^3 - x^2y - 3x[x^2 - 3xy + y^2] + 4xy^2\} - 3xy^2 \quad \text{combining terms in brackets}$$

$$= 3x^3 - \{2x^3 - x^2y - 3x^3 + 9x^2y - 3xy^2 + 4xy^2\} - 3xy^2 \quad \text{removing brackets by distributive axiom}$$

$$= 3x^3 - \{-x^3 + 8x^2y + xy^2\} - 3xy^2 \quad \text{combining terms in braces}$$

$$= 3x^3 + x^3 - 8x^2y - xy^2 - 3xy^2 \quad \text{removing braces by use of (2.23)}$$

$$= 4x^3 - 8x^2y - 4xy^2 \quad \text{combining similar terms}$$

EXERCISE 2.3
Products of Monomials and Polynomials

Find the products in Probs. 1 to 12.

1 $(5x^3y^2)(3x^6y^4)$

2 $(4a^4b^3)(6a^2b^5)$

3 $(3c^2d)(7cd^2)$

4 $(8h^5k^2)(2h^3k^4)$

5 $(7x^2y^3z^4)(-3x^3y^5z^7)$

6 $(-4a^5b^3c^2)(5a^3b^6c^9)$

7 $(-8cd^2e^3)(3c^4d^2e)$

8 $(7x^7y^3z^2)(-5x^2y^5z^3)$

9 $(3a^2b^3)(2ab^2)(4a^4b^4)$

10 $(4xy^2z^4)(5x^3y^3z)(-2xy^4z^2)$

11 $(-7c^3d^4e^5)(-2c^2d^6e)(-3cd^3e^2)$

12 $(5a^4bc^2)(-2ab^3c^4)(-4a^3b^2c^5)$

Find the required powers in Probs. 13 to 20.

13 $(3a^2)^2$

14 $(3x^2y^3)^3$

15 $(-3x^2y^4)^5$

16 $(-2a^2b^3)^4$

17 $(5a^2b^3z)^3$

18 $(-2x^3y^2)^5$

19 $(-3cd^3)^4$

20 $(-5r^2s^3t^5)^2$

Find the products in Probs. 21 to 32. In Probs. 29 to 32, combine similar terms after multiplying.

21 $3x^2(2x^4 - 3x^2)$

22 $2a^3(3a^4 - 5a^3)$

23 $3x^2y(6x^3y + 4xy^2)$

24 $2a^3b(3ab^2 + 5a^3b)$

25 $2xy^2z^3(3x^2y - 2x^3y^2 + 5x^3y^4)$

26 $(-4ab^2c^2 + 3a^3c - 2b^4c^5)(2a^3bc^4)$

27 $-6x^3y^3z^3(-3yz^4 + 5xy^4z^2 - 2x^2z^3)$

28 $(2x^2y - 3x^3y^2 + 4y^3z^4)(3xy^2z^3)$

29 $3x^4y(2x^3y^2z^2 - 5x^4y^4z^3) - 5x^3y^2z(x^4yz - 3x^5y^3z^2)$

30 $3a^3b^4c^2(6a^2b^3 - 5a^4c^5 + 4b^2c^3) - 6a^2b^4c(3a^3b^3c + 2ab^2c^4)$

31 $8p^3q^4r^5(4p^2q^3 - 3q^5r^6) - 4p^3q^7r^4(8p^2r - 3p^3q - 6q^2r^7)$

32 $4a^3b^2c^4(2 - 3a^2b^5) + 3a^5b^6c^4(4b + 3a^3c) - 2a^3bc^2(4bc^2 - 3a^5b^5c^3)$

Remove the symbols of grouping in the following problems, and combine similar terms.

33 $2a + 4[3a - 2(b - a)] + 2b$

34 $x + 3(y - z) + 5[x - 3(y - 2z)]$

35 $a[a - 2(b + 3c) + 7] + a[2b - 3(a + c) - 5]$

36 $2x[8x - 3x(x + 3) + 2x^2] - 3x^2[2 - x(2 + 3x) - x^2 + 3x]$

37 $5\{2x - 3[2x - 3(2x - 3) + 2x] + 3\}$

38 $3x\{x^2 - x[2x - 3(x + 3) + 2] + x^2\}$

39 $2a - 3\{3a - 2[3(3a - 2) - (4a - 3) + (a - 3)] - 12\} + 9$

40 $3x - \{5x + 2[3x + 3(2x + 3y) - (x - y) - 7(x - y) + 3] - (8x + 7y)\} + 5y$

2.10
THE PRODUCT OF TWO POLYNOMIALS

We obtain the product of two polynomials by repeated applications of the distributive axiom and the commutative axiom. The method is illustrated in the following example, in which we obtain the product of $3x^3 - 4y^3 - 6x^2y + 2xy^2$ and $2xy - 5x^2 + 3y^2$:

$$(3x^3 - 4y^3 - 6x^2y + 2xy^2)(2xy - 5x^2 + 3y^2)$$

$$= (3x^3 - 4y^3 - 6x^2y + 2xy^2)(2xy) + (3x^3 - 4y^3 - 6x^2y + 2xy^2)(-5x^2)$$

$$+ (3x^3 - 4y^3 - 6x^2y + 2xy^2)(3y^2) \qquad \text{by left-hand distributive axiom (2.16)}$$

$$= 6x^4y - 8xy^4 - 12x^3y^2 + 4x^2y^3 - 15x^5$$

$$+ 20x^2y^3 + 30x^4y - 10x^3y^2 + 9x^3y^2 - 12y^5 - 18x^2y^3 + 6xy^4$$

$$\text{by right-hand distributive axiom (2.15)}$$

$$= -15x^5 + 30x^4y + 6x^4y - 10x^3y^2 - 12x^3y^2$$

$$+ 9x^3y^2 + 20x^2y^3 + 4x^2y^3 - 18x^2y^3 - 8xy^4 + 6xy^4 - 12y^5$$

$$\text{by commutative axiom for addition (2.13)}$$

$$= -15x^5 + 36x^4y - 13x^3y^2 + 6x^2y^3 - 2xy^4 - 12y^5 \qquad \text{combining similar terms}$$

The above process can be abbreviated considerably by use of the method that we shall next explain. We first arrange the terms in each polynomial so that the exponents of one of the literal numbers are in descending numerical order. In this case we shall base our arrangement on the exponents of x and shall proceed as follows:

$$3x^3 - 6x^2y + 2xy^2 - 4y^3$$
$$\underline{- 5x^2 + 2xy + 3y^2}$$
$$-15x^5 + 30x^4y - 10x^3y^2 + 20x^2y^3 \qquad\qquad -5x^2(3x^3 - 6x^2y + 2xy^2 - 4y^3)$$
$$+ 6x^4y - 12x^3y^2 + 4x^2y^3 - 8xy^4 \qquad\quad\; 2xy(3x^3 - 6x^2y + 2xy - 4y^3)$$
$$9x^3y^2 - 18x^2y^3 + 6xy^4 - 12y^5 \quad 3y^2(3x^3 - 6x^2y + 2xy - 4y^3)$$
$$\overline{-15x^5 + 36x^4y - 13x^3y^2 + 6x^2y^3 - 2xy^4 - 12y^5}$$

For convenience in adding, the terms in the partial products are arranged so that similar terms form columns.

EXERCISE 2.4
Multiplication of Polynomials

Find the products in Probs. 1 to 28.

1 $(a + b)(a - b)$ 2 $(3x + y)(3x - y)$

3 $(2a + 4b)(2a - 4b)$ 4 $(5c + 6d)(5c - 6d)$

5 $(x + y)^2$ 6 $(a^2 + 2b^2)^2$

7 $(3y^2 + 2z^3)^2$ 8 $(4a^4 + 3a^2)^2$

9 $(a - b)^2$ 10 $(2x - 3y)^2$

11 $(4a^2 - 5b^2)^2$

12 $(6x^3 - 7y^2)^2$

13 $(3x + 5y)(2x + 3y)$

14 $(4a + 3b)(2a + 5b)$

15 $(5c + 3d)(2c - d)$

16 $(6x - 5y)(3x + y)$

17 $(7p - 2q)(4p - 3q)$

18 $(9a - 2b)(7a - 3b)$

19 $(8x^2 - y^2)(7x^2 + y^2)$

20 $(3a^3 - 2b^2)(5a^3 + 3b^2)$

21 $(x + y)(x^2 - xy + y^2)$

22 $(a - 2)(a^2 + 2a + 4)$

23 $(5a^3 - b^3)(25a^6 + 5a^3b^3 + b^6$

24 $(2y^2 + 3z^2)(4y^4 - 6y^2z^2 + 9z^4)$

25 $(2x - 3y)(3x^2 + 2xy + 10y^2)$

26 $(4a + 5b)(2a^2 - 3ab + b^2)$

27 $(x^3 + 2y^3)(2x^6 + 4x^3y^3 + 4y^6)$

28 $(3c^2 - 7d^2)(2c^4 - c^2d^2 - 5d^4)$

In Probs. 29 to 40, multiply the first polynomial by the second.

29 $2x^2 - 3xy + y^2$
$4x^2 + xy - 2y^2$

30 $4a^4 - 3a^2 + 5$
$2a^4 + 4a^2 - 1$

31 $3x^3y - 5x^2y^2 - 2xy^3$
$x^2y + 3xy^2 - 4y^3$

32 $7r^5s + 5r^3s^3 + 3rs^5$
$4r^4s^2 - 2r^2s^4 - s^6$

33 $2a + 3b - 2c$
$4a - 2b + c$

34 $4x^2 - 2y^2 + z^2$
$3x^2 + 5y^2 - 2z^2$

35 $3ab - 2ac + 4bc$
$2ab + 5ac - 8bc$

36 $4x^2y + 3xy^2 - 4y^3$
$2x^2y - xy^2 + 3y^3$

37 $a^4 - a^3 + a^2 - 1$
$a + 1$

38 $x^4 - 2x^3 + 4x^2 - 8x + 16$
$x + 2$

39 $x^3 - x^2y + 2xy^2 + 2y^3$
$x^2 + xy - y^2$

40 $8x^3 - 4x^2 + 2x - 1$
$4x^2 + 2x + 1$

2.11
AXIOMS AND THEOREMS OF DIVISION

Quotient

We stated earlier that the quotient of a and b is indicated by $a \div b$, $\dfrac{a}{b}$, or
a/b. We define the *quotient* of a and b as the number x such that $bx = a$, or
more briefly,

■ $\dfrac{a}{b} = x$ if and only if $bx = a$.   (2.34)

The number a is the *dividend*, and b is the *divisor*.

Reciprocal

If $a \neq 0$, we define the quotient $1/a$ as the number such that

■ $a \cdot \dfrac{1}{a} = \dfrac{1}{a} \cdot a = 1$ (2.35)

The number $1/a$ is called the *reciprocal* of a.

We use this definition to prove two very important theorems.

Cancellation Theorem for Multiplication

■ If $a \neq 0$ and $ab = ac$, then $b = c$ (2.36)

PROOF

$$ab = ac \qquad \text{given}$$

$$\frac{1}{a}(ab) = \frac{1}{a}(ac) \qquad \text{by multiplicativity axiom for equality (2.5)}$$

$$\left(\frac{1}{a} \cdot a\right)b = \left(\frac{1}{a} \cdot a\right)c \qquad \text{by associative axiom (2.29)}$$

$$1 \cdot b = 1 \cdot c \qquad \text{by (2.35)}$$

$$b = c \qquad \text{since } 1 \cdot b = b \text{ and } 1 \cdot c = c$$

■ If $ab = 0$ and $a \neq 0$, then $b = 0$ (2.37)

PROOF

$$a \cdot b = 0 \qquad \text{given}$$

$$a \cdot 0 = 0 \qquad \text{by (2.26)}$$

$$a \cdot b = a \cdot 0 \qquad \text{since each is equal to zero}$$

$$b = 0 \qquad \text{by cancellation theorem (2.36)}$$

Zero in Division

We are now in a position to discuss the role of zero in division. Using the definition $a/b = x$ if and only if $bx = a$, we have the following possibilities:

1. If $a = 0$ and $b \neq 0$, then $bx = 0$. Hence, by (2.36), $x = 0$. Therefore,

■ $\dfrac{0}{b} = 0 \qquad b \neq 0$ (2.38)

2. If $a \neq 0$ and $b = 0$, then $bx = a$ becomes $0 \cdot x = a$ and there is no replacement for x for which this statement is true, since $0 \cdot x = 0$. Hence, if the divisor is zero and the dividend is not zero, the quotient does not exist.

3. If $a = 0$ and $b = 0$, then $bx = 0$ becomes $0 \cdot x = 0$ and this statement is true for every replacement for x, and therefore the quotient does not exist as a *unique* number.

Therefore, division is not an admissible operation if the divisor is zero. Since $a \cdot 1 = 1 \cdot a$, it follows that

■ $\quad \dfrac{a}{a} = 1 \qquad a \neq 0$ $\hfill$ (2.39)

and

■ $\quad \dfrac{a}{1} = a$ $\hfill$ (2.40)

We next prove two theorems that are very useful in the division of monomials and in fractions.

■ $\quad \dfrac{a}{b} = a\!\left(\dfrac{1}{b}\right) = \dfrac{1}{b}(a)$ $\hfill$ (2.41)

PROOF Let

$$\frac{a}{b} = x$$

then,

$a = bx \qquad$ **by (2.34)**

Also,

$\dfrac{1}{b}(a) = \dfrac{1}{b}(bx) \qquad$ **by multiplicativity axiom for equalities (2.5)**

$\quad = \left(\dfrac{1}{b} \cdot b\right)x \qquad$ **by associative axiom**

$\quad = 1 \cdot x \qquad$ **by (2.35)**

$\quad = x$

Hence,

$$\frac{1}{b}(a) = \frac{a}{b}$$

since each is equal to x.

■ $\quad \dfrac{ab}{cd} = \dfrac{a}{c} \cdot \dfrac{b}{d}$ $\hfill$ (2.42)

PROOF Let

$$\frac{a}{c} = x \qquad \text{and} \qquad \frac{b}{d} = y$$

Then,

$$a = cx \qquad \text{and} \qquad b = dy$$

Furthermore, we have

$$\frac{ab}{cd} = \frac{1}{cd}(ab) \qquad\qquad \text{by (2.41)}$$

$$= \frac{1}{cd}(cx \cdot dy) \qquad \text{replacing } a \text{ by } cx \text{ and } b \text{ by } dy$$

$$= \left(\frac{1}{cd} \cdot cd\right)xy \qquad \text{by commutative and associative axioms}$$

$$= xy \qquad\qquad\qquad \text{by (2.35) and (2.19)}$$

$$= \frac{a}{c} \cdot \frac{b}{d} \qquad\qquad \text{replacing } x \text{ and } y \text{ by } a/b \text{ and } c/d, \text{ respectively}$$

Since $cd = dc$, we can by a similar argument show that

$$\frac{ab}{cd} = \frac{a}{d} \cdot \frac{b}{c}$$

Law of Signs for Division

We next derive the law of signs for division. We use the equality $bx = a$ and recall that x is the quotient of a and b. We have the following possibilities:

b	x	a
positive	positive	positive
positive	negative	negative
negative	negative	positive
negative	positive	negative

From the tabulation we see that if a and b are both positive or both negative, the quotient x is positive. Furthermore, if a is positive and b is negative, or if a is negative and b is positive, then the quotient x is negative. Hence, we have the following law.

■ The quotient of two positive numbers or of two negative numbers is positive. The quotient of a positive number and a negative number, or of a negative number and a positive number, is negative.

The final theorem that we prove in this section is called the *law of exponents for division*. If in the definition of a quotient (2.34) we replace a by a^m and b by a^n, where m and n are positive integers with $m > n$, we have

$$\frac{a^m}{a^n} = x \text{ if and only if } a^n x = a^m$$

and we shall determine x. Since $m > n$ and m and n are positive integers, $m - n$ is a positive integer and can be used as an exponent. Now if we replace x in $a^n x$ by a^{m-n}, we get

$$a^n a^{m-n} = a^{n+m-n} \qquad \text{by law of exponents for multiplication}$$

$$= a^{m+n-n} \qquad \text{by commutative axiom}$$

$$= a^m \qquad \text{since } n - n = 0$$

Hence, if $x = a^{m-n}$, then $a^n x = a^m$. Therefore, we have the law of exponents for division:

$$\blacksquare \quad \frac{a^m}{a^n} = a^{m-n} \qquad m > n \tag{2.43}$$

We shall later prove that this law holds if $m < n$, and now we examine the situation if $m = n$. In this case, (2.43) becomes

$$\frac{a^n}{a^n} = a^{n-n} = a^0$$

Since our definition of a^n requires that n be a positive integer, a^0 has no meaning. By (2.39), however,

$$\frac{a^n}{a^n} = 1$$

and it is logical to define a^0 as the number 1. Hence, by this definition, we have:

Zero as an Exponent

$$\blacksquare \quad a^0 = 1 \qquad a \neq 0 \tag{2.44}$$

EXAMPLE

$$\frac{x^6}{x^2} = x^{6-2} = x^4$$

2.12

MONOMIAL DIVISORS

We obtain the quotient of two monomials that have powers of the same variable by use of (2.42), as illustrated now by Example 1.

EXAMPLE 1

$$\frac{48a^4b^2}{12a^2b^2} = \frac{48}{12} \cdot \frac{a^4}{a^2} \cdot \frac{b^2}{b^2} \qquad \text{by (2.42)}$$

$$= 4a^{4-2}b^{2-2} \qquad \text{by law of exponents}$$

$$= 4a^2b^0$$

$$= 4a^2 \qquad \text{since } b^0 = 1$$

Division of a Polynomial by a Monomial

If the dividend is a polynomial and the divisor is a monomial, we use the following formula as a first step in obtaining the quotient:

$$\blacksquare \quad \frac{a + b - c}{d} = \frac{a}{d} + \frac{b}{d} - \frac{c}{d} \tag{2.45}$$

PROOF

$$\frac{a + b - c}{d} = \frac{1}{d}(a + b - c) \qquad \text{by (2.41)}$$

$$= \frac{a}{d} + \frac{b}{d} - \frac{c}{d} \qquad \text{by distributive axiom and (2.41)}$$

Therefore, if the dividend is a polynomial and the divisor is a monomial, the quotient is the algebraic sum of the quotients obtained by dividing each term in the dividend by the divisor.

EXAMPLE 2

Divide $6x^4 + 4x^3y^3 - 3x^2y^2 - 2x^2$ by $3x^2$.

SOLUTION

$$\frac{6x^4 + 4x^3y^3 - 3x^2y^2 - 2x^2}{3x^2} = \frac{6x^4}{3x^2} + \frac{4x^3y^3}{3x^2} - \frac{3x^2y^2}{3x^2} - \frac{2x^2}{3x^2}$$

$$= 2x^2 + \frac{4}{3}xy^3 - x^0y^2 - \frac{2}{3}x^0$$

$$= 2x^2 + \frac{4}{3}xy^3 - y^2 - \frac{2}{3} \qquad \text{since } x^0 = 1$$

Degree of a Polynomial

In this section, we shall discuss the procedure for obtaining the quotient of two polynomials. We shall discuss only polynomials in which each term is the product of an integer and one or more integral powers of a variable. The *degree* of a polynomial in any variable is the number that is equal to the greatest exponent of that variable in the polynomial. For example, $3x^4 + 2x^3y + 5x^2y^2 + xy^3$ is a polynomial of degree 4 in x and of degree 3 in y.

Before discussing the procedure for dividing one polynomial by another, we shall consider the quotient of 23 and 5. By methods of arithmetic, we have

$$\tfrac{23}{5} = 4\tfrac{3}{5} = 4 + \tfrac{3}{5}$$

Hence, if we divide 23 by 5, we obtain the integer 4 and the fraction $\tfrac{3}{5}$. We call the mixed number $4\tfrac{3}{5}$ the *complete quotient*, the integer 4 the *partial quotient*, and the numerator of the fraction $\tfrac{3}{5}$ the *remainder*.

Likewise,

$$\frac{6x^2 + 4}{3x} = \frac{6x^2}{3x} + \frac{4}{3x} = 2x + \frac{4}{3x}$$

and we have the complete quotient $2x + 4/3x$, the partial quotient $2x$, and the remainder 4.

We may readily verify that in each of the above examples, the following relation is satisfied:

Dividend = (divisor)(partial quotient) + remainder

since

$$23 = 5(4) + 3 \qquad \text{and} \qquad 6x^2 + 4 = 3x(2x) + 4$$

Hereafter in this section, we shall use the word "quotient" to refer to the partial quotient.

In order to divide one polynomial by another, we first arrange the terms in each polynomial so that they are in the order of the descending powers of some variable that appears in each. Then we seek the quotient that is a polynomial and that satisfies the relation

Dividend = (divisor)(quotient) + remainder

where the degree of the remainder in the variable chosen as the basis of the arrangement of terms is less than the degree of the divisor in that variable. We shall illustrate the procedure by the following example.

EXAMPLE 1

Find the quotient obtained by dividing $6x^2 + 5x - 1$ by $2x - 1$.

SOLUTION   Here the dividend is $6x^2 + 5x - 1$, the divisor is $2x - 1$, and we seek the quotient that satisfies the relation

$$6x^2 + 5x - 1 = (2x - 1)(\text{quotient}) + \text{remainder} \tag{1}$$

Since the degree of the dividend is 2, the degree of the divisor is 1, and the degree of the remainder is less than 1, it follows that the degree of the quotient is 1. Hence we write the quotient in the form $ax + b$, substitute this expression in Eq. (1), and get

$$6x^2 + 5x - 1 = (2x - 1)(ax + b) + \text{remainder} \tag{2}$$

from which we can determine the values of a and b by the following procedure. We first perform the indicated multiplication in Eq. (2) and obtain

$$6x^2 + 5x - 1 = (2x - 1)ax + (2x - 1)b + \text{remainder} \tag{3}$$

By inspection we see that the only terms in the left and right members of Eq. (3) that involve x^2 are $6x^2$ and $(2x)(ax) = 2ax^2$, respectively. Hence, $2ax^2 = 6x^2$ and it follows that $a = 3$. Now, we substitute 3 for a in Eq. (3) and subtract $(2x - 1)3x = 6x^2 - 3x$ from each member and get

$$8x - 1 = (2x - 1)b + \text{remainder} \tag{4}$$

Again by inspection, we see that the only terms in Eq. (4) that include x are $8x$ and $2bx$, respectively. Consequently, $2bx = 8x$, and therefore, $b = 4$. Finally, we substitute 4 for b in Eq. (4) and subtract $(2x - 1)4 = 8x - 4$ from each member and get $3 = \text{remainder}$. Hence,

$$6x^2 + 5x - 1 = (2x - 1)(3x + 4) + 3 \tag{5}$$

Therefore, the quotient is $3x + 4$, and the remainder is 3.

We shall now divide $6x^2 + 5x - 1$ by $2x - 1$ by use of the usual long-division process. If this process is examined closely, it may be seen that it is a condensation of the above procedure.

```
divisor   3x + 4      quotient
2x − 1)6x² + 5x − 1         dividend
        6x² − 3x           (2x − 1)(3x)
             8x − 1        subtracting
             8x − 4        (2x − 1)4
                  3        remainder
```

The formal steps in the process of dividing one polynomial by another are the following:

1. Arrange the terms in both the dividend and divisor in the order of the descending powers of a variable that appears in each.
2. Divide the first term in the dividend by the first term in the divisor to get the first term in the quotient.
3. Multiply the divisor by the first term in the quotient, and subtract the product from the dividend.
4. Treat the remainder obtained in step 3 as a new dividend, and repeat steps 2 and 3.
5. Continue this process until a remainder is obtained that is of lower degree than the divisor in the variable that is chosen as a basis for the arrangement in step 1.

The quotient can be checked by the use of the relation

$$\text{Dividend} = (\text{divisor})(\text{quotient}) + \text{remainder}$$

We shall further illustrate the process by the following example.

EXAMPLE 2

Divide $6x^4 - 6x^2y^2 - 3y^4 + 5xy^3 - x^3y$ by $-2y^2 + 2x^2 + xy$.

SOLUTION We shall arrange the terms in the dividend and divisor in the order of the descending powers of x and proceed as follows:

$$
\begin{array}{l}
\underline{\;3x^2 - 2xy + y^2}\quad \text{quotient}\\
2x^2 + xy - 2y^2\,\overline{)6x^4 - x^3y - 6x^2y^2 + 5xy^3 - 3y^4}\quad\text{dividend}\\
\underline{6x^4 + 3x^3y - 6x^2y^2}\quad (2x^2 + xy - 2y^2)3x^2\\
- 4x^3y+ 5xy^3 - 3y^4\quad\text{subtracting}\\
\underline{- 4x^3y - 2x^2y^2 + 4xy^3}\quad (2x^2 + xy - 2y^2)(-2xy)\\
2x^2y^2 + xy^3 - 3y^4\quad\text{subtracting}\\
\underline{2x^2y^2 + xy^3 - 2y^4}\quad (2x^2 + xy - 2y^2)y^2\\
- y^4\quad\text{remainder}
\end{array}
$$

EXERCISE 2.5
Division

Find the quotients in Probs. 1 to 24.

1. x^6/x^2
2. a^8/a^3
3. $-z^7/z^4$
4. $b^6/-b^2$
5. $8a^4/2a^3$
6. $27x^5y^2/9x^3y^2$
7. $-32a^2b^3c^4/4ab^3c^2$
8. $-16x^3y^5z^3/-4x^2y^2z^2$
9. $45c^4d^5e^3/-9c^4d^2e^3$
10. $-36r^7s^5t^3/9r^5s^2t$
11. $-42a^{10}b^8c^5/-7a^3b^5c^2$
12. $56x^{12}y^{10}z^8/-8x^8y^6z^2$
13. $(6a^6 + 8a^4)/2a^2$
14. $(9b^7 - 6b^5)/3b^3$
15. $(12x^8 - 9x^4)/-3x^4$
16. $(-24a^6 - 16a^4)/-4a^3$

17 $(6x^4y^3 - 4x^3y^4 + 2xy^5)/2xy^3$

18 $(12a^{12}b^6 - 8a^8b^4 - 4a^4b^2)/4a^4b^2$

19 $(8x^3y^2z^5 - 16x^4y^3z^4 - 24x^5y^5z^3)/-4x^3y^2z^3$

20 $(9a^4b^3z^5 + 12a^5b^4z^4 - 27a^6b^5z^3)/3a^4b^2z^3$

21 $(20r^7s^5t^3 - 25r^4s^4t^4 - 35rs^2t^5)/-5rs^2t^3$

22 $(-12x^8y^2z^5 - 18x^6y^3z^4 + 24x^2y^4z^5)/-6x^2y^2z^4$

23 $(14p^{14}q^7r^5 - 21p^9q^8r^6 + 35p^7q^9r^4)/7p^7q^3r^2$

24 $(12a^{10}b^8c^6 - 16a^7b^6c^4 - 24a^4b^4c^2)/-4a^4b^2c$

In Probs. 25 to 36, divide the first expression by the second.

25 $2c^2 - 5c + 3, \quad c - 1$

26 $6x^2 + xy - 2y^2, \quad 2x - y$

27 $6a^2 - ab - 2b^2, \quad 2a + b$

28 $3a^2 + 10ab - 8b^2, \quad a + 4b$

29 $6x^3 - 13x^2y + 8xy^2 - 3y^3, \quad 2x - 3y$

30 $6a^3 - a^2b + ab^2 - 2b^3, \quad 3a - 2b$

31 $6x^3 - 13x^2y + 8xy^2 - 3y^3, \quad 2x - 3y$

32 $5w^3 + 23w^2z + 14wz^2 + 8z^3, \quad w + 4z$

33 $4x^4 + x^3 - 4x^2 + 6x - 3, \quad x^2 + x - 1$

34 $x^3 + 5x^2y + 5xy^2 - 3y^3, \quad x^2 + 2xy - y^2$

35 $12x^3y^3 - 4x^4y^2 + x^6 - 9x^2y^4, \quad 2x^2y - 3xy^2 + x^3$

36 $6c^3 - 11c^2d + 7cd^2 - 6d^3, \quad 3c^2 - cd + 2d^2$

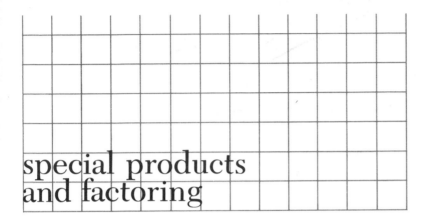

special products and factoring

In this chapter we shall present methods which contribute to speed and accuracy in computation. First we shall discuss certain products in which the computation can be performed mentally. We shall also present methods for factoring certain types of expressions. The process of factoring is necessary for dealing with fractions, is useful in solving equations, and is often helpful in simplifying complicated expressions.

3.1
THE PRODUCT OF TWO BINOMIALS WITH
CORRESPONDING TERMS SIMILAR

The corresponding terms of the binomials $ax + by$ and $cx + dy$ are similar. We obtain the product of these two binomials by the procedure shown here in which we make use of (2.15), the right-hand distributive axiom with $c = cx + dy$.

$$(ax + by)(cx + dy) = ax(cx + dy) + by(cx + dy)$$

$$= acx^2 + adxy + bcxy + bdy^2 \qquad \text{by left-hand distributive and commutative axioms (2.16) and (2.28)}$$

$$= acx^2 + (ad + bc)xy + bdy^2 \qquad \text{since } adxy + bcxy = (ad + bc)xy$$

Hence, we have

■ $(ax + by)(cx + dy) = acx^2 + (ad + bc)xy + bdy^2$ \hfill (3.1)

We see, by observing the product on the right, that we obtain the product of two binomials with corresponding terms similar by performing the following steps:

1. Multiply the first terms in the binomials to obtain the first term in the product.
2. Add the products obtained by multiplying the first term in each binomial by the second term in the other. This yields the second term in the product.
3. Multiply the second terms in the binomials to get the third term in the product.

Ordinarily, the computation required by these three steps can be done mentally, and the result can be written with no intermediate steps. This is illustrated by the following example.

EXAMPLE 1

Obtain the product of $2x - 5y$ and $4x + 3y$.

SOLUTION We write the product as shown here, proceed as directed following the problem, and record the results in the positions indicated by the flow lines.

$$(2x - 5y)(4x + 3y) = 8x^2 - 14xy - 15y^2$$

Get these products mentally:

1. $2x \cdot 4x = $ —————
2. $(2x \cdot 3y) + (-5y \cdot 4x) = 6xy - 20xy = $ —
3. $-5y \cdot 3y = $ —————

The Square of the Sum or of the Difference of Two Numbers

The square of the sum of two numbers x and y is expressed as $(x + y)^2$. Since $(x + y)^2 = (x + y)(x + y)$, we may use Formula (3.1) and get

$$(x + y)^2 = (x + y)(x + y)$$

$$= x^2 + (xy + xy) + y^2 \qquad \text{by (3.1)}$$

$$= x^2 + 2xy + y^2$$

Consequently,

■ $(x + y)^2 = x^2 + 2xy + y^2$ (3.2)

Similarly,

■ $(x - y)^2 = x^2 - 2xy + y^2$ (3.3)

Therefore we have the following rule:

■ The square of the sum or of the difference of two numbers is the square of the first term, plus or minus twice the product of the first term and the second term, plus the square of the second term.

The Product of the Sum and the Difference of the Same Two Numbers

The product of the sum and the difference of the numbers a and b is expressed as $(a + b)(a - b)$. If we apply Formula (3.1) to this product, we get

$$(a + b)(a - b) = a^2 + ab - ab - b^2 = a^2 - b^2$$

Consequently,

■ $(a + b)(a - b) = a^2 - b^2$ (3.4)

Therefore we have the following rule:

■ The product of the sum and the difference of the same two numbers is equal to the difference of their squares.

We shall illustrate the application of Formulas (3.2), (3.3), and (3.4) with two examples.

EXAMPLE 2

By use of Formulas (3.2) and (3.3), obtain the square of $2a + 5b$ and the square of $3x - 4y$.

SOLUTION

$(2a + 5b)^2 = (2a)^2 + 2(2a)(5b) + (5b)^2$ **by (3.2)**

$\qquad = 4a^2 + 20ab + 25b^2$ **by (2.32) and commutative and associative axioms, (2.28) and (2.29)**

$(3x - 4y)^2 = (3x)^2 - 2(3x)(4y) + (4y)^2$ **by (3.3)**

$\qquad = 9x^2 - 24xy + 16y^2$

EXAMPLE 3

By use of Formula (3.4), obtain the product of $3x + 5y$ and $3x - 5y$ and also the product of 104 and 96.

SOLUTION

$(3x + 5y)(3x - 5y) = (3x)^2 - (5y)^2$ **by (3.4)**

$\qquad = 9x^2 - 25y^2$

To get the product of 104 and 96, we use the fact that $104 = 100 + 4$ and $96 = 100 - 4$. Therefore,

$$(104)(96) = (100 + 4)(100 - 4)$$

$$= 100^2 - 4^2 \qquad \text{by (3.4)}$$

$$= 10,000 - 16$$

$$= 9984$$

Products Involving Trinomials

The square of a trinomial may be obtained by suitably grouping the terms and then applying (3.2) or (3.3). We shall illustrate the procedure with two examples.

EXAMPLE 4

Obtain the square of $y + z + w$ by use of (3.2).

SOLUTION We shall treat $z + w$ as a single number and indicate our intention by the use of parentheses thus, $y + (z + w)$. Then $[y + (z + w)]^2$ is the square of the sum of two numbers, and we may obtain the square by two applications of (3.2) as now shown.

$$(y + z + w)^2 = [y + (z + w)]^2$$

$$= y^2 + 2y(z + w) + (z + w)^2 \qquad \text{by (3.2)}$$

$$= y^2 + 2yz + 2yw + z^2 + 2zw + w^2$$

$$= y^2 + z^2 + w^2 + 2yz + 2yw + 2zw$$

EXAMPLE 5

Obtain the square of $2a - 3b - 5c$.

SOLUTION In this problem we shall enclose the first two terms in parentheses and then apply (3.3) to obtain

$$(2a - 3b - 5c)^2 = [(2a - 3b) - 5c]^2$$

$$= (2a - 3b)^2 - 2(2a - 3b)(5c) + (-5c)^2 \qquad \text{by (3.3)}$$

$$= 4a^2 - 12ab + 9b^2 - 20ac + 30bc + 25c^2$$

In some cases it is possible to group the terms in two trinomials so that one of them is the sum of two numbers and the other is the difference of the same two numbers. Then the product can be obtained by the use of (3.4) and (3.2) or (3.3). The following examples illustrate such situations.

EXAMPLE 6

57

3.2 The Square of
a Polynomial

Obtain the product of $3x + 2y + 5z$ and $3x + 2y - 5z$.

SOLUTION   If we enclose the first two terms in each trinomial in parentheses, we obtain

$$(3x + 2y) + 5z \quad \text{and} \quad (3x + 2y) - 5z$$

and we have the sum and the difference of the same two numbers. Hence, we can obtain the product by first using (3.4) and then complete the problem by using (3.2). Thus, we get

$$[(3x + 2y) + 5z]\,[(3x + 2y) - 5z]$$
$$= (3x + 2y)^2 - (5z)^2 \qquad \text{by (3.4)}$$
$$= 9x^2 + 12xy + 4y^2 - 25z^2$$

EXAMPLE 7

Obtain the product of $(3a + 4b + c)(3a - 4b - c)$.

SOLUTION   We first notice that if we group the first two terms in each trinomial together, we do not obtain the product of the sum and difference of the same two numbers. If, however, we group the terms in this way $[3a + (4b + c)]\,[3a - (4b + c)]$, we see that the expressions in the first and second brackets are, respectively, the sum and the difference of the same two numbers. Note that the parentheses in the second trinomial were inserted after a minus sign, and so the signs of all enclosed terms were changed. We may now complete the solution as indicated here.

$$(3a + 4b + c)(3a - 4b - c) = [3a + (4b + c)]\,[3a - (4b + c)]$$
$$= (3a)^2 - (4b + c)^2 \qquad \text{by (3.4)}$$
$$= 9a^2 - (16b^2 + 8bc + c^2) \qquad \text{by (3.2)}$$
$$= 9a^2 - 16b^2 - 8bc - c^2 \qquad \text{removing parentheses}$$

3.2
THE SQUARE OF A POLYNOMIAL

We may use (3.2) and the result obtained in Example 4 of the previous section to obtain the square of a polynomial containing four terms. The method is illustrated in Example 1.

EXAMPLE 1

Obtain the square of $x + y + z + w$.

SOLUTION

$$(x + y + z + w)^2 = [x + (y + z + w)]^2$$

$$= x^2 + 2x(y + z + w) + (y + z + w)^2 \qquad \text{by (3.2)}$$

$$= x^2 + 2xy + 2xz + 2xw + y^2 + z^2 \qquad \text{by (2.16) and}$$

$$+ w^2 + 2yz + 2yw + 2zw \qquad \text{Example 4, Sec. 3.1}$$

$$= x^2 + y^2 + z^2 + w^2 + 2xy + 2xz$$

$$+ 2xw + 2yz + 2yw + 2zw \qquad \text{by (2.13)}$$

The preceding example and Example 4, Sec. 3.1, illustrate the following rule for obtaining the square of a polynomial:

■   The square of a polynomial is equal to the sum of the squares of the separate terms increased by twice the product of each term and the sum of all terms that follow it.

At present we are not in a position to prove that this rule is true for polynomials containing more than four terms. The usefulness of the rule, however, justifies its inclusion here. We shall illustrate the application of the rule with the following example.

EXAMPLE 2

Obtain the square of $2x + 3y - 4z - 2w$.

SOLUTION

$$(2x + 3y - 4z - 2w)^2 = (2x)^2 + (3y)^2 + (-4z)^2 + (-2w)^2 + 2(2x)(3y - 4z - 2w)$$

$$+ 2(3y)(-4z - 2w) + 2(-4z)(-2w)$$

$$= 4x^2 + 9y^2 + 16z^2 + 4w^2 + 12xy - 16xz - 8xw - 24yz$$

$$- 12yw + 16zw$$

EXERCISE 3.1
Special Products

Find the product in each of the following problems.

1	$(3x + 1)(x + 5)$	**2**	$(4c + d)(6c + d)$
3	$(2y + 1)(y + 3)$	**4**	$(5z + 1)(z + 4)$
5	$(6x + 3)(x + 1)$	**6**	$(3x + 2)(2x + 1)$
7	$(5c + d)(6c + d)$	**8**	$(7x + 2y)(3x + 4y)$
9	$(c - 2d)(2c - 3d)$	**10**	$(7r - 2t)(2r - t)$
11	$(7x - 9y)(2x - 7y)$	**12**	$(5a - 7b)(8a - 2b)$

13 $(8k - 3m)(9k - 5m)$

14 $(6c - 11d)(2c - 5d)$

15 $(x - 6y)(8x - 5y)$

16 $(5c - 6d)(2c - 3d)$

17 $(3x + 2y)(x - y)$

18 $(7x - 3y)(x + 4y)$

19 $(3a - 10b)(4a + 7b)$

20 $(7h + 3k)(4h - 7k)$

21 $(2c + 3d)(7c - 9d)$

22 $(8m - 3p)(5m + 2p)$

23 $(5a - 11b)(3a + 5b)$

24 $(5h + 8k)(4h - 7k)$

25 $(a + 2b)^2$

26 $(3c + d)^2$

27 $(2x + 3)^2$

28 $(3r + 2s)^2$

29 $(3r + t)^2$

30 $(9w + 2z)^2$

31 $(5m + 2n)^2$

32 $(4x + 6)^2$

33 $(5c - 2d)^2$

34 $(11x - 3y)^2$

35 $(9p - 7q)^2$

36 $(4x^2 - 3x)^2$

37 $(8x - 3y)^2$

38 $(3a - 9b)^2$

39 $(12x - 5y)^2$

40 $(3r - 5s)^2$

41 $(a + 3)(a - 3)$

42 $(r + s)(r - s)$

43 $(a + 5b)(a - 5b)$

44 $(x + 7)(x - 7)$

45 $(9w + z)(9w - z)$

46 $(3m + 2n)(3m - 2n)$

47 $(6x + 5y)(6x - 5y)$

48 $(7u + 2v)(7u - 2v)$

49 $(101)(99) = (100 + 1)(100 - 1)$

50 $(32)(28) = (30 + 2)(30 - 2)$

51 $(49)(51)$ **52** $(26)(34)$ **53** $(23)(17)$

54 $(25)(15)$ **55** $(42)(38)$ **56** $(66)(74)$

57 $(a + b - c)^2$

58 $(x + y + 1)^2$

59 $(x + 2y + 4z)^2$

60 $(c - 3d - 4e)^2$

61 $(x + y - z - w)^2$

62 $(a - 2b + c - 2d)^2$

63 $(2r - s - 3t - u)^2$

64 $(3x - 2y + z - w)^2$

65 $(3a + 4b + c)(3a + 4b - c) = [(3a + 4b) + c][(3a + 4b) - c]$

66 $(3m - q + 2z)(3m - q - 2z)$

67 $(4a - 5b - 2c)(4a - 5b + 2c)$

68 $(3x + 5y - z)(3x - 5y + z) = [3x + (5y - z)][3x - (5y - z)]$

69 $(x^2 + x + 1)(x^2 - x - 1)$

70 $(4a + b + 1)(4a - b - 1)$

71 $(a + b + c + d)(a + b - c - d) = [(a + b) + (c + d)][(a + b) - (c + d)]$

72 $(x + 2z + y - 4)(x + 2z - y + 4)$

73 $(2a - 5b + 3c)(2a - 5b + 2c) = [(2a - 5b) + 3c][(2a - 5b) + 2c]$

74 $(c - d + 3e)(c - d - 5e)$

75 $(2x - y - 4z)(2x - y + 6z)$

76 $(6a - 4b - c)(3a - 2b + 2c)$

77 $(3x^2 - x + 2y)(5x^2 + x - 2y) = [3x^2 - (x - 2y)][5x^2 + (x - 2y)]$

78 $(5a + 2c - 2b)(2a - 3c + 3b)$ **79** $(2x + y - z)(x - y + z)$

80 $(4r - s - t)(3r + s + t)$

3.3
FACTORING

A number is *factored* if it is expressed as the product of two or more other numbers. Several such expressions may be possible. For example, $6 = 6 \cdot 1 = 3 \cdot 2 = 9 \cdot \frac{2}{3}$. In this section, however, we shall consider only *prime* factors. A *prime number* is an integer greater than 1 that has no factors except itself and 1. Therefore, the only prime factors of 6 are 3 and 2. We shall limit our discussion in this section to polynomials in which the numerical coefficients are integers. Examples of such polynomials are $3x^2 + 2x + 1$, $4x^3 + 2x^2y - xy^2 + 2y^3$, and $4ab - 3bc + 3ac + 4cd$. A polynomial with integral coefficients is said to be *prime* or *irreducible* if it has no factors of the same type except itself and 1.

3.4
FACTORS OF A QUADRATIC TRINOMIAL

A trinomial of the type $ax^2 + bxy + cy^2$ where a, b, and c are integers is a *quadratic trinomial with integral coefficients*. In this section, we shall discuss the method for finding the two binomial factors of such a trinomial if the factors exist. Since we shall use (3.1) for this purpose, we shall rewrite it here with the members interchanged.

■ $acx^2 + (ad + bc)xy + bdy^2 = (ax + by)(cx + dy)$ (3.5)

To use (3.5) in factoring $3x^2 - 10xy - 8y^2$, we must find four numbers a, b, c, and d such that $ac = 3$, $bd = -8$, and $ad + bc = -10$. The only possibilities for a and c are ± 3† and ± 1, and these numbers must have the same sign since $ac > 0$. The possibilities for b and d are ± 4 and ∓ 2 or ± 8 and ∓ 1, where the double sign indicates that if one of the two numbers is positive, the other must be negative. If we let $a = 3$ and $c = 1$, then $ad + bc = 3d + b = -10$, and this is true only when $d = -4$ and $b = 2$. Therefore,

$$3x^2 - 10xy - 8y^2 = (3x + 2y)(x - 4y)$$

We call attention to the fact that $(-3x - 2y)$ and $(-x + 4y)$ are also factors of $3x^2 - 10xy - 8y^2$, as the reader may verify. By the distributive and commutative axioms, however,

$$(-3x - 2y)(-x + 4y) = -1\,(3x + 2y)(-1)(x - 4y)$$
$$= (-1)(-1)(3x + 2y)(x - 4y)$$
$$= (3x + 2y)(x - 4y)$$

† This symbol means $+3$ or -3.

Hence, if the first term of the trinomial is positive, we choose positive values for a and c. Furthermore, if we had let $a = 1$ and $c = 3$, then $ad + bc = -10$ would have been $d + 3b = -10$, which is satisfied by $b = -4$ and $d = 2$; so the factors would be $(x - 4y)(3x + 2y)$, which are the same as obtained above except in reverse order.

Usually, if a trinomial is factorable, the factors may be found after relatively few trials. If the first and last terms of the trinomial can be factored in more than one way, several combinations may be tried before the correct one is found. If the correct factors are not readily seen, it is advisable to list the possible corresponding values of a and c and also of b and d, and then systematically try each possibility until the correct combination is found. We shall further illustrate the method with two examples.

EXAMPLE 1

Factor $6x^2 + 47xy + 15y^2$.

SOLUTION   We refer to (3.5) and see that a, c, b, and d must have values that satisfy $ac = 6$, $bd = 15$, and $ad + bc = 47$. The corresponding values of a and c are

a	6	3
c	1	2

and of b and d,

b	± 3	± 5	± 15	± 1
d	± 5	± 3	± 1	± 15

where the corresponding values must have the same sign. Since, however, $ad + bc = 47$ is positive, and a and b are also positive, we may rule out the negative signs for b and d. Now, using $a = 6$ and $c = 1$, we have $ad + bc = 6d + b = 47$. We readily verify that no one of the above pairs of corresponding values of b and d satisfies $6d + b = 47$. Hence, we use $a = 3$ and $b = 2$ and have $3d + 2c = 47$, and we see at once that $d = 15$ and $c = 1$ satisfy this equation. Consequently, we have

$$6x^2 + 47xy + 15y^2 = (3x + y)(2x + 15y)$$

EXAMPLE 2

Factor $12x^2 + 71xy - 60y^2$.

SOLUTION   Referring to (3.5), we see that a, c, b, and d must be chosen so that

$$ac = 12, \quad bd = -60, \quad \text{and} \quad ad + bc = 71$$

Hence, we have the following possibilities:

a	12	6	4
c	1	2	3

b	± 60	± 30	± 20	± 15	± 12	± 10	± 6	± 5	± 4	± 3	± 2	± 1
d	∓ 1	∓ 2	∓ 3	∓ 4	∓ 5	∓ 6	∓ 10	∓ 12	∓ 15	∓ 20	∓ 30	∓ 60

Note that since $bd = -60$, b and d must have opposite signs. Now, using the three possible pairs of values for a and c, we have the following equations for determining b and d:

$$a = 12 \qquad c = 1 \qquad 12d + b = 71$$

$$a = 6 \qquad c = 2 \qquad 6d + 2b = 71$$

$$a = 4 \qquad c = 3 \qquad 4d + 3b = 71$$

We can eliminate the second equation at once since $6d + 2b = 2(3d + b)$ is an even number and 71 is an odd number. We readily see that the first equation cannot be satisfied if d equals $|60|$, $|30|$, $|20|$, $|15|$, $|12|$, or $|10|$ since in each case $|12d| \geq 120$ and $|b| \leq 6$. Furthermore, with a little pencil work we can verify that the other remaining pairs of values of b and d will not satisfy $12d + b = 71$. Consequently, if the trinomial is factorable, we must find the values of d and b that satisfy the third equation. An examination of the possibilities reveals that the values $d = 20$ and $b = -3$ satisfy the condition since

$$4(20) + 3(-3) = 80 - 9 = 71$$

Consequently,

$$12x^2 + 71xy - 60y^2 = (4x - 3y)(3x + 20y)$$

After sufficient practice, one can mentally perform most of the steps illustrated in the above examples and quickly arrive at the proper combination.

3.5
TRINOMIALS THAT ARE PERFECT SQUARES

If a trinomial is the square of a binomial, we know by Formulas (3.2) and (3.3) that two of its terms are perfect squares and hence are positive, and that the third term is twice the product of the square roots of these two. Furthermore, such a trinomial is the square of a binomial composed of the square roots of the two perfect-square terms of the trinomial connected by the sign of the other term.

EXAMPLE Factor

(1): $4x^2 - 12xy + 9y^2$ (2): $9a^2 + 24ab + 16b^2$ and

(3): $(2a - 3b)^2 - 8(2a - 3b) + 16$

SOLUTION Since in (1)

$$4x^2 = (2x)^2, \qquad 9y^2 = (3y)^2, \qquad \text{and} \qquad 12xy = 2(2x)(3y)$$

we have

$$4x^2 - 12xy + 9y^2 = (2x)^2 - 2(2x)(3y) + (3y)^2 = (2x - 3y)(2x - 3y)$$

$$= (2x - 3y)^2 \tag{1}$$

$$9a^2 + 24ab + 16b^2 = (3a)^2 + 2(3a)(4b) + (4b)^2 = (3a + 4b)^2 \tag{2}$$

$$(2a - 3b)^2 - 8(2a - 3b) + 16 = (2a - 3b)^2 + 2(2a - 3b)4 + 4^2$$

$$= [(2a - 3b) - 4]^2$$

$$= (2a - 3b - 4)^2 \tag{3}$$

Note A trinomial that is a perfect square can be factored by the method given in Sec. 3.4. However, if a trinomial is recognized as a perfect square, it can be factored more quickly by this method.

EXERCISE 3.2
Factoring Trinomials

Factor each of the following expressions.

1 $x^2 + x - 12$	**2** $b^2 - b - 6$	**3** $y^2 + y - 2$
4 $a^2 - a - 20$	**5** $z^2 + z - 56$	**6** $x^2 - x - 72$
7 $b^2 + b - 90$	**8** $c^2 - c - 42$	**9** $a^2 + 8ab + 7b^2$
10 $x^2 + 7xy + 12y^2$	**11** $c^2 + 9cd + 14d^2$	**12** $r^2 + 13rs + 36s^2$

13 $a^2 - 13ab + 22b^2$ **14** $p^2 - 9pq + 20q^2$

15 $x^2 - 7xy + 6y^2$ **16** $c^2 - 8cd + 15d^2$

17 $6a^2 + 8a - 8$ **18** $6x^2 - x - 15$

19 $6a^2 + a - 5$ **20** $8x^2 + 2xy - 6y^2$

21 $8c^2 - cd - 7d^2$ **22** $12r^2 + rs - 20s^2$

23 $12a^2 - 4ab - 21b^2$ **24** $9x^2 - 6xy - 8y^2$

25 $x^2 + 2x + 1$ **26** $a^2 - 6a + 9$

27 $c^2 - 10c + 25$ **28** $y^2 + 4y + 49$

29 $4x^2 + 4x + 1$ **30** $9a^2 + 6a + 1$

31 $25x^2 - 10x + 1$ **32** $4y^2 + 12xy + 9x^2$

33 $9a^2 + 24ab + 16b^2$

34 $25a^2 - 20ab + 4b^2$

35 $49c^2 + 42cd + 9d^2$

36 $25d^2 - 40df + 16f^2$

37 $100x^2 + 220xy + 121y^2$

38 $81a^2 - 198ab + 121b^2$

39 $400r^2 + 280rs + 49s^2$

40 $900c^2 - 780cd + 169d^2$

41 $12x^2 + 7xy - 12y^2$

42 $12a^2 - 8ab - 15b^2$

43 $24y^2 - 50yz - 9z^2$

44 $18x^2 + 19xz - 12z^2$

45 $18a^2 + 9ab - 20b^2$

46 $28c^2 + 5cd - 12d^2$

47 $15x^2 - 44xy - 20y^2$

48 $24a^2 - 11ab - 18b^2$

49 $12x^2 + 43x + 36$

50 $12a^2 + 56a + 9$

51 $8a^2 + 99ab + 36b^2$

52 $8x^2 + 41xy + 36y^2$

53 $5c^2 - 36cd + 36d^2$

54 $9r^2 - 17rs + 8s^2$

55 $9a^2 - 18ab + 8b^2$

56 $6x^2 - 25xy + 21y^2$

57 $x^2 + y^2 + 1 + 2xy + 2x + 2y = (x + y)^2 + 2(x + y) + 1$

58 $a^2 + 2ab + b^2 + 2c(a + b) + c^2$ **59** $a^2 + b^2 + 1 - 2ab + 2a - 2b$

60 $x^2 + y^2 + 1 - 2xy - 2x + 2y$ **61** $(x - 3)^2 + 3(x - 3) + 2$

62 $(y - 4)^2 - 5(y - 4) + 6$ **63** $(b + 1)^2 - 3(b + 1) - 10$

64 $(a + 3)^2 - 2(a + 3) - 3$ **65** $(a + b)^2 + (a + b) - 2$

66 $(x + y)^2 + 3z(x + y) - 10z^2$ **67** $(a - 2b)^2 - 2c(a - 2b) - 8c^2$

68 $(x + 3y)^2 + z(x + 3y) - 12z^2$

3.6
FACTORS OF A BINOMIAL

The Difference of Two Squares

If we interchange the members of Formula (3.4), we obtain

■ $a^2 - b^2 = (a + b)(a - b)$ ⠀⠀⠀⠀⠀⠀⠀⠀⠀⠀⠀⠀⠀⠀⠀⠀⠀⠀⠀⠀(3.6)

Consequently we have the following rule for factoring the difference of the squares of two numbers:

■ The difference of the squares of two numbers is equal to the product of the sum and the difference of the two numbers.

We shall illustrate the application of this rule with the following example.

EXAMPLE 1   Factor

(1): $49a^2 - 16b^2$ ⠀⠀(2): $(a + 3b)^2 - 4$ ⠀⠀and⠀⠀   (3): $x^2 - (y + z)^2$

SOLUTION

$$49a^2 - 16b^2 = (7a)^2 - (4b)^2$$

$$= (7a + 4b)(7a - 4b) \tag{1}$$

$$(a + 3b)^2 - 4 = (a + 3b)^2 - 2^2$$
$$= (a + 3b + 2)(a + 3b - 2) \tag{2}$$

$$x^2 - (y + z)^2 = [x + (y + z)][x - (y + z)]$$
$$= (x + y + z)(x - y - z) \tag{3}$$

The Sum and Difference of Two Cubes

The sum and difference of two cubes can be expressed as $x^3 + y^3$ and $x^3 - y^3$, respectively. If we divide $x^3 + y^3$ by $x + y$ by the method of Sec. 2.13, we get $x^2 - xy + y^3$ as a quotient. Hence,

■ $\quad x^3 + y^3 = (x + y)(x^2 - xy + y^2) \tag{3.7}$

Similarly,

■ $\quad x^3 - y^3 = (x - y)(x^2 + xy + y^2) \tag{3.8}$

Hence, we have the following two rules:

■ If a binomial is expressed as the sum of the cubes of two numbers, one factor is the sum of the two numbers. The other factor is the square of the first number minus the product of the two numbers plus the square of the second number.

■ If a binomial is expressed as the difference of the cubes of two numbers, one factor is the difference of the two numbers. The other factor is the square of the first number plus the product of the two numbers plus the square of the second number.

EXAMPLE 2 Factor

(1): $8x^3 + 27y^3$ (2): $27a^3 - 64b^3$

SOLUTION

$$8x^3 + 27y^3 = (2x)^3 + (3y)^3$$
$$= (2x + 3y)[(2x)^2 - (2x)(3y) + (3y)^2]$$
$$= (2x + 3y)(4x^2 - 6xy + 9y^2) \tag{1}$$

$$27a^3 - 64b^6 = (3a)^3 - (4b^2)^3$$
$$= (3a - 4b^2)[(3a)^2 + (3a)(4b^2) + (4b^2)^2]$$
$$= (3a - 4b^2)(9a^2 + 12ab^2 + 16b^4) \tag{2}$$

Note 1 In the case of the sum of two cubes, the sign between the two terms of the first factor is plus, and the sign of the middle term of the second factor is minus.

Note 2   In the case of the difference of two cubes, the sign between the two terms of the first factor is minus, and the sign of the middle term of the second factor is plus.

Note 3   In each case, the middle term of the second factor is the product of the two terms of the first factor (not twice the product).

Frequently, the factors obtained by use of Formulas (3.6), (3.7), and (3.8) can be further factored by a repeated application of one or more of these formulas.

EXAMPLE 3   Factor

$$(1):\quad x^6 - y^6 \qquad (2):\quad a^8 - y^8$$

SOLUTION

$$
\begin{aligned}
x^6 - y^6 &= (x^3)^2 - (y^3)^2 \\
&= (x^3 - y^3)(x^3 + y^3) && \text{by (3.6)} \\
&= (x - y)(x^2 + xy + y^2)(x + y)(x^2 - xy + y^2) && \text{by (3.8) and (3.7)} \qquad (1)
\end{aligned}
$$

$$
\begin{aligned}
x^8 - y^8 &= (x^4)^2 - (y^4)^2 \\
&= (x^4 - y^4)(x^4 + y^4) && \text{by (3.6)} \\
&= (x^2 - y^2)(x^2 + y^2)(x^4 + y^4) && \text{by (3.6)} \\
&= (x - y)(x + y)(x^2 + y^2)(x^4 + y^4) && \text{by (3.6)} \qquad (2)
\end{aligned}
$$

EXERCISE 3.3
Factoring Binomials

Find the factors of the binomials in Probs. 1 to 20 by use of (3.6).

1 $a^2 - 9$	**2** $x^2 - 16$	**3** $c^2 - 36$
4 $y^2 - 49$	**5** $4a^2 - 9b^2$	**6** $9x^2 - 16y^2$
7 $25c^2 - 49d^2$	**8** $4h^2 - 81k^2$	**9** $9x^4 - 64y^2$
10 $16a^6 - 25b^4$	**11** $49c^8 - 4c^6$	**12** $36x^{10} - 25y^4$
13 $121h^{12} - 4k^6$	**14** $100r^{10} - 49r^4$	**15** $64c^8 - 25d^{10}$
16 $144a^{12} - 121b^6$	**17** $81x^{18} - 16y^8$	**18** $625h^{10} - 16k^2$
19 $36r^6 - 121s^8$	**20** $64a^{16} - 49b^{14}$	

Factor the binomials in Probs. 21 to 40 by use of (3.7) and (3.8).

21 $a^3 - 1$	**22** $b^3 + 1$	**23** $c^3 - 64$
24 $x^3 + 27$	**25** $8a^3 + b^3$	**26** $8x^3 - 27y^3$
27 $27c^3 + 64d^3$	**28** $125a^3 - b^3$	**29** $x^6 + 1$
30 $8a^6 - 1$	**31** $c^3 - 8d^9$	**32** $x^3 + 27y^{12}$

33 $8a^6 - 27b^{12}$ **34** $125r^{15} + 8s^6$ **35** $64h^{12} - 27k^9$

36 $216x^{18} + y^3$ **37** $64a^{15} - 125b^3$ **38** $27c^{12} + 243b^9$

39 $512r^{24} - 27s^6$ **40** $343x^{15} + 216y^9$

Factor the expressions in Probs. 41 to 52.

41 $x^2 + 2xy + y^2 - 4$ **42** $a^2 + 2ab + b^2 - 9$

43 $b^2 + 4b + 4 - c^2$ **44** $4x^2 + 4x + 1 - y^2$

45 $x^2 - 6xy + 9 - z^2$ **46** $4a^2 - 12ab + 9b^2 - 4c^2$

47 $25c^2 - 20cb + 4b^2 - 9d^2$ **48** $36x^2 - 12xy + y^2 - 16z^2$

49 $9 - a^2 - 2ab - b^2$ **50** $4 - r^2 + 2rs - s^2$

51 $x^2 - y^2 - 4yz - 4z^2$ **52** $c^2 - 4a^2 + 4ab - b^2$

Factor the following binomials by repeated applications of (3.6), (3.7), and (3.8).

53 $a^4 - b^4$ **54** $16c^4 - 1$ **55** $81x^4 - 16y^4$

56 $16a^8 - 1$ **57** $x^6 - 1$ **58** $a^9 + b^9$

59 $x^{12} - 1$ **60** $a^{12} - b^6$

3.7
COMMON FACTORS

Each term of a polynomial may be divisible by the same monomial. This monomial is called the *common factor*. Such a polynomial can be factored by expressing it as the product of the common factor and the sum of the quotients obtained by dividing each term of the polynomial by the common factor. This procedure is justified by the distributive axiom. If the factors thus obtained are not prime. we continue factoring by use of one or more of the methods previously discussed. For example,

$$ax + ay - az = a(x + y - z)$$

$$6a^3b + 3a^2b^2 - 18ab^3 = 3ab(2a^2 + ab - 6b^2)$$
$$= 3ab(2a - 3b)(a + 2b)$$

This method can be extended to include polynomials in which the terms are polynomials that have a common factor that is not a monomial. For example,

$$(a + b)(a - b) + 2(a + b) = (a + b)(a - b + 2)$$

$$(x - 1)(x + 2) - (x - 1)(2x - 3) = (x - 1)[(x + 2) - (2x - 3)]$$
$$= (x - 1)(x + 2 - 2x + 3)$$
$$= (x - 1)(-x + 5)$$

3.8

FACTORING BY GROUPING

Frequently, the terms of a polynomial can be grouped in such a way that each group has a common factor, and then the method of common factors can be applied. We shall illustrate the method by three examples.

EXAMPLE 1

Factor $ax + bx - ay - by$.

SOLUTION We notice that the first two terms have the common factor x and the third and fourth have the common factor y. Hence, we group the terms in this way,

$$(ax + bx) - (ay + by)$$

and then proceed as follows:

$$ax + bx - ay - by = (ax + bx) - (ay + by)$$
$$= x(a + b) - y(a + b)$$
$$= (a + b)(x - y)$$

with x as common factor of first group, y as common factor of second group and with $a + b$ as common factor of $x(a + b)$ and $y(a + b)$

EXAMPLE 2

Factor $a^2 + ab - 2b^2 + 2a - 2b$.

SOLUTION Since

$$a^2 + ab - 2b^2 = (a + 2b)(a - b) \qquad \text{and} \qquad 2a - 2b = 2(a - b)$$

we proceed as indicated here:

$$a^2 + ab - 2b^2 + 2a - 2b = (a^2 + ab - 2b^2) + (2a - 2b)$$
$$= (a + 2b)(a - b) + 2(a - b)$$
$$= (a - b)(a + 2b + 2) \qquad \text{with } a - b \text{ as common factor}$$

EXAMPLE 3

Factor $4c^2 - a^2 + 2ab - b^2$.

SOLUTION

$$4c^2 - a^2 + 2ab - b^2 = 4c^2 - (a^2 - 2ab + b^2)$$
$$= (2c)^2 - (a - b)^2 \qquad \text{by (3.3)}$$
$$= [2c + (a - b)][2c - (a - b)] \qquad \text{by (3.6)}$$
$$= (2c + a - b)(2c - a + b)$$

TRINOMIALS REDUCIBLE TO THE DIFFERENCE OF TWO SQUARES

Frequently, it is possible to convert a trinomial into the difference of two squares by adding and subtracting a monomial that is a perfect square. For example, if we add $4a^2b^2$ to $a^4 + 2a^2b^2 + 9b^4$ and then subtract $4a^2b^2$, we have

$$a^4 + 2a^2b^2 + 9b^4 = a^4 + 2a^2b^2 + 9b^4 + 4a^2b^2 - 4a^2b^2$$

$$= a^4 + 6a^2b^2 + 9b^4 - 4a^2b^2$$

$$= (a^2 + 3b^2)^2 - (2ab)^2$$

This process is possible only when the trinomial becomes a perfect square when a perfect-square monomial is *added* to it.

After a trinomial has been converted into the difference of two squares, it can be factored by use of (3.6).

EXAMPLE
Factor $4x^4 - 21x^2y^2 + 9y^4$.

SOLUTION   In this trinomial, $4x^4$ and $9y^4$ are perfect squares, and twice the product of their square roots is

$$2(2x^2)(3y^2) = 12x^2y^2$$

Hence, the trinomial becomes a perfect square if we add $9x^2y^2$ to it. Then we must also subtract $9x^2y^2$ and get

$$4x^4 - 21x^2y^2 + 9y^4 = 4x^4 - 21x^2y^2 + 9y^4 + 9x^2y^2 - 9x^2y^2$$

$$= 4x^4 - 12x^2y^2 + 9y^4 - 9x^2y^2$$

$$= (2x^2 - 3y^2)^2 - (3xy)^2$$

$$= [(2x^2 - 3y^2) + 3xy][(2x^2 - 3y^2) - 3xy] \quad \text{by (3.6)}$$

$$= (2x^2 + 3xy - 3y^2)(2x^2 - 3xy - 3y^2)$$

EXERCISE 3.4
Factoring by Grouping

Factor the expressions in Probs. 1 to 60 by the methods of Secs. 3.7 and 3.8.

1 $3a + 12$ **2** $6x - 9y$ **3** $5a - 10b$

4 $7x - 21y$ **5** $a^2 + 3ab$ **6** $8xy - 10xz$

7 $x^2y - xy^2 + xyz$ **8** $2a^2b - 8ab^2 + 6abc$

9 $3a^2 - 3b^2$

10 $5x^2 - 20y^2$

11 $7c^2 - 63d^2$

12 $2x^4 - 2$

13 $a^3 - 4ab^2$

14 $2x^3 - 18xy^2$

15 $12c^4 - 3c^2d^2$

16 $x^4y - x^2y^3$

17 $2a^3 - 16b^3$

18 $81 + 3x^3$

19 $a^4 - ab^3$

20 $x^4y + xy^4$

21 $3x^2 + 3xy - 6y^2$

22 $10a^2 + 15ab - 10b^2$

23 $6x^2 + 24xy + 24y^2$

24 $20a^2 - 60ab + 45b^2$

25 $8c^3 - 2c^2d - cd^2$

26 $9x^3 + 3x^2y - 30xy^2$

27 $20c^2d + 4cd^2 - 24d^3$

28 $24a^3 - 2a^2d - 2ad^2$

29 $ab + ac + bd + cd$

30 $xy + 2x + 3y + 6$

31 $cx + cy + 2dx + 2dy$

32 $3xw + 3xz + 2yw + 2yz$

33 $4ac + 8ad - bc - 2bd$

34 $15rt + 10ru - 3st - 6su$

35 $6cd - 15ce + 8fd - 20fe$

36 $12ax + 10ay - 6bx - 5by$

37 $x^2 - xz - xy + yz$

38 $2a^2 - 4ac - ab + 2bc$

39 $3c^2 - 3ce - 4dc + 4de$

40 $10x^2 - 5xz - 4xy + 2yz$

41 $a^2 - b^2 + a - b$

42 $x^2 - z^2 - 2x - 2z$

43 $x^2 - 4y^2 + x + 2y$

44 $4c^2 - d^2 - 6c + 3d$

45 $a^2 - ab - 2b^2 + 2a + 2b$

46 $x^2 + 2xy - 3y^2 - x + y$

47 $c^2 - 2cd - 3d^2 - 3c - 3d$

48 $a^2 - 2ac - 3c^2 + ad - 3cd$

Convert each of the following trinomials to the difference of two squares, and then obtain the factors.

49 $a^4 - 3a^2 + 1$

50 $x^4 - 5x^2 + 4$

51 $y^4 + 5y^2 + 9$

52 $b^4 + 4$

53 $a^4 + a^2b^2 + b^4$

54 $x^4 - 8x^2y^2 + 4y^4$

55 $c^4 + 3c^2d^2 + 4d^4$

56 $4b^4 - 8b^2c^2 + c^4$

57 $4x^4 + 11x^2y^2 + 9y^4$

58 $9m^4 - 7m^2n^2 + n^4$

59 $a^4 - 12a^2b^2 + 16b^4$

60 $4c^4 - 16c^2d^2 + 9d^4$

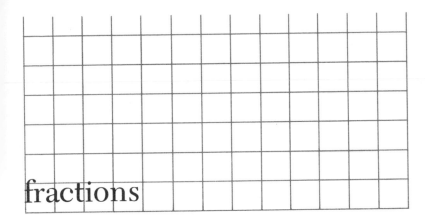

fractions

Fractions such as $\frac{1}{2}, \frac{3}{4}, \frac{2}{3}$, and $\frac{5}{7}$ commonly occur in arithmetic and are used constantly in everyday living. Algebraic fractions are equally important in mathematics and in all fields in which algebra is applied. Skill in the operations that involve fractions is essential for progress in any of these fields. In this chapter we consider the basic operations dealing with fractions.

4.1
DEFINITIONS AND FUNDAMENTAL PRINCIPLE

Fraction, Numerator, Denominator

In Chap. 2 we defined the number a/b as the quotient of a and b. We now define a/b as a fraction:

■ A *fraction* is a number of the type a/b, where $b \neq 0$. The number a is the *numerator* of the fraction, and b is the *denominator*.

We frequently refer to the numerator and denominator as *members* of the fraction.

The letters a and b in the definition may stand for integers, irrational numbers, monomials, or other algebraic expressions. Hence,

$$\frac{4a}{7}, \quad \frac{3x^2}{2xy}, \quad \frac{c^2 - d^2}{c^2 + d^2}, \quad \text{and} \quad \frac{(x+y)(x^2 + xy + y^4)}{(x-y)(x^2 + x^2y + y^2)}$$

are examples of fractions.

In Chap. 2 we proved theorems (2.41) and (2.42), which we shall use repeatedly in this chapter:

■ $\quad \dfrac{a}{b} = a\left(\dfrac{1}{b}\right) = \dfrac{1}{b}(a) \qquad b \neq 0$ (2.41)

■ $\quad \dfrac{ab}{cd} = \dfrac{a}{c} \cdot \dfrac{b}{d} \qquad c \neq 0, d \neq 0$ (2.42)

Equality of Fractions

We also defined the reciprocal of the nonzero number a as the number $1/a$ such that

■ $\quad a \cdot \dfrac{1}{a} = \dfrac{1}{a} \cdot a = 1$ (2.35)

We now state and prove a theorem that establishes the condition under which two fractions are equal.

■ If $n \neq 0$ and $q \neq 0$, $\quad \dfrac{m}{n} = \dfrac{p}{q} \quad$ if and only if $\quad mq = np$ (4.1)

PROOF We first assume that $m/n = p/q$ and prove that $mq = np$.

$$\frac{m}{n} = \frac{p}{q} \qquad \text{assumed}$$

$$n \cdot q \cdot \frac{m}{n} = n \cdot q \cdot \frac{p}{q} \qquad \text{by multiplicativity axiom for equalities (2.5)}$$

$$n \cdot q \cdot m \cdot \frac{1}{n} = n \cdot q \cdot p \cdot \frac{1}{q} \qquad \text{by (2.41)}$$

$$n \cdot \frac{1}{n} \cdot q \cdot m = q \cdot \frac{1}{q} \cdot n \cdot p \qquad \text{by commutative axiom (2.28)}$$

$$\left(n \cdot \frac{1}{n}\right)(q \cdot m) = \left(q \cdot \frac{1}{q}\right)(n \cdot p) \qquad \text{by associative axiom (2.29)}$$

$$1 \cdot qm = 1 \cdot np \qquad \text{since } n \cdot \frac{1}{n} = q \cdot \frac{1}{q} = 1 \text{ by (2.35)}$$

$$qm = np \qquad \text{since } 1 \cdot qm = qm \text{ and } 1 \cdot np = np$$

Now we assume that $mq = np$ and prove that $m/n = p/q$.

$$mq = np \qquad \text{assumed}$$

$$\frac{1}{nq}(mq) = \frac{1}{nq}(np) \qquad \text{by multiplicative axiom for equalities}$$

$$\frac{mq}{nq} = \frac{np}{nq} \qquad \text{by (2.41)}$$

$$\frac{m}{n} \cdot \frac{q}{q} = \frac{n}{n} \cdot \frac{p}{q} \qquad \text{by (2.42)}$$

$$\frac{m}{n} = \frac{p}{q} \qquad \text{since } q/q = n/n = 1$$

Fundamental Principal of Fractions

We are now in position to prove the fundametal principle of fractions that is stated as follows:

■   If the numerator and denominator of a fraction are multiplied or divided by the same nonzero number, the resulting fraction is equal to the given fraction.

If we express this statement as a formula, we have

$$\frac{m}{n} = \frac{mf}{nf} = \frac{m/d}{n/d} \qquad f \neq 0, \quad n \neq 0, \quad d \neq 0 \tag{4.2}$$

PROOF   We consider the fraction m/n, multiply the numerator and denominator by the same nonzero number f, and prove that

$$\frac{m}{n} = \frac{mf}{nf} \tag{1}$$

If in (4.1) we replace p by mf and q by nf, we have $m/n = mf/nf$ if and only if $mnf = nmf$. Furthermore, the products mnf and nmf are equal by the commutative axiom. Hence statement (1) is true.

If we divide m and n by d, we have $\dfrac{m/d}{n/d}$. Now, if in (1) we replace f by $1/d$, we have

$$\frac{m}{n} = \frac{m \cdot \dfrac{1}{d}}{n \cdot \dfrac{1}{d}} = \frac{\dfrac{m}{d}}{\dfrac{n}{d}} \qquad \text{by (2.41)}$$

This completes the proof.

EXAMPLE 1

$$\frac{3x}{4y} = \frac{3x \cdot 2x^2 y^3}{4y \cdot 2x^2 y^3} = \frac{6x^3 y^3}{8x^2 y^4}$$

EXAMPLE 2

$$\frac{a}{b} = \frac{a(x+y)}{b(x+y)}$$

EXAMPLE 3

$$\frac{15x^3 y^7}{25x^5 y^2} = \frac{15x^3 y^7 \div 5x^3 y^2}{25x^5 y^2 \div 5x^3 y^2} = \frac{3x^{3-3} y^{7-2}}{5x^{5-3} y^{2-2}}$$

$$= \frac{3x^0 y^5}{5x^2 y^0}$$

$$= \frac{3y^5}{5x^2} \qquad \text{since } x^0 = y^0 = 1$$

There are three signs associated with a fraction: the sign preceding the fraction, the sign preceding the numerator, and the sign preceding the denominator. For example,

$$+\frac{+n}{+d} \qquad -\frac{+n}{-d} \qquad +\frac{-n}{-d}$$

We shall prove that if two of these signs in a given fraction are changed, the resulting fraction is equal to the given fraction.

PROOF

$$\frac{n}{d} = \frac{-1 \cdot n}{-1 \cdot d} = \frac{-n}{-d} \qquad \text{by (4.2)}$$

Furthermore,

$$\frac{n}{d} = -(-1)\frac{n}{d} \qquad \text{since } -(-1) = 1 \text{ by (2.22)}$$

$$= -\left(\frac{-1}{1} \cdot \frac{n}{d}\right) \qquad \text{since } -1/1 = -1 \text{ by (2.40)}$$

$$= -\frac{-n}{d}$$

Similarly, we can prove that $n/d = -n/-d$. Consequently, we have

$$\blacksquare \qquad \frac{n}{d} = \frac{-n}{-d} = -\frac{-n}{d} = -\frac{n}{-d} \qquad (4.3)$$

EXAMPLE 4

75

4.2 Conversion of
Fractions

$$\frac{-x}{y-x} = \frac{-(-x)}{-(y-x)} = \frac{x}{x-y}$$

EXAMPLE 5

$$\frac{y^3 - x^3}{x - y} = -\frac{-(y^3 - x^3)}{x - y} = -\frac{x^3 - y^3}{x - y}$$

EXAMPLE 6

$$\frac{x^2 - y^2}{y - x} = -\frac{x^2 - y^2}{-(y - x)} = -\frac{x^2 - y^2}{x - y}$$

4.2
CONVERSION OF FRACTIONS

Lowest Terms

Usually, fractions that are obtained as the result of operations are expressed in *lowest terms*. A fraction is said to be in lowest terms if the members have no common factor. For the present we shall require that the common factor be an integer, a monomial with an integral numerical coefficient, or a polynomial with integral numerical coefficients. In many operations we are required to convert a fraction to an equal fraction with a denominator in a specified form. We accomplish these conversions by use of the fundamental principle of fractions.

EXAMPLE 1   Convert

$$\frac{a + b}{a - b}$$

to an equal fraction with $a^2 - b^2$ as the denominator.

SOLUTION   We multiply each member by $a + b$ and obtain

$$\frac{a + b}{a - b} = \frac{(a + b)(a + b)}{(a - b)(a + b)} = \frac{a^2 + 2ab + b^2}{a^2 - b^2}$$

EXAMPLE 2   Convert each of the fractions

$$\frac{a}{xy}, \quad \frac{3b}{x^2y}, \quad \text{and} \quad \frac{2c}{xy^2}$$

to an equal fraction having x^2y^2 as the denominator.

SOLUTION   We multiply the members of the first by xy, the members of the second by y, and the members of the third by x and obtain

$$\frac{axy}{x^2y^2}, \quad \frac{3by}{x^2y^2}, \quad \text{and} \quad \frac{2cx}{x^2y^2}$$

To reduce a fraction to lowest terms, we divide the numerator and denominator by every factor that is common to both. If the members of the fractions are polynomials, it is advisable to factor each as a first step in the reduction.

EXAMPLE 3   Reduce to lowest terms

$$\frac{27a^2bc^4}{36a^3b^2c^3}$$

SOLUTION   We divide each member by $9a^2bc^3$, the factor that is common to both members, and get

$$\frac{27a^2bc^4}{36a^3b^2c^3} = \frac{3c}{4ab}$$

EXAMPLE 4   Reduce to lowest terms

$$\frac{x^3 - 2x^2y + xy^2}{x^3 - xy^2}$$

SOLUTION

$$\frac{x^3 - 2x^2y + xy^2}{x^3 - xy^2} = \frac{x(x-y)(x-y)}{x(x-y)(x+y)} \qquad \text{factoring numerator and denominator}$$

$$= \frac{x-y}{x+y} \qquad \text{dividing each member by } x(x-y)$$

EXERCISE 4.1
Conversion of Fractions

Express the fraction in each of Probs. 1 to 28 as an equal fraction that has the second expression as a denominator.

1   $\dfrac{4}{2-x}, \ x-2$

2   $\dfrac{x+y}{y-x}, \ x-y$

3   $\dfrac{y-3x}{x-3y}, \ 3y-x$

4   $\dfrac{-cd}{d-c}, \ c-d$

5   $\dfrac{x+y}{-x(x-y)}, \ x^2-xy$

6   $\dfrac{a-2b}{-a(a-b)}, \ a^2-ab$

7   $\dfrac{a+b}{b^2-a^2},\ a^2-b^2$ 8 $\dfrac{x-2}{8-x^3},\ x^3-8$

9   $\dfrac{2a}{3b^2},\ 15b^2$ 10 $\dfrac{3a}{4b^2},\ 12cb^3$

11   $\dfrac{4x}{5y^2},\ 20x^2y^3$ 12 $\dfrac{3ab}{7cd},\ 21ac^2d^2$

13   $\dfrac{a+b}{a-b},\ a^2-b^2$ 14 $\dfrac{2a+b}{a+3b},\ a^2-9b^2$

15   $\dfrac{x-6}{x+3},\ x^2-9$ 16 $\dfrac{3c-d}{2c-d},\ 4c^2-d^2$

17   $\dfrac{a-b}{a+b},\ a^2+2ab+b^2$ 18 $\dfrac{x+2y}{x-2y},\ x^2-4xy+4y^2$

19   $\dfrac{x-y}{x^2+xy+y^2},\ x^3-y^3$ 20 $\dfrac{a-1}{a^2-a+1},\ a^3+1$

21   $\dfrac{15xy^2}{18x^3y^4},\ 6x^2y^2$ 22 $\dfrac{12a^2b}{18a^4b^2},\ 3a^2b$

23   $\dfrac{32x^2y^3z^4}{48x^5y^4z^5},\ 3x^3yz$ 24 $\dfrac{35a^2bc}{56a^4b^2c},\ 8a^2b$

25   $\dfrac{(x-3)(x+5)}{(x-3)(x+6)},\ x+6$ 26 $\dfrac{(2a-5b)(a+b)}{(a-b)(2a-5b)},\ a-b$

27   $\dfrac{x^2-4}{x^2-x-2},\ x+1$ 28 $\dfrac{a^2-4b^2}{a^2-ab-2b^2},\ a+b$

Reduce the following fractions to lowest terms.

29   $\dfrac{16x^2y^2}{24x^3y}$ 30 $\dfrac{14x^3y}{21xy^3}$ 31 $\dfrac{36c^4d^3e}{48c^5d^2e^3}$

32   $\dfrac{24p^3qr^3}{84p^3q^2r^2}$ 33 $\dfrac{xy-x}{y^2-1}$ 34 $\dfrac{ab-3a}{b^2-9}$

35   $\dfrac{xyz-2xy}{z^2-4}$ 36 $\dfrac{a^2bc-3a^2b}{c^2-9}$ 37 $\dfrac{2a-b}{2a^2+ab-b^2}$

38   $\dfrac{2xy-x^2}{x^2-4xy+4y^2}$ 39 $\dfrac{3x^2+xy-2y^2}{x^2+xy}$ 40 $\dfrac{c^2-5c+6}{c^2-4}$

41   $\dfrac{x^2+x-6}{x^2+5x+6}$ 42 $\dfrac{x^2+2x-8}{x^2+6x+8}$

43   $\dfrac{a^2+ab-2b^2}{2a^2-ab-b^2}$ 44 $\dfrac{a^2-ab-12b^2}{a^2-3ab-4b^2}$

45   $\dfrac{(r-s)(2r^2+3rs+s^2)}{(2r+s)(r^2+rs-2s^2)}$ 46 $\dfrac{(x-2y)(2x^2+9xy+4y^2)}{(x+4y)(x^2-5xy+6y^2)}$

47   $\dfrac{(m+3n)(m^2-mn-2n^2)}{(m+n)(2m^2+7mn+3n^2)}$ 48 $\dfrac{(a-b)(a^2+7ab+10b^2)}{(a+5b)(a^2+2ab-3b^2)}$

49   $\dfrac{2ax+2ay+bx+by}{2ay-2ax+by-bx}$ 50 $\dfrac{2ac+bc-6ad-3bd}{4ac+2bc+2ad+bd}$

51 $\dfrac{3wy - 6wz - xy + 2xz}{6wy - 2xy + 3wz - xz}$

52 $\dfrac{rt - ru + 2st - 2su}{3rt - st + su - 3ru}$

53 $\dfrac{x^3 + y^3}{x^2 - y^2}$

54 $\dfrac{x^6 - y^6}{x^4 - y^4}$

55 $\dfrac{a^2 - ab + b^2}{a^3 + b^3}$

56 $\dfrac{c^2 + 2c + 4}{c^3 - 8}$

57 $\dfrac{a - 1}{(2a - 1)a - 1}$

58 $\dfrac{x + 2}{(x + 3)x + 2}$

59 $\dfrac{b - 2}{(3b - 5)b - 2}$

60 $\dfrac{y - 3}{(2y - 5)y - 3}$

In each of the following problems, write three more fractions that are equal to the given fraction by making only sign changes.

61 $\dfrac{x + 3}{x - 4}$

62 $\dfrac{a + 1}{a - 3}$

63 $\dfrac{a + b}{b - a}$

64 $\dfrac{x^2 - x + 2}{2x^2 + x - 1}$

4.3
MULTIPLICATION OF FRACTIONS

When we read Theorem (2.42) from right to left, we have the following rule for obtaining the product of two fractions:

Product of Two or More Fractions

■ $\quad \dfrac{a}{c} \cdot \dfrac{b}{d} = \dfrac{ab}{cd}$ (4.4)

By using the associative axiom (2.29) and two applications of (4.4), we can obtain the product of three or more fractions as follows:

$$\frac{a}{b} \cdot \frac{n}{d} \cdot \frac{r}{s} = \left(\frac{a}{b} \cdot \frac{n}{d} \right) \cdot \frac{r}{s} \qquad \text{by associative axiom (2.29)}$$

$$= \left(\frac{an}{bd} \right) \cdot \frac{r}{s} \qquad \text{by (4.4)}$$

$$= \frac{anr}{bds} \qquad \text{by (4.4)}$$

This rule may be extended by a similar argument to include the product of any given number of fractions; the extended rule is stated as follows:

■ The product of two or more fractions is a fraction whose numerator is the product of the numerators and whose denominator is the product of the denominators.

The application of this rule is illustrated by the following example.

EXAMPLE 1 Obtain the product of

$$\frac{a}{b}, \quad \frac{c}{d}, \quad \text{and} \quad \frac{x-y}{x+y}$$

SOLUTION

$$\frac{a}{b} \cdot \frac{c}{d} \cdot \frac{x-y}{x+y} = \frac{ac(x-y)}{bd(x+y)}$$

Frequently, the numerator and denominator of a product have a common factor. In such cases the fraction should be reduced to lowest terms by dividing both members by the common factor. It is advisable to factor the numerators and denominators of the fractions, if possible, before the final product is written; then the factors that are common to the numerator and denominator of the product can be more easily detected.

EXAMPLE 2 Obtain the product of

$$\frac{2}{3}, \quad \frac{6}{7}, \quad \text{and} \quad \frac{21}{8}$$

SOLUTION

$$\frac{2}{3} \cdot \frac{6}{7} \cdot \frac{21}{8} = \frac{2}{3} \cdot \frac{3 \cdot 2}{7} \cdot \frac{3 \cdot 7}{2 \cdot 2 \cdot 2} \qquad \text{factoring 6, 21, and 8}$$

$$= \frac{2 \cdot 2 \cdot 3 \cdot 3 \cdot 7}{2 \cdot 2 \cdot 2 \cdot 3 \cdot 7} \qquad \text{by (2.28) and (4.4)}$$

$$= \frac{3}{2} \qquad \qquad \text{dividing numerator and denominator by } 2 \cdot 2 \cdot 3 \cdot 7$$

EXAMPLE 3 Obtain the product of

$$\frac{a^2 - 4b^2}{2a^2 - 7ab + 3b^2}, \quad \frac{6a - 3b}{2a + 4b}, \quad \text{and} \quad \frac{a^2 - 4ab + 3b^2}{a^2 - ab - 2b^2}$$

SOLUTION

$$\frac{a^2 - 4b^2}{2a^2 - 7ab + 3b^2} \cdot \frac{6a - 3b}{2a + 4b} \cdot \frac{a^2 - 4ab + 3b^2}{a^2 - ab - 2b^2}$$

$$= \frac{(a - 2b)(a + 2b)}{(2a - b)(a - 3b)} \cdot \frac{3(2a - b)}{2(a + 2b)} \cdot \frac{(a - b)(a - 3b)}{(a + b)(a - 2b)} \qquad \text{factoring}$$

$$= \frac{3(a - 2b)(a + 2b)(2a - b)(a - 3b)(a - b)}{2(a - 2b)(a + 2b)(2a - b)(a - 3b)(a + b)} \qquad \text{by (4.4) and commutative axiom (2.28)}$$

$$= \frac{3(a - b)}{2(a + b)} \qquad \qquad \text{dividing numerator and denominator by } (a - 2b)(a + 2b)(2a - b)(a - 3b)$$

4.4

DIVISION OF FRACTIONS

By the definition of the quotient in Sec. 2.11, $a \div b = x$ if $bx = a$. Now, if $a = n/d$ and $b = q/p$, then $n/d \div q/p$ is the number x such that

$$\frac{q}{p}(x) = \frac{n}{d}$$

Since

$$\frac{q}{p} \cdot \frac{p}{q} = \frac{qp}{pq} = 1$$

we may determine x by use of the multiplicative axiom for equalities (2.5) with $a = q/p$, $b = n/d$, and $c = p/q$ and get

$$\frac{q}{p} \cdot \frac{p}{q}(x) = \frac{n}{d} \cdot \frac{p}{q} \qquad \text{by commutative axioms (2.28) and (2.5)}$$

$$x = \frac{n}{d} \cdot \frac{p}{q} \qquad \text{since } (q/p)(p/q) = 1$$

Consequently, since $n/d \div q/p = x$, we have

■ $\quad \dfrac{n}{d} \div \dfrac{q}{p} = \dfrac{n}{d} \cdot \dfrac{p}{q}$ $\hfill$ (4.5)

Therefore, to obtain the quotient of two fractions, multiply the dividend by the reciprocal of the divisor.

EXAMPLE 1 Divide

$$\frac{3x^2}{4a} \qquad \text{by} \qquad \frac{6x^3}{5a^2}$$

SOLUTION

$$\frac{3x^2}{4a} \div \frac{6x^3}{5a^2} = \frac{3x^2}{4a} \cdot \frac{5a^2}{6x^3} = \frac{15a^2x^2}{24ax^3} = \frac{5a}{8x}$$

EXAMPLE 2 Divide

$$\frac{x^2 - y^2}{x + 3y} \qquad \text{by} \qquad \frac{x - y}{x^2 + 3xy}$$

$$\frac{x^2 - y^2}{x + 3y} \div \frac{x - y}{x^2 + 3xy} = \frac{x^2 - y^2}{x + 3y} \cdot \frac{x^2 + 3xy}{x - y}$$ inverting divisor and multiplying

$$= \frac{(x^2 - y^2)(x^2 + 3xy)}{(x + 3y)(x - y)}$$ by (4.4)

$$= \frac{(x - y)(x + y)(x)(x + 3y)}{(x + 3y)(x - y)}$$ factoring

$$= (x + y)x$$

$$= x^2 + xy$$ dividing numerator and
denominator by $(x + 3y)(x - y)$

EXERCISE 4.2
Multiplication and Division of Fractions

Perform the operations indicated in the following problems.

1 $\frac{3}{4} \cdot \frac{8}{9} \cdot \frac{2}{3}$

2 $\frac{4}{5} \cdot \frac{10}{3} \cdot \frac{9}{8}$

3 $\frac{2}{7} \cdot \frac{21}{4} \cdot \frac{5}{6}$

4 $\frac{8}{3} \cdot \frac{12}{5} \cdot \frac{5}{24}$

5 $\frac{3}{7} \cdot \frac{28}{15} \div \frac{3}{5}$

6 $\frac{12}{5} \cdot \frac{15}{2} \div \frac{10}{3}$

7 $\frac{16}{3} \cdot \frac{9}{64} \cdot \frac{12}{5} \div \frac{6}{5}$

8 $\frac{8}{7} \cdot \frac{5}{4} \cdot \frac{7}{15} \div \frac{2}{3}$

9 $\frac{2a^3 b}{3ab^3} \cdot \frac{5a^2 b^4}{2a^4 b^2}$

10 $\frac{3x^3 y^5}{10x^4 y} \cdot \frac{5x^2 y^4}{9xy^3}$

11 $\frac{4c^2 d^3}{3c^4 d} \cdot \frac{6c^5 d^2}{5cd^4}$

12 $\frac{4ab^3 c}{21a^5 b^2 c^2} \cdot \frac{7a^2 bc^3}{12a^3 b^2 c^4}$

13 $\frac{2x^2}{3y^2} \cdot \frac{6xy}{7} \cdot \frac{14y^3}{5x^3}$

14 $\frac{3a^3}{14b^3} \cdot \frac{7a^4}{12b^4} \cdot \frac{8b^6}{3a^8}$

15 $\frac{5a^5}{7b^3} \cdot \frac{14b^7}{10a^2} \cdot \frac{2a^4}{9b^2}$

16 $\frac{2c^2 d^3}{3ef^2} \cdot \frac{4ef^5}{3c^4 d} \cdot \frac{3c^3 e^2}{4df^2}$

17 $\frac{5x^4 y^3}{4xy^6} \div \frac{10x^7 y^5}{12xy^4}$

18 $\frac{3a^4 b}{7a^5 b^6} \div \frac{9a^6 b^5}{14a^7 b^3}$

19 $\frac{12b^3 c^5}{11a^2 b^4 c} \div \frac{15a^3 bc^6}{22ab^3}$

20 $\frac{6x^2 yz^4}{7x^3 y^2 z} \div \frac{18x^4 y^4 z}{21y^6 z^4}$

21 $\frac{3x^3}{10y^2} \cdot \frac{5y}{6z^3} \div \frac{9x^2}{4z^2}$

22 $\frac{25a^2 b^2}{12c^2} \cdot \frac{36bc^3}{5a^3} \div \frac{15b^3}{7ac}$

23 $\frac{5a^2 b}{6c^3} \cdot \frac{8bc^2}{15a} \div \frac{14b^3}{9ac}$

24 $\frac{48x^3 y^2}{25z^2} \cdot \frac{15z^3}{16x^2 y} \div \frac{81xz^2}{35y^3}$

25 $\dfrac{4w + 2z}{xy - y^2} \cdot \dfrac{wx - wy}{4xw + 2xz}$

26 $\dfrac{3a + 9b}{c + 2d^2} \cdot \dfrac{ac + 2ad^2}{2ac + 6bc}$

27 $\dfrac{4cd + 2d^2}{9c^2 + 3cd} \cdot \dfrac{9c + 3d}{8cd + 4d^2}$

28 $\dfrac{4x - 2y}{5x + 10y} \cdot \dfrac{x^2 + 2xy}{2xy - y^2}$

29 $\dfrac{4a - 2b}{5a + 10b} \div \dfrac{2ab - b^2}{a^2 + 2ab}$

30 $\dfrac{cd + 3ce}{d^2 - 2de} \div \dfrac{3cd + 9ce}{d^3 - 2d^2e}$

31 $\dfrac{mn - 2n^2}{9m - 3n} \div \dfrac{m^2 - 2mn}{6m - 2n}$

32 $\dfrac{ac^2 - bc^2}{cd + d^2} \div \dfrac{ab - b^2}{2cd + 2d^2}$

33 $\dfrac{3bc + 3c^2}{2ab - 2ac} \cdot \dfrac{b^2 - c^2}{b^2 + 3bc + 2c^2}$

34 $\dfrac{x^2 - 4y^2}{x - 2y} \cdot \dfrac{x^2 + xy - 6y^2}{x^3y - 4xy^3}$

35 $\dfrac{2ab + 2ac}{3b^2 - 3bc} \cdot \dfrac{b^2 - c^2}{b^2 + 2bc + c^2}$

36 $\dfrac{x^2 + xy - 2y^2}{x^2 - 4y^2} \cdot \dfrac{x^2 + 4xy + 3y^2}{x^2 + 2xy - 3y^2}$

37 $\dfrac{x^3}{x + y} \cdot \dfrac{x^2 - y^2}{x^2} \cdot \dfrac{x}{x - y}$

38 $\dfrac{a^2 - ab}{ab + b^2} \cdot \dfrac{a}{ab - b^2} \cdot \dfrac{a^2 + ab}{b}$

39 $\dfrac{c^3}{c + 2d} \cdot \dfrac{c^2 - 4d^2}{d^3} \cdot \dfrac{d^4}{c - 2d}$

40 $\dfrac{p^2 - q^2}{pq^2} \cdot \dfrac{p^2}{pq + q^2} \cdot \dfrac{q^4}{p^2 - pq}$

41 $\dfrac{a^3}{b^3 + ab^2} \cdot \dfrac{b^2 - a^2}{a^2b} \div \dfrac{a^2b - a^3}{b}$

42 $\dfrac{a^3}{a^2b - b^3} \cdot \dfrac{a^3 + a^2b}{b^3} \div \dfrac{a^4}{ab^2 - b^3}$

43 $\dfrac{x^3}{x + 2y} \cdot \dfrac{x^2 - 4y^2}{y^3} \div \dfrac{x - 2y}{y^4}$

44 $\dfrac{3c^2d}{c^2 - d^2} \cdot \dfrac{c^2 + cd}{2d^2} \div \dfrac{3c^3}{2cd - 2d^2}$

45 $\dfrac{a - b}{a + b} \cdot \dfrac{a^2 + 2ab}{ab - 3b^2} \cdot \dfrac{a^2 - 2ab - 3b^2}{a^2 + ab - 2b^2}$

46 $\dfrac{2x^2 + xy - y^2}{2x^2 - xy - y^2} \cdot \dfrac{2x^2 + xy}{xy + 2y^2} \cdot \dfrac{x - y}{x + y}$

47 $\dfrac{6c^2 - 4cd}{6bc + 9bd} \cdot \dfrac{2c^2 + 5cd + 3d^2}{c^2 + cd - 2d^2} \cdot \dfrac{c + 2d}{3c - 2d}$

48 $\dfrac{4r^2 + rs - 3s^2}{2r^2 + 3rs - 2s^2} \cdot \dfrac{2r - s}{2r + s} \cdot \dfrac{3rs + 6s^2}{4r^2 - 3rs}$

49 $\dfrac{5a^2 + 4ab - b^2}{a^2 - 7ab + 10b^2} \cdot \dfrac{5a^2 - 25ab}{3ab^2 + 3b^3} \div \dfrac{5a - b}{a - 2b}$

50 $\dfrac{5x^2 + 7xy + 2y^2}{2x^2 - 5xy + 3y^2} \cdot \dfrac{2x - 3y}{3x - 5y} \div \dfrac{30xy + 12y^2}{6x^2y - 10xy^2}$

51 $\dfrac{4c^2d + 8cd^2}{14c^3 + 2c^2d} \cdot \dfrac{3c - d}{c - 2d} \div \dfrac{3c^2 + 5cd - 2d^2}{7c^2 - 13cd - 2d^2}$

52 $\dfrac{6d - 5e}{2d + 3e} \cdot \dfrac{9d^3e - 18d^2e^2}{48d^2e^2 - 60de^3} \div \dfrac{6d^2 - 17de + 10e^2}{8d^2 + 2de - 15e^2}$

53 $\dfrac{2x^2 + 3xy + y^2}{x^2 - 4y^2} \cdot \dfrac{2x^2 + 3xy - 2y^2}{x^2 - 2xy + y^2} \cdot \dfrac{x^2 - 3xy + 2y^2}{4x^2 - y^2}$

54 $\dfrac{2a^2 - ab - 3b^2}{3a^2 - 2ab - b^2} \cdot \dfrac{2a^2 - ab - b^2}{a^2 - 2ab - 3b^2} \cdot \dfrac{3a^2 - 8ab - 3b^2}{2a^2 - 7ab + 6b^2}$

55 $\dfrac{4c^2 + 11cd - 3d^2}{c^2 + 3cd - 4d^2} \cdot \dfrac{3c^2 - cd - 4d^2}{4c^2 - 17cd + 4d^2} \div \dfrac{3c^2 + 5cd - 12d^2}{c^2 - 16d^2}$

56 $\dfrac{15p^2 - pr - 2r^2}{p^2 + 5pr + 4r^2} \cdot \dfrac{3p^2 + 11pr - 4r^2}{5p^2 + 8pr - 4r^2} \div \dfrac{9p^2 - r^2}{p^2 + pr - 2r^2}$

57 $\dfrac{(x+2)x-3}{(x-3)x+2} \cdot \dfrac{(x-1)x-2}{(x+4)x+3} \cdot \dfrac{(x+4)x+4}{(x-1)x-6}$

58 $\dfrac{(a+1)a-2}{(a-5)a+4} \cdot \dfrac{(a+4)a+3}{(a+1)a-6} \cdot \dfrac{(a-6)a+8}{(a-2)a+1}$

59 $\dfrac{(x-3)x-4}{(x-2)x-3} \cdot \dfrac{(x-2)x+(x-2)}{(x-4)x+2(x-4)} \div \dfrac{(x+1)x-6}{(x+1)x-2}$

60 $\dfrac{(2c-3)c-2}{(2c+3)c-9} \cdot \dfrac{(2c-1)c-3}{(2c-3)c-2} \div \dfrac{(2c+3)c+1}{(2c-5)c-2(c-3)}$

4.5
THE LOWEST COMMON MULTIPLE

As we shall see in the next section, we can find the sum of two or more fractions only when the denominators of the fractions are equal. Hence, if the denominators of the fractions to be added are different, we must change each fraction to an equal fraction with the denominators of the resulting fractions all the same. This denominator should be the lowest common multiple (LCM) of the denominators; it is called the lowest common denominator (LCD).

A common multiple of a set of integers is an integer that is divisible by each integer in the set. For example, 48 is a common multiple of 2, 3, and 8. Also, 24 is a common multiple of 2, 3, and 8, and there is no integer less than 24 that is divisible by each of these three numbers. Hence, 24 is the least common multiple of 2, 3, and 8. In arithmetic the LCM of a set of integers is defined to be the least positive integer that is divisible by each integer in the set. In this chapter, however, we shall be dealing with polynomials, and we cannot define the LCM of a set of polynomials in this way, since the adjective "least" has no meaning when applied to polynomials. Therefore, we shall define the LCM of a set of polynomials in this way:

■ The lowest common multiple (LCM) of a set of polynomials with integral coefficients is the polynomial P such that P is divisible by each polynomial in the set, and furthermore, every polynomial that is divisible by each member of the set is also divisible by P.

We stated above that the LCM of 2, 3, and 8 is 24. We shall now show that 24 satisfies the LCM definition. By definition of a multiple, any multiple of 8 can be expressed in the form $8n$, where n is an integer. If this multiple is divisible by 3, then n must be divisible by 3; hence $n = 3m$, where m is an integer. Consequently, $8n = 24m$. Furthermore, since $24m$ is divisible by 2, $24m$ is a common multiple of 2, 3, and 8. The least value of $24m$ is obtained when we let $m = 1$. Also, since $24m$ is divisible by 24, 24 is the LCM of 2, 3, and 8.

We shall further illustrate the definition of the LCM in Example 1.

EXAMPLE 1
Find the LCM of $(a-1)^2$, $(a+1)^2$, and $(a-2)(a-1)$.

SOLUTION In explaining the method for solving this problem, we shall use the letters p, q, and r to stand for polynomials in which the coefficients are integers. We first notice that any multiple of $(a-1)^2$ can be expressed in the form $p(a-1)^2$. Second, since $(a-1)^2$ is not divisible by $(a+1)^2$, then $p(a-1)^2$ is divisible by $(a+1)^2$ only if $p = q(a+1)^2$. Therefore a common multiple of $(a-1)^2$ and $(a+1)^2$ is $q(a+1)^2(a-1)^2$. Third, since $q(a+1)^2(a-1)^2$ is divisible by $a-1$ and neither of the last two factors is divisible by $a-2$, then $q(a+1)^2(a-1)^2$ is divisible by $(a-2)(a-1)$ only if $q = r(a-2)$. Now we replace q by $r(a-2)$ and get $r(a-2)(a+1)^2(a-1)^2$. This is clearly a multiple of $(a-1)^2$, $(a+1)^2$, and $(a-2)(a-1)$. Fourth, we let $r = 1$ and get $(a-2)(a+1)^2(a-1)^2$. This is the required LCM, since it is divisible by each of the three given expressions. Furthermore $r(a-2)(a+1)^2(a-1)^2$ is divisible by the LCM if r is any polynomial of the specified type.

In order to obtain the LCM of a set of polynomials, we first express each polynomial of the set as the product of powers of its prime factors. Then by definition, the following satement is true.

■   The LCM of a set of polynomials must have as a factor the highest power of each prime factor that appears in any polynomial of the set and must have no other factor.

EXAMPLE 2

Find the LCM of $x^2 - 2xy + y^2$, $x^2 + 2xy + y^2$, $x^2 - y^2$, $x^2 - 3xy + 2y^2$, and $2x^2 + 3xy + y^2$.

SOLUTION We first write each of these polynomials in the factored form shown here:

$$x^2 - 2xy + y^2 = (x-y)^2$$

$$x^2 + 2xy + y^2 = (x+y)^2$$

$$x^2 - y^2 = (x-y)(x+y)$$

$$x^2 - 3xy + 2y^2 = (x-2y)(x-y)$$

$$2x^2 + 3xy + y^2 = (2x+y)(x+y)$$

The prime factors which appear are $(x-y)$, $(x+y)$, $(x-2y)$, and $(2x+y)$. However, $(x-y)$ and $(x+y)$ have exponents 2 in the first and the second polynomials, respectively. Hence, the LCM is $(x-y)^2(x+y)^2(x-2y)(2x+y)$.

4.6
ADDITION OF FRACTIONS

By use of the distributive axiom (2.16) and of (2.41), we have

$$\frac{a}{n} + \frac{b}{n} + \frac{c}{n} + \cdots + \frac{r}{n} = \frac{1}{n}(a+b+c+\cdots+r) = \frac{a+b+c+\cdots+r}{n}$$

Hence, the sum of two or more fractions with identical denominators is the fraction that has the sum of the given numerators as the numerator and the common denominator as the denominator. For example,

$$\frac{3a}{2xy} + \frac{5a}{2xy} - \frac{c}{2xy} = \frac{3a + 5a - c}{2xy} = \frac{8a - c}{2xy}$$

and

$$\frac{x+y}{x+3y} + \frac{x-y}{x+3y} - \frac{2x+y}{x+3y} = \frac{(x+y) + (x-y) - (2x+y)}{x+3y}$$

$$= \frac{x+y+x-y-2x-y}{x+3y} \qquad \text{removing parentheses}$$

$$= \frac{-y}{x+3y}$$

If the denominators of the fractions to be added are different, we convert each fraction to an equal fraction with the LCD as the denominator and proceed as in the above examples.

EXAMPLE 1 Express

$$\frac{1}{6x} + \frac{1}{3y} - \frac{3x + 2y}{12xy}$$

as a single fraction.

SOLUTION The LCD of the given fractions is $12xy$. To convert the given fractions to equal fractions with $12xy$ as a denominator, we multiply each member of the first fraction by $2y$ and each member of the second by $4x$. We thereby obtain

$$\frac{1}{6x} + \frac{1}{3y} - \frac{3x + 2y}{12xy} = \frac{2y}{12xy} + \frac{4x}{12xy} - \frac{3x + 2y}{12xy}$$

$$= \frac{2y + 4x - (3x + 2y)}{12xy}$$

$$= \frac{2y + 4x - 3x - 2y}{12xy} \qquad \text{removing parentheses}$$

$$= \frac{x}{12xy} \qquad \text{combining similar terms}$$

$$= \frac{1}{12y} \qquad \begin{array}{l} \text{dividing numerator} \\ \text{and denominator by } x \end{array}$$

EXAMPLE 2 Combine

$$\frac{3x + y}{x^2 - y^2} - \frac{2y}{x(x - y)} - \frac{1}{x + y}$$

into a single fraction.

SOLUTION The denominators are $x^2 - y^2 = (x + y)(x - y)$; $x(x - y)$; and $x + y$. Therefore the LCD is $x(x + y)(x - y)$. Consequently, we multiply the numerator and denominator of the first, the second, and the third fraction by x, $x + y$, and $x(x - y)$ respectively, and complete the computation as:

$$\frac{3x + y}{(x + y)(x - y)} - \frac{2y}{x(x - y)} - \frac{1}{x + y}$$

$$= \frac{x(3x + y)}{x(x + y)(x - y)} - \frac{2y(x + y)}{x(x + y)(x - y)} - \frac{x(x - y)}{x(x + y)(x - y)}$$

$$= \frac{x(3x + y) - 2y(x + y) - x(x - y)}{x(x + y)(x - y)}$$

$$-\frac{3x^2 + xy - 2xy - 2y^2 - x^2 + xy}{x(x + y)(x - y)}$$

$$= \frac{2x^2 - 2y^2}{x(x + y)(x - y)}$$

$$= \frac{2(x + y)(x - y)}{x(x + y)(x - y)}$$

$$= \frac{2}{x}$$

EXERCISE 4.3
Addition of Fractions

Combine the fractions in the following problems into a single fraction.

1 $\frac{3}{4} - \frac{1}{3} - \frac{1}{6}$ **2** $\frac{1}{2} + \frac{1}{6} - \frac{1}{9}$

3 $\frac{1}{6} + \frac{1}{3} - \frac{1}{5}$ **4** $\frac{5}{9} + \frac{1}{2} - \frac{4}{3}$

5 $\frac{5}{3} + \frac{1}{12} - \frac{5}{4} + \frac{3}{2}$ **6** $\frac{2}{5} - \frac{4}{3} + \frac{7}{6} + \frac{1}{10}$

7 $\frac{3}{7} - \frac{10}{21} + \frac{3}{14} - \frac{2}{3}$ **8** $\frac{1}{4} - \frac{7}{6} - \frac{1}{3} + \frac{4}{5}$

9 $\dfrac{x}{6yz} + \dfrac{3y}{10xz} - \dfrac{2z}{15xy}$

10 $\dfrac{3c}{8ab} + \dfrac{a}{6bc} - \dfrac{5b}{12ac}$

11 $\dfrac{3c}{7de} + \dfrac{5d}{14ec} - \dfrac{e}{4cd}$

12 $\dfrac{5x}{6yz} - \dfrac{7y}{12xz} - \dfrac{3z}{16xy}$

13 $\dfrac{a+5}{6} + \dfrac{3a+1}{9} - \dfrac{a+9}{2}$

14 $\dfrac{x+3}{6} - \dfrac{2x-4}{8} + \dfrac{x-3}{3}$

15 $\dfrac{3x^2 - 2y}{xy} + \dfrac{2}{x} - \dfrac{12x}{5y}$

16 $\dfrac{2a - 3b}{ab} + \dfrac{a + 3ab}{a^2 b} - \dfrac{a+b}{ab^2}$

17 $\dfrac{3c - d}{cd} - \dfrac{3c^2 + d}{c^2 d} + \dfrac{d^2 + 1}{cd^2}$

18 $\dfrac{2p - r}{p^2 r} + \dfrac{3p - 4r}{2pr^2} - \dfrac{p^2 - 3r^2}{3p^2 r^2}$

19 $\dfrac{x - 4y}{2y^3} + \dfrac{6x - y}{3xy^2} - \dfrac{y - 2x}{6x^2 y}$

20 $\dfrac{5a - 2b}{4b^2} + \dfrac{3a + 2b}{6ab} - \dfrac{a + 5b}{3a^2}$

21 $\dfrac{x - y}{x + 3y} + \dfrac{y}{x}$ **22** $\dfrac{6r + s}{3r - 2s} + \dfrac{3r}{s}$

23 $\dfrac{3a}{2b} - \dfrac{3a + 4b}{5a + 2b}$ **24** $\dfrac{2c + 5d}{6c + 4d} - \dfrac{c}{2d}$

25 $\dfrac{4x^2 - 6xy + 3y^2}{2x(x + y)} - \dfrac{2x - 3y}{x + y}$

26 $\dfrac{a^2 - 2ab - b^2}{b(a - b)} + \dfrac{a + b}{a - b}$

27 $\dfrac{6z^2 + 3zw - w^2}{3z(z - w)} - \dfrac{2z + w}{z - w}$

28 $\dfrac{2r^2 - 7rs - 12s^2}{2r(3r - 4s)} + \dfrac{2r + 4s}{3r - 4s}$

29 $\dfrac{2a^2 - 6ab}{a^2 - 2ab - 3b^2} + \dfrac{a + 2b}{a + b}$

30 $\dfrac{b^2}{a^2 - 3ab + 2b^2} - \dfrac{a - b}{a - 2b}$

31 $\dfrac{x-y}{x+y} - \dfrac{2y^2 - 4xy}{x^2 - xy - 2y^2}$

32 $\dfrac{3z^2 + 4zw}{z^2 + zw - 6z^2} - \dfrac{z+2w}{z-2w}$

33 $\dfrac{2}{b} - \dfrac{1}{a+b} + \dfrac{1}{a-b}$

34 $\dfrac{3}{3c-2d} + \dfrac{7}{c+2d} - \dfrac{2}{2c+d}$

35 $\dfrac{1}{r-2t} - \dfrac{4}{r+t} + \dfrac{3}{r+2t}$

36 $\dfrac{3}{x-y} - \dfrac{1}{x+y} - \dfrac{4}{2x-y}$

37 $\dfrac{1}{p+r} - \dfrac{4}{2p-r} + \dfrac{1}{p-2r}$

38 $\dfrac{3}{h+1} - \dfrac{8}{2h+1} + \dfrac{1}{h-1}$

39 $\dfrac{1}{a-b} - \dfrac{2}{2a+b} + \dfrac{3}{a+b}$

40 $\dfrac{9}{x+3y} - \dfrac{5}{x-y} + \dfrac{16}{x-2y}$

41 $\dfrac{4c+d}{c^2 - cd - 2d^2} - \dfrac{1}{c+d} - \dfrac{2}{c-2d}$

42 $\dfrac{3}{x+2y} + \dfrac{x+6y}{x^2 - 4y^2} - \dfrac{2}{x-2y}$

43 $\dfrac{a-b}{a+b} - \dfrac{6b^2}{a^2 - ab - 2b^2} + \dfrac{a}{a-2b}$

44 $\dfrac{14pr + 10r^2}{p^2 + 7pr + 10r^2} + \dfrac{4p}{p+5r} - \dfrac{3p}{p+2r}$

45 $\dfrac{r-2t}{r+2t} - \dfrac{r-2t}{r-t} + \dfrac{2r^2 - 5t^2}{r^2 + rt - 2t^2}$

46 $\dfrac{2z^2 - 3wz}{2w^2 - 5wz - 3z^2} - \dfrac{3w+z}{2w+z} + \dfrac{w-2z}{w-3z}$

47 $\dfrac{5ab - 2a^2}{a^2 + ab - 2b^2} + \dfrac{a-2b}{a-b} - \dfrac{a+5b}{a+2b}$

48 $\dfrac{z+w}{2z-w} - \dfrac{7zw - w^2}{2z^2 + 3zw - 2w^2} - \dfrac{z-w}{z+2w}$

49 $\dfrac{2x}{x^2 - y^2} - \dfrac{1}{x+y} - \dfrac{y}{x^2 - xy}$

50 $\dfrac{c}{2cd + 3d^2} - \dfrac{9d}{4c^2 - 9d^2} + \dfrac{3}{2c-3d}$

51 $\dfrac{6}{b^2 - 1} - \dfrac{3}{b^2 - b} + \dfrac{3}{b + 1}$

52 $\dfrac{3a}{a^2 + ab - 2b^2} + \dfrac{2b}{a^2 + 3ab + 2b^2} - \dfrac{2}{a + b}$

53 $\dfrac{x - y}{2x^2 + 3xy + y^2} - \dfrac{x + y}{2x^2 - xy - y^2} + \dfrac{8}{3(2x + y)}$

54 $\dfrac{3a + b}{4(2a + b)(a + b)} - \dfrac{5(a + b)}{8(3a - b)(2a + b)} + \dfrac{1}{8(2a + b)}$

55 $\dfrac{1}{3c - 2d} + \dfrac{3c + 2d}{12(3c - 2d)(c - d)} - \dfrac{3c - 2d}{12(3c + 2d)(c - d)}$

56 $\dfrac{f}{(f + g)(f + 3g)} - \dfrac{1}{f - g} + \dfrac{f}{(f + g)(f - g)}$

57 $\dfrac{3w^2}{(2w + z)(w - z)} + \dfrac{w^2 + wz - z^2}{(2w - z)(w - z)} - \dfrac{4w^2}{(2w + z)(2w - z)}$

58 $\dfrac{11h + 2k}{2(3h + k)(2h - k)} - \dfrac{7h}{2(3h + k)(2h + 3k)} - \dfrac{6k}{(2h + 3k)(2h - k)}$

59 $\dfrac{32r - 2s}{6r^2 - 19rs + 10s^2} + \dfrac{r - 3s}{6r^2 - rs - 2s^2} - \dfrac{20r + 4s}{4r^2 - 8rs - 5s^2}$

60 $\dfrac{a^2 + b^2}{2a^2 + 5ab + 3b^2} - \dfrac{a^2 - 11b^2}{2a^2 - ab - 3b^2} + \dfrac{10ab - 24b^2}{4a^2 - 9b^2}$

4.7
COMPLEX FRACTIONS

A complex fraction is a fraction in which the numerator, the denominator, or both contain fractions. For example,

$$\dfrac{1 + \frac{1}{3}}{1 - \frac{1}{4}}, \qquad \dfrac{\frac{2x - y}{y}}{x + \frac{2}{y}}, \qquad \text{and} \qquad \dfrac{\frac{a + b}{a - b} - \frac{a}{2b}}{2 - \frac{1}{a^2 b^2}}$$

are complex fractions.

We simplify a complex fraction by converting it to an equal fraction that has no fractions in either the numerator or the denominator. A complex fraction can be simplified by first simplifying the numerator and the denominator and then by finding their quotient. Usually, however, the most efficient method consists of the following steps:

1 Find the LCM of the denominators of the fractions that appear in the complex fraction.
2 Multiply the members of the complex fraction by the LCM found in step 1.
3 Simplify the result obtained in step 2.

EXAMPLE 1 Simplify

$$\frac{4 - \dfrac{1}{x}}{16 - \dfrac{1}{x^2}}$$

SOLUTION Since the LCM of the denominators is x^2, we proceed as follows:

$$\frac{4 - \dfrac{1}{x}}{16 - \dfrac{1}{x^2}} = \frac{x^2\left(4 - \dfrac{1}{x}\right)}{x^2\left(16 - \dfrac{1}{x^2}\right)} \qquad \text{multiplying numerator and denominator by } x^2$$

$$= \frac{4x^2 - x}{16x^2 - 1} \qquad \text{performing indicated operations}$$

$$= \frac{x(4x - 1)}{(4x + 1)(4x - 1)} \qquad \text{factoring}$$

$$= \frac{x}{4x + 1} \qquad \text{dividing numerator and denominator by } 4x - 1$$

EXAMPLE 2 Simplify

$$\frac{2 - \dfrac{3}{a + 2}}{\dfrac{1}{a - 1} + \dfrac{1}{a + 2}}$$

SOLUTION The LCM of the denominators is $(a - 1)(a + 2)$. Therefore, we proceed as follows:

$$\frac{2 - \dfrac{3}{a + 2}}{\dfrac{1}{a - 1} + \dfrac{1}{a + 2}} = \frac{(a - 1)(a + 2)\left(2 - \dfrac{3}{a + 2}\right)}{(a - 1)(a + 2)\left(\dfrac{1}{a - 1} + \dfrac{1}{a + 2}\right)} \qquad \text{multiplying each member by } (a - 1)(a + 2)$$

$$= \frac{2(a-1)(a+2) - 3(a-1)}{a+2+a-1}$$ performing indicated operations

$$= \frac{2(a^2 + a - 2) - 3(a-1)}{2a+1}$$

$$= \frac{2a^2 + 2a - 4 - 3a + 3}{2a+1}$$ removing parentheses

$$= \frac{2a^2 - a - 1}{2a+1}$$ combining terms

$$= \frac{(2a+1)(a-1)}{2a+1}$$ factoring

$$= a - 1$$ dividing numerator and denominator by $2a + 1$

In the complex fraction

$$\frac{1 - \dfrac{1}{1 - \dfrac{1}{a}}}{\dfrac{1}{1 - \dfrac{a}{3}} - 1}$$

the numerator and denominator are themselves complex fractions. In such cases the member or members that contain complex fractions should be simplified first. If the fraction thus obtained is complex, the process is continued until a fraction is obtained in which no fractions appear in either the numerator or the denominator. As an example we shall simplify the above complex fraction.

$$\frac{1 - \dfrac{a}{a-1}}{\dfrac{3}{3-a} - 1}$$ multiplying members of complex fraction in numerator by a and members of complex fraction in denominator by 3

$$= \frac{\dfrac{a-1-a}{a-1}}{\dfrac{3-3+a}{3-a}}$$ adding fractions in numerator and denominator

$$= \frac{\dfrac{-1}{a-1}}{\dfrac{a}{3-a}} \qquad \text{combining terms}$$

$$= \frac{a-3}{a(a-1)} \qquad \text{multiplying numerator and denominator by } (a-1)(-a+3)$$

EXERCISE 4.4
Complex Fractions

Simplify the following complex fractions.

1 $\dfrac{1}{1-\frac{3}{5}}$ **2** $\dfrac{3}{2+\frac{1}{4}}$

3 $\dfrac{3}{1+\frac{5}{6}}$ **4** $\dfrac{5}{3+\frac{3}{4}}$

5 $\dfrac{\frac{1}{2}+\frac{1}{3}}{\frac{3}{4}-\frac{7}{6}}$ **6** $\dfrac{\frac{2}{3}+\frac{1}{2}}{\frac{1}{2}+\frac{5}{6}}$

7 $\dfrac{5-\frac{5}{6}}{4+\frac{4}{9}}$ **8** $\dfrac{2+\frac{1}{2}}{3-\frac{6}{7}}$

9 $\dfrac{2+\dfrac{1}{x}}{4-\dfrac{1}{x^2}}$ **10** $\dfrac{a-\dfrac{1}{a}}{1-\dfrac{1}{a^2}}$

11 $\dfrac{3+\dfrac{1}{a}}{9-\dfrac{1}{a^2}}$ **12** $\dfrac{x-\dfrac{1}{x}}{1-\dfrac{1}{x}}$

13 $\dfrac{3+\dfrac{5}{x}-\dfrac{2}{x^2}}{1-\dfrac{4}{x^2}}$ **14** $\dfrac{x+\dfrac{8}{x^2}}{x-2+\dfrac{4}{x}}$

15 $\dfrac{2-\dfrac{7}{x}+\dfrac{3}{x^2}}{2+\dfrac{3}{x}-\dfrac{2}{x^2}}$ **16** $\dfrac{a+\dfrac{1}{a^2}}{a-1+\dfrac{1}{a}}$

17 $\dfrac{x-1}{1-\dfrac{x}{x+1}}$

18 $\dfrac{a-3}{1-\dfrac{5}{a+2}}$

19 $\dfrac{x+5}{1+\dfrac{2}{x+3}}$

20 $\dfrac{a+3}{1+\dfrac{6}{a-3}}$

21 $\dfrac{r+\dfrac{2rt}{r-3t}}{r-t}$

22 $\dfrac{1-\dfrac{3e}{d+e}}{d-2e}$

23 $\dfrac{3+\dfrac{3y}{x-2y}}{x-y}$

24 $\dfrac{r+\dfrac{2rt}{r-3t}}{r-t}$

25 $\dfrac{1-\dfrac{b}{a-b}}{1-\dfrac{3b}{a+b}}$

26 $\dfrac{r-\dfrac{rt}{r+2t}}{r+\dfrac{t^2}{r+2t}}$

27 $\dfrac{2+\dfrac{3y}{x-y}}{2-\dfrac{3y}{x+2y}}$

28 $\dfrac{a-\dfrac{ab}{a-b}}{a-\dfrac{2b^2}{a-b}}$

29 $\dfrac{1-\dfrac{c}{b-c}}{1-\dfrac{3c}{b+c}}$

30 $\dfrac{1-\dfrac{6d}{2x+3d}}{2+\dfrac{d}{c-2d}}$

31 $\dfrac{2-\dfrac{3b}{a+b}}{\dfrac{5b}{a+2b}-2}$

32 $\dfrac{\dfrac{3x}{x+3y}-3}{2-\dfrac{2x}{x-3y}}$

33 $\dfrac{\dfrac{1}{x+1}+\dfrac{2}{x^2-1}}{\dfrac{2}{x-1}+1}$

34 $\dfrac{\dfrac{w}{w+1}-\dfrac{w^2}{w^2-1}}{\dfrac{1}{w-1}+1}$

35 $\dfrac{1+\dfrac{x+y}{x}}{\dfrac{1}{x-y}+\dfrac{x}{x^2-y^2}}$

36 $\dfrac{\dfrac{2a+3}{a+2}-\dfrac{2a}{a+1}}{\dfrac{a}{a+2}-1}$

37 $\dfrac{1}{1 - \dfrac{1}{1 - \dfrac{1}{a}}}$

38 $\dfrac{a}{\dfrac{1}{1 - \dfrac{a}{3}} - 1}$

39 $\dfrac{1}{1 - \dfrac{1}{1 - \dfrac{1}{x + 1}}}$

40 $\dfrac{1 - \dfrac{1}{a}}{1 + \dfrac{1}{1 + \dfrac{2}{a - 2}}}$

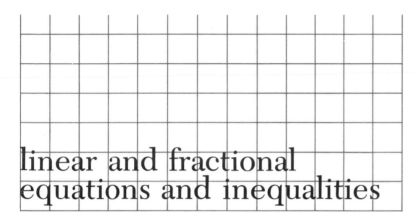

linear and fractional equations and inequalities

Chapter

5

Heretofore, we have been concerned with formal operations that followed prescribed rules of procedure. In this chapter we shall investigate conditions under which two algebraic expressions are equal. For example, we shall explain methods for finding the replacement for x so that a statement such as $\frac{1}{2}(x + 3) = \frac{1}{4}(x - 2)$ is true. A statement of this type is called an *equation*. The equation is a powerful tool in mathematics and is essential in the development and understanding of the problems in the physical sciences and engineering.

5.1
OPEN SENTENCES

In this section we shall consider statements of the type of

$$5 + x = 9 \tag{1}$$

x is an integer between 2 and 6 $\hfill (2)$

x is a color in the United States flag $\hfill (3)$

No one of these statements is true as it stands. However, (1) is true if x is replaced by 4, (2) is true if x is replaced by an element of $\{3, 4, 5\}$, and (3) is true if x is replaced by a color of the set $\{red, white, blue\}$. Furthermore, each of the statements (1), (2), and (3) is false if x is replaced by any number or word other than the replacement specified above.

Variable and Replacement Set

In statements (1) and (2), x stands for a number, and in (3), x stands for a word. Under such circumstances, the letter x is called a *variable* and illustrates the following definition:

■ A *variable* is a symbol, usually a letter, that stands for one or more elements of a specified set. The specified set is called the *replacement set*.

Open Sentence

The assertions (1), (2), and (3) are called *open sentences* and illustrate the following definition:

■ An *open sentence* is an assertion containing a variable that is neither true nor false but that becomes a true or false statement if the variable is replaced by an element chosen from an appropriate set.

Truth Set

■ If an open sentence becomes a true statement when the variable is replaced by each element of a set T, then T is called the *truth set* for the open sentence.

According to this definition, the truth sets for the open sentences (1), (2), and (3) are $\{4\}$, $\{3, 4, 5\}$, and $\{red, white, blue\}$, respectively.

Other examples of open sentences and their truth sets follow:

Open Sentence	Truth Set
x is a state of the United States larger than California.	$\{Alaska, Texas\}$
x is a color in a rainbow.	$\{violet, indigo, blue, green, yellow, orange, red\}$
x is a woman who has been president of the United States.	$\varnothing$
x is a positive odd integer less than 10.	$\{1, 3, 5, 7, 9\}$
x is an element of $\{a, b, c, d\} \cap \{a, c, d, e, f\}$	$\{a, c, d\}$
$x - 8 = 2$	$\{10\}$

Open Sentences

Find the truth set for each of the following open sentences.

1 X is an American state that borders on the Pacific Ocean.

2 X is the name of a former president of the United States, the name of an American city, and the name of an American state.

3 New x is the name of an American state.

4 X is the last name of two presidents of the United States.

5 New x is the name of a large American city and the name of an American state.

6 X is an American state admitted to the Union since 1950.

7 X is the name of one of the three largest automobile manufacturers in the United States.

8 X is a United States copper-nickel coin now widely circulated.

9 X is a country bordered by the Rio Grande and New x is an American state.

10 X-ball is one of three popular intercollegiate sports.

11 X-ball is a game played in a gymnasium.

12 South x is an American state.

13 X is the name of a large American river and of an American state.

14 X Sox is the name of a major league baseball team.

15 X is the name of a bird and of a major league baseball team.

16 X is a five-letter word that is the name of a color and also a familiar family name.

17 X is a two-digit number, and the sum of the digits is 4.

18 X is a number containing not more than three digits, and the sum of the digits is 3.

19 X is a prime number less than 20.

20 X is a two-digit number, the sum of whose digits is 7.

21 X is a positive two-digit number divisible by 3 such that the first digit minus the second is 2.

22 X is a three-digit number in which the first and third digits are equal, and the sum of the three digits is 9.

23 X is positive, less than 100, and the cube of an integer.

24 X is a three-digit number with the third digit 5 and the sum of the first two digits 5.

25 $x = \{1, 3, 5, 7\} \cap \{1, 2, 5, 6, 7\}$.

26 $x = \{y \mid y$ is divisible by 3$\} \cap \{y \mid y < 10\}$.

27 $x = \{y \mid y$ is a perfect square$\} \cap \{y \mid y < 50\}$.

28 X is an element of $\{x \mid x$ is a positive integer divisible by 2 and $x \le 10\} \cup \{x \mid x$ is a positive integer divisible by 3 and $0 < x \le 10\}$.

29 X is an element of $\{x \mid 0 < x \le 25$ and x is a perfect square$\} \cup \{x \mid 0 < x \le 27$ and x is divisible by $6\}$.

30 $x = \{x \mid 0 < x < 16$ and x is divisible by $2\} - \{y \mid 0 < y \le 15$ and y is divisible by $3\}$.

31 $x = \{1, 2, 3\} \cup \{4, 5, 6\} \cap \{3, 5, 7\}$.

32 $x = (\{2, 4, 6, 8, 10\} - \{2, 3, 6, 9\}) \cap \{6, 8, 10\}$.

5.2
EQUATIONS

Open sentences of the type

$$3x - 5 = 4 + 2x \tag{1}$$

$$\frac{2x - 3}{x + 2} = \frac{x - 1}{4} \tag{2}$$

$$x + 1 = x + 3 \tag{3}$$

$$\frac{x - 1}{3} + \frac{x + 1}{2} = \frac{5x + 1}{6} \tag{4}$$

are called equations and illustrate the following definition.

Equations Defined

■ An *equation* is an open sentence which states that two expressions are equal. Each of the two expressions is called a *member* of the equation.

If an equation is a true statement after the variable is replaced by a specific number, then that number is called a *root* of the equation and is said to *satisfy* it.

For example, 9 is a root of Eq. (1) since $3(9) - 5 = 4 + 2(9)$. Furthermore, 2 and 5 are roots of Eq. (2), since each member of (2) is equal to $\frac{1}{4}$ if x is replaced by 2, and each member is equal to 1 if x is replaced by 5.

Solution Set

■ The set of all roots of an equation is called the *solution set* of the equation.

The procedure for finding the solution set of an equation is called *solving* the equation.

As stated above, the solution sets of Eqs. (1) and (2) are $\{9\}$ and $\{2, 5\}$, respectively. Equation (3), however, is a false statement for every replacement for x, since the sum of a number and 1 is not equal to the sum of the same number and 3. Therefore, the solution set of Eq. (3) is the empty set $\varnothing$.

Equation (4), however, is a true statement for every replacement for x, since if we add the fractions at the left of the equality sign, we get the fraction on the right. Therefore, the solution set of Eq. (4) is the set of all numbers. Equations (1), (2), and (3) are called *conditional equations*, and Eq. (4) is an *identity*. These illustrate the following definitions:

Conditional Equation

■ A *conditional algebraic equation* is an equation whose solution set contains a finite number of elements, or an equation whose solution set is $\varnothing$.

Identity

■ An *identity* is an equation whose solution set is the set of all permissible numbers. A permissible number is a replacement for the variable for which each member of the equation is a number.

For example, 1 is not a permissible number for the equation

$$\frac{x-2}{x-1} + x = \frac{x^2 - 2}{x-1}$$

since if x is replaced by 1, we have $-1/0 + 1 = -1/0$, and $-1/0$ is not a number.

We use set notation to state that $\{9\}$ is the solution set of $3x - 5 = 4 + 2x$ in this way:

$$\{x \mid 3x - 5 = 4 + 2x\} = \{9\}$$

Similarly, for Eq. (2) we have

$$\left\{ x \,\middle|\, \frac{2x-3}{x+2} = \frac{x-1}{4} \right\} = \{2, 5\}$$

We now introduce the notation $f(x)$, read "f of x," which is very important in mathematics. In this chapter we shall use $f(x)$ to stand for an algebraic expression in x. This notation will be discussed more fully in a later chapter where it will be given a broader interpretation.

Now if, in a given discussion, we let $f(x) = 3x^2 - 2x + 1$, then throughout the discussion, $f(x)$ stands for the trinomial $3x^2 - 2x + 1$. It should be noted that the variable that appears in the parentheses in $f(x)$ is the same as the variable appearing in the expression for which $f(x)$ stands. Therefore, if

$$f(x) = 3x^2 - 2x + 1$$

then,

$$f(t) = 3t^2 - 2t + 1$$
$$f(t + 1) = 3(t + 1)^2 - 2(t + 1) + 1$$

and

$$f(4) = 3(4^2) - 2(4) + 1 = 41$$

As illustrated by the last equation, if x is replaced by 4, then $f(x)$ becomes $f(4)$ and is a constant.

The notations $h(x)$, $g(x)$, and $F(x)$, or any other letter followed by a second letter enclosed in parentheses are also used to name algebraic expressions. In this sense then, each of the following statements is an equation:

$$f(x) = c \qquad \text{where } c \text{ is a constant} \tag{5}$$

$$f(x) = h(x) \tag{6}$$

$$f(x) + k(x) = g(x) \tag{7}$$

If r, r', and r'' are the respective roots of these equations, then

$$f(r) = c \qquad \text{from (5)}$$

$$f(r') = h(r') \qquad \text{from (6)}$$

$$f(r'') + k(r'') = g(r'') \qquad \text{from (7)}$$

Hereafter in this chapter, we shall deal only with conditional equations and the methods for solving them. The variable in an equation is often called the *unknown*, and we shall frequently refer to it in this way.

5.3
EQUIVALENT EQUATIONS

The objective in solving an equation is to find a replacement for the variable that satisfies the equation. The simpler the equation is in form, the easier it is to solve. For example, consider the equations

$$7x - 45 = 5x - 43 \tag{1}$$

$$2x = 2 \tag{2}$$

At this stage, the only way that we can find a root of Eq. (1) is to guess at a number, substitute it for x, and see if it satisfies the equation. In Eq. (2), however, it is obvious that the root is 1. Now, if we substitute 1 for x in Eq. (1) and combine terms, we get $-38 = -38$. Therefore 1 is also a root of Eq. (1). Even though Eqs. (1) and (2) are different statements, each statement is true if x is replaced by 1. Two equations of this type are *equivalent* and illustrate the following definition:

■ Two equations are *equivalent* if their solution sets are equal.

The procedure for solving a given equation is to obtain a succession of equivalent equations, each simpler in form than the preceding, until we ultimately obtain one whose solution set can be easily found. The following theorem is essential for this purpose.

■ If $f(x)$, $g(x)$, and $h(x)$ are polynomials, then the two equations
$f(x) = g(x)$ and $f(x) + h(x) = g(x) + h(x)$ are equivalent   (5.1)

To illustrate this theorem, we let $f(x) = 5x + 3$, $g(x) = 2x - 9$, and $h(x) = -2x - 3$. Then $f(x) = g(x)$ is the equation

$$5x + 3 = 2x - 9 \tag{3}$$

and

$$f(x) + h(x) = g(x) + h(x)$$

becomes

$$5x + 3 - 2x - 3 = 2x - 9 - 2x - 3$$

and by combining similar terms in each member, we have

$$3x = -12 \tag{4}$$

Now, it is obvious that -4 is a root of (4) since $3(-4) = -12$, and since $5(-4) + 3 = 2(-4) - 9$, then -4 is a root of (3).

We shall now prove Theorem (5.1).

PROOF If r is a root of

$$f(x) = g(x)$$

then

$$f(r) = g(r)$$

Furthermore, $h(r)$ is a constant, and therefore by the additivity axiom for equalities (2.4),

$$f(r) + h(r) = g(r) + h(r)$$

Hence, r is a root of

$$f(x) + h(x) = g(x) + h(x)$$

Conversely, if r is a root of

$$f(x) + h(x) = g(x) + h(x)$$

then

$$f(r) + h(r) = g(r) + h(r)$$

and it follows by (2.4) that

$$f(r) + h(r) - h(r) = g(r) + h(r) - h(r)$$

Therefore, since

$$h(r) - h(r) = 0$$

we have

$$f(r) = g(r)$$

and therefore r is a root of

$$f(x) = g(x)$$

Another useful theorem in solving equations follows:

■ If k is a nonzero constant, then the equations $f(x) = g(x)$ and $k \cdot f(x) = k \cdot g(x)$ are equivalent   (5.2)

The proof of this theorem depends on the multiplicative axiom of equalities (2.5) and will be left as an exercise for the student.

We use this theorem when one or more of the coefficients or constant terms in $f(x) = g(x)$ are fractions. If k is the LCM of the denominators of the fractions in the equation, then $k \cdot f(x) = k \cdot g(x)$ will contain no fractions.

We illustrate the use of Theorems (5.1) and (5.2) in Example 1.

EXAMPLE 1

Solve the equation $\frac{1}{2}x + \frac{2}{3} = \frac{1}{4}x - \frac{1}{6}$.

SOLUTION The LCM of the denominators in the equation is 12. Consequently, we proceed as follows:

$\frac{1}{2}x + \frac{2}{3} = \frac{1}{4}x - \frac{1}{6}$	given equation
$12(\frac{1}{2}x + \frac{2}{3}) = 12(\frac{1}{4}x - \frac{1}{6})$	by (5.2) with $k = 12$
$6x + 8 = 3x - 2$	multiplying
$6x + 8 - 3x - 8 = 3x - 2 - 3x - 8$	by (5.1) with $h(x) = -3x - 8$
$3x = -10$	combining similar terms
$x = -\frac{10}{3}$	by (5.2) with $k = \frac{1}{3}$

Hence, the solution set of the given equation is $\{-\frac{10}{3}\}$. We verify this fact by replacing x in the given equation with $-\frac{10}{3}$ and finding that each member is equal to -1.

LINEAR EQUATIONS IN ONE VARIABLE

The equation $f(x) = g(x)$ is a linear equation in x if $f(x)$ and $g(x)$ are of degree 1 in x, or if one of them is of degree 1 and the other is a constant.

Examples of linear equations in one variable are

$$ax = b, \quad 3y + 5 = 7, \quad \text{and} \quad 2x - 7 = 8 - 4x$$

The steps in the solution of a linear equation in one variable are illustrated in Example 1 of Sec. 5.3. We shall now list those steps.

1   If one or more of the coefficients or constant terms in the equation $f(x) = g(x)$ are fractions, multiply each member of the equation by the LCM of the denominators and equate the products. The resulting equation is $k \cdot f(x) = k \cdot g(x)$.
2   Formulate the polynomial $h(x)$ whose terms are the negatives of the constant terms in $k \cdot f(x)$, and the negatives of the terms that involve x in $k \cdot g(x)$. Then write the equation $k \cdot f(x) + h(x) = k \cdot g(x) + h(x)$.
3   Combine similar terms in each member of the equation obtained in step 2, and obtain an equation of the type of $ax = b$, with $a \neq 0$.
4   Multiply each member of $ax = b$ by $1/a$, and obtain $x = b/a$. The number b/a is a root of the given equation since the equation obtained in each step is equivalent to the given equation.
5   Finally, replace x in the given equation by the number obtained in step 4 in order to verify that it is a root.

EXAMPLE 1

Find the solution set of $\dfrac{2x}{3} - \dfrac{3}{4} = \dfrac{5}{6} - \dfrac{x}{8}$.

SOLUTION   The LCM of the denominators is 24. Hence we proceed as follows:

$$24\left(\frac{2x}{3} - \frac{3}{4}\right) = 24\left(\frac{5}{6} - \frac{x}{8}\right) \qquad \text{multiplying each member by 24}$$

$$16x - 18 = 20 - 3x$$

The negative of the constant term in the left member is 18 and the negative of the term involving x in the right member is $3x$. Hence,

$$h(x) = 18 + 3x$$

and if we add $18 + 3x$ to each member, we get

$$16x - 18 + 18 + 3x = 20 - 3x + 18 + 3x$$

$$19x = 38 \qquad \text{combining similar terms}$$

$$x = 2 \qquad \text{multiplying each member by } \tfrac{1}{19}$$

Therefore, a possible set is $\{2\}$.

Verification If we replace x by 2 in the given equation, we get

$$\frac{2}{3}(2) - \frac{3}{4} = \frac{4}{3} - \frac{3}{4} = \frac{16 - 9}{12} = \frac{7}{12} \qquad \text{for left member}$$

and

$$\frac{5}{6} - \frac{2}{8} = \frac{20 - 6}{24} = \frac{14}{24} = \frac{7}{12} \qquad \text{for right member}$$

Hence, $\{2\}$ is the solution set of the given equation.

EXAMPLE 2

Find the set $\{x \mid 4x - 1 = 7\}$.

SOLUTION The required set is the solution set of

$$4x - 1 = 7$$

This equation has no fractions, so step 1 is not necessary. To apply step 2, we note that there is no term with x in the right member and that the negative of the constant term in the left is 1. Therefore, $h(x) = 1$. Hence we proceed as follows:

$$4x - 1 + 1 = 7 + 1 \qquad \text{adding } h(x) = 1 \text{ to each member}$$

$$4x = 8 \qquad \text{combining terms}$$

$$x = 2 \qquad \text{multiplying each member by } \tfrac{1}{4}$$

Verification Replacing x by 2 in $4x - 1 = 7$, we have $8 - 1 = 7$. Therefore,

$$\{x \mid 4x - 1 = 7\} = \{2\}$$

EXERCISE 5.2
Solution of Linear Equations

Find the solution set of the equation in each of Probs. 1 to 32.

1	$3x - 5 = 1$	**2**	$7 = 2x - 3$
3	$5x - 8 = 3x$	**4**	$4x + 9 = 7x$
5	$4x - 5 = 2x + 9$	**6**	$7x - 5 = 4x + 22$
7	$5x + 3 = 8x - 15$	**8**	$6x + 5 = 8x - 11$
9	$3(2x - 7) = 5(x - 2)$		
10	$3(3x - 7) = 8(x - 1)$		

11 $2(5x + 6) = 7(3x - 14)$

12 $5(7x - 12) = 9(3x + 4)$

13 $2(3x + 4) = 3(3x + 2)$

14 $5(4x + 1) = 3(8x + 1)$

15 $7(4x - 1) = 2(8x + 1)$

16 $3(5x + 3) = 5(5x + 1)$

17 $3 - \dfrac{x}{4} = 2\left(5 - \dfrac{x}{8}\right)$

18 $\dfrac{1}{3}\left(\dfrac{x}{2} + 3\right) = \dfrac{1}{2}\left(\dfrac{x}{3} - 5\right)$

19 $\dfrac{1}{3}\left(\dfrac{x}{4} - 7\right) = \dfrac{1}{4}\left(\dfrac{x}{3} + 5\right)$

20 $\dfrac{1}{4}\left(\dfrac{x}{3} + 8\right) = \dfrac{1}{2}\left(\dfrac{x}{6} - 4\right)$

21 $\frac{2}{3}x + \frac{3}{4} - \frac{1}{4}x = \frac{1}{2}x - \frac{1}{4}$

22 $\frac{5}{6}x + \frac{5}{9} - \frac{2}{9}x = \frac{2}{3}x - \frac{4}{9}$

23 $\frac{5}{6}x + \frac{7}{9} - \frac{3}{4}x = \frac{1}{9}x - \frac{2}{9}$

24 $\frac{3}{8}x - \frac{5}{12} + \frac{1}{6}x = \frac{1}{2}x + \frac{7}{12}$

25 $\frac{4}{5}x + 7 = \frac{2}{3}x + 9$

26 $\frac{5}{6}x - 8 = \frac{2}{5}x + 5$

27 $\frac{5}{9}x + 5 = \frac{4}{5}x - 6$

28 $\frac{5}{8}x - 6 = \frac{1}{6}x + 5$

29 $bx + 3 = \dfrac{3b}{a} + ax$

30 $b(x - 1) = b(a - x) - 2b$

31 $a(x - 1) + b(x - 1) = ax - 2b$

32 $a(x - 1) + b(x + 2) = b$

33 Find the set $\left\{x \,\middle|\, \dfrac{2x + 5}{3} = 3x - 10\right\}$

34 Find the set $\left\{x \,\middle|\, \dfrac{3x - 1}{5} = 11 - x\right\}$

35 Find the set $\left\{x \,\middle|\, \dfrac{2x + 3}{7} = 2x - 15\right\}$

36 Find the set $\left\{x \,\middle|\, \dfrac{3x - 2}{2} = 2x - 5\right\}$

37 Find the set $\left\{x \,\middle|\, \dfrac{2x - 3}{3} + 5 = \dfrac{3x - 2}{2}\right\}$

38 Find the set $\left\{x \,\middle|\, \dfrac{3x + 4}{7} + 1 = \dfrac{x + 7}{3}\right\}$

39 Find the set $\left\{x \,\middle|\, \dfrac{2x + 5}{5} + 2 = \dfrac{x + 4}{2}\right\}$

40 Find the set $\left\{x \,\middle|\, \dfrac{3x - 4}{8} - 1 = \dfrac{x + 3}{5}\right\}$

41 Find the set $\left\{ x \left| \dfrac{2x+1}{3} + \dfrac{1}{3} = \dfrac{x+3}{2} \right. \right\}$

42 Find the set $\left\{ x \left| \dfrac{3x+2}{6} + \dfrac{1}{6} = \dfrac{2x+6}{5} \right. \right\}$

43 Find the set $\left\{ x \left| \dfrac{4x-7}{5} - \dfrac{2}{5} = \dfrac{x+3}{2} \right. \right\}$

44 Find the set $\left\{ x \left| \dfrac{4x+3}{7} + \dfrac{3}{7} = \dfrac{5x+3}{8} \right. \right\}$

45 Find the set $\left\{ x \left| \dfrac{5x+2}{4} - \dfrac{2x+2}{7} = x \right. \right\}$

46 Find the set $\left\{ x \left| \dfrac{2x-1}{5} + \dfrac{5x+7}{8} = x+1 \right. \right\}$

47 Find the set $\left\{ x \left| \dfrac{3x+15}{5} + \dfrac{x+1}{2} = 2x-10 \right. \right\}$

48 Find the set $\left\{ x \left| \dfrac{6x+3}{5} - \dfrac{4x-3}{9} = x-2 \right. \right\}$

5.5
FRACTIONAL EQUATIONS

If at least one fraction with the variable in the denominator appears in an equation, then the equation is a *fractional equation*. For example,

$$\frac{x}{x+1} + \frac{5}{8} = \frac{5}{2(x+1)} + \frac{3}{4}$$

is a fractional equation, but

$$\frac{x+1}{2} - 5x = 7$$

is not.

We use the following theorem in solving a fractional equation:

■ If $k(x)$ is a polynomial, then each root of $f(x) = g(x)$ is also a root of
$k(x) \cdot f(x) = k(x) \cdot g(x)$ (5.3)

PROOF If r is a root of $f(x) = g(x)$, then $f(r) = g(r)$. Furthermore, $k(r)$ is a constant. Hence, by the multiplicative axiom for equalities (2.5), we have

$$k(r) \cdot f(r) = k(r) \cdot g(r)$$

Hence r is a root of

$$k(x) \cdot f(x) = k(x) \cdot g(x)$$

As an example, we let

$$f(x) = 3x - 2, \quad g(x) = 5x + 8, \quad \text{and} \quad k(x) = x - 3$$

Then $f(x) = g(x)$ becomes

$$3x - 2 = 5x + 8 \tag{1}$$

and

$$k(x) \cdot f(x) = k(x) \cdot g(x)$$

is

$$(x - 3)(3x - 2) = (x - 3)(5x + 8) \tag{2}$$

Now, if x is replaced by -5, the members of Eq. (1) are equal since each is equal to -17, and the members of Eq. (2) are equal, both being $(-8)(-17)$. Hence, -5 is a root of each equation.

The converse of Theorem (5.3) is not true. For example, 3 is not a root of $3x - 2 = 5x + 8$, since $7 \neq 23$. However, 3 is a root of $(x - 3)(3x - 2) = (x - 3)(5x + 8)$, since if x is replaced by 3, each member of the latter equation is equal to 0.

If $f(x) = g(x)$ is a fractional equation, and $k(x)$ is the LCM of the denominators, then $k(x) \cdot f(x) = k(x) \cdot g(x)$ will contain no fractions. If the latter equation is linear, we can solve it by the methods in Sec. 5.4.

EXAMPLE 1 Solve the equation

$$\frac{x}{x + 1} + \frac{5}{8} = \frac{5}{2(x + 1)} + \frac{3}{4} \tag{3}$$

SOLUTION We use Theorem (5.3) as a first step in the solution. Since the LCM of the denominators is $8(x + 1)$, we let $k(x) = 8(x + 1)$, multiply each member of Eq. (3) by $8(x + 1)$, and then proceed as indicated:

$$8(x + 1)\left(\frac{x}{x + 1} + \frac{5}{8}\right) = 8(x + 1)\left[\frac{5}{2(x + 1)} + \frac{3}{4}\right] \tag{4}$$

$$8x + 5(x + 1) = 4(5) + 6(x + 1) \qquad \text{performing indicated multiplication}$$

$$8x + 5x + 5 = 20 + 6x + 6 \qquad \text{by left-hand distributive axiom}$$

$$8x + 5x + 5 - 6x - 5 = 20 + 6x + 6 - 6x - 5 \qquad \text{adding } -6x - 5 \text{ to each member}$$

$$7x = 21 \qquad \text{combining terms}$$

$$x = 3 \qquad \text{multiplying by } \tfrac{1}{7}$$

By Theorem (5.3) we know that each root of Eq. (3) is a root of Eq. (4), but we do not know that the converse is true. Hence, we must replace x by 3 in

Eq. (3) and see if it is a root. When this is done, we see that each member is equal to $\frac{11}{8}$. Hence, the root of Eq. (3) is 3, and we have

$$\left\{ x \,\middle|\, \frac{x}{x+1} + \frac{5}{8} = \frac{5}{2(x+1)} + \frac{3}{4} \right\} = \{3\}$$

EXAMPLE 2 Find

$$\left\{ x \,\middle|\, \frac{2}{x+1} - 3 = \frac{4x+6}{x+1} \right\}$$

The required set is the solution set of

$$\frac{2}{x+1} - 3 = \frac{4x+6}{x+1} \tag{5}$$

We first use Theorem (5.3), multiply each member by $x + 1$ and get

$$2 - 3x - 3 = 4x + 6 \tag{6}$$

Then proceed as follows:

$$-3x - 1 = 4x + 6 \qquad \text{combining terms}$$
$$-3x - 1 + 1 - 4x = 4x + 6 + 1 - 4x \qquad \text{adding } 1 - 4x \text{ to each member}$$
$$-7x = 7 \qquad \text{combining terms}$$
$$x = -1$$

Now if we replace x by -1 in Eq. (5), the left member becomes $(\frac{2}{0}) - 3$, which is not a number. Furthermore, the right member becomes $\frac{2}{0}$, which also is not a number. Hence, neither member of Eq. (5) is defined when $x = -1$, and we therefore cannot accept -1 as a root. Furthermore, by Theorem (5.3) each root of Eq. (5) is a root of Eq. (6), and the only root of Eq. (6) is -1. Hence, we conclude that Eq. (5) has no roots, and that

$$\left\{ x \,\middle|\, \frac{2}{x+1} - 3 = \frac{4x+6}{x+1} \right\} = \varnothing$$

where $\varnothing$ is the empty set.

EXERCISE 5.3
Fractional Equations

Find the solution set for the equation in each of Probs. 1 to 28.

1 $\dfrac{x+1}{x-2} = \dfrac{x+3}{x-1}$

2 $\dfrac{x+5}{x+1} = \dfrac{x+1}{x-1}$

3 $\dfrac{x+5}{x-3} = \dfrac{x+2}{x-4}$

4 $\dfrac{x+8}{x+1} = \dfrac{x-2}{x-4}$

5 $\dfrac{2x-1}{3x-4} = \dfrac{6x+3}{9x-4}$

6 $\dfrac{3x+4}{4x+4} = \dfrac{6x+4}{8x+3}$

7 $\dfrac{2x-3}{5x-3} = \dfrac{4x-1}{10x+8}$

8 $\dfrac{4x+5}{3x+2} = \dfrac{12x+1}{9x-4}$

9 $\dfrac{4}{x-1} - \dfrac{3}{x+2} = \dfrac{18}{(x+2)(x-1)}$

10 $\dfrac{5}{x+4} + \dfrac{1}{x-2} = \dfrac{30}{(x+4)(x-2)}$

11 $\dfrac{6}{x+3} - \dfrac{2}{x-1} = \dfrac{8}{(x+3)(x-1)}$

12 $\dfrac{6}{x+6} - \dfrac{2}{x+3} = \dfrac{18}{(x+6)(x+3)}$

13 $\dfrac{5}{x-5} - \dfrac{6}{x+2} = \dfrac{28}{x^2-3x-10}$

14 $\dfrac{3}{x-9} - \dfrac{4}{x-4} = \dfrac{10}{x^2-13x+36}$

15 $\dfrac{8}{x-3} - \dfrac{2}{x-6} = \dfrac{24}{x^2-9x+18}$

16 $\dfrac{6}{x-4} - \dfrac{5}{x+2} = \dfrac{45}{x^2-2x-8}$

17 $\dfrac{5}{x+6} - \dfrac{2}{3x-4} = \dfrac{20}{3x^2-14x-24}$

18 $\dfrac{5}{2x-3} - \dfrac{2}{x+3} = \dfrac{30}{2x^2+3x-9}$

19 $\dfrac{3}{2x+1} + \dfrac{2}{x+3} = \dfrac{60}{2x^2+7x+3}$

20 $\dfrac{5}{3x+5} - \dfrac{1}{x+3} = \dfrac{20}{3x^2+14x+15}$

21 $\dfrac{7}{2x-1} - \dfrac{35}{2x^2-3x+1} = \dfrac{5}{3x-3}$

22 $\dfrac{4}{3x+1} = \dfrac{20}{3x^2-8x-3} + \dfrac{1}{2x-6}$

23 $\dfrac{7}{4x+3} - \dfrac{2}{3x-4} = \dfrac{70}{12x^2-7x-12}$

24 $\dfrac{9}{4x-2} = \dfrac{3}{2x+8} + \dfrac{81}{4x^2+14x-8}$

25 $\dfrac{4x-4}{2x+1} - \dfrac{2x-2}{x+2} - \dfrac{3x-2}{(x+1)(x+2)} = 0$

26 $\dfrac{x+1}{x+3} - \dfrac{x-4}{x-2} - \dfrac{15}{(2x+1)(x-2)} = 0$

27 $\dfrac{x-5}{x-1} - \dfrac{x-6}{x+3} = \dfrac{15x-79}{(3x+1)(x-1)}$

28 $\dfrac{x+1}{x+4} - \dfrac{x-1}{x+7} = \dfrac{10x-41}{(2x-7)(x+4)}$

29 Find the set $\left\{ x \,\middle|\, \dfrac{2}{x+2} + \dfrac{3}{x+6} = \dfrac{5}{x+4} \right\}$

30 Find the set $\left\{ x \,\middle|\, \dfrac{2}{x+6} + \dfrac{2}{2x-9} = \dfrac{3}{x-1} \right\}$

31 Find the set $\left\{ x \,\middle|\, \dfrac{1}{x+5} + \dfrac{3}{3x-1} = \dfrac{2}{x+1} \right\}$

32 Find the set $\left\{ x \,\middle|\, \dfrac{2}{2x+1} - \dfrac{1}{3x-6} = \dfrac{2}{3x+9} \right\}$

33 Find the set $\left\{ x \,\middle|\, \dfrac{2}{x+5} + \dfrac{6}{3x+5} = \dfrac{4}{x+3} \right\}$

34 Find the set $\left\{ x \,\middle|\, \dfrac{3}{3x+7} - \dfrac{1}{2x+18} = \dfrac{1}{2x-2} \right\}$

35 Find the set $\left\{ x \,\middle|\, \dfrac{3}{3x-4} - \dfrac{1}{2x-6} = \dfrac{1}{2x+4} \right\}$

36 Find the set $\left\{ x \,\middle|\, \dfrac{1}{x-4} - \dfrac{1}{2x-6} = \dfrac{1}{2x-9} \right\}$

37 Find the set $\left\{ x \,\middle|\, \dfrac{2}{2x-1} - \dfrac{1}{3x-9} = \dfrac{2}{3x+6} \right\}$

38 Find the set $\left\{ x \,\middle|\, \dfrac{2}{x+4} + \dfrac{6}{3x-8} = \dfrac{4}{x-1} \right\}$

39 Find the set $\left\{ x \,\middle|\, \dfrac{3}{3x-11} - \dfrac{1}{2x-6} = \dfrac{1}{2x-8} \right\}$

40 Find the set $\left\{ x \,\middle|\, \dfrac{1}{x+4} + \dfrac{3}{3x-4} = \dfrac{2}{x} \right\}$

41 Find the set $\left\{ x \,\middle|\, \dfrac{1}{x-5} - \dfrac{1}{2x-8} = \dfrac{1}{2x-11} \right\}$

42 Find the set $\left\{ x \,\middle|\, \dfrac{2}{x+8} + \dfrac{2}{2x+1} = \dfrac{3}{x+3} \right\}$

43 Find the set $\left\{ x \,\middle|\, \dfrac{2}{2x-1} - \dfrac{1}{3x-10} = \dfrac{2}{3x+7} \right\}$

44 Find the set $\left\{ x \,\middle|\, \dfrac{2}{x-8} + \dfrac{1}{x-10} = \dfrac{3}{x-9} \right\}$

Show that the equation in each of Probs. 45 to 48 has no roots.

45 $\dfrac{x-6}{x-7} = 2 + \dfrac{1}{x-7}$

46 $\dfrac{x+2}{x-9} = 5 + \dfrac{11}{x-9}$

47 $\dfrac{6}{x-5} - \dfrac{1}{x+5} = \dfrac{2x+20}{x^2-25}$

48 $\dfrac{5}{x+3} - \dfrac{2}{x-2} = \dfrac{x-12}{x^2+x-6}$

5.6
INEQUALITIES

The following theorems dealing with the order relation will be used in this section:

■   If $a < c$ and $b > 0$, then, $ab < cb$ (5.4)

PROOF

$a < c$	given
$c > a$	by (2.7)
$cb > ab$	by (2.33)
$ab < cb$	by (2.7)

$\cdot /$

■   If $a > b$ and $k < 0$, then $ak < kb$ (5.5)

PROOF

$a > b$	given
$a - b > 0$	by (2.24)
$(a-b)k < 0$	since product of negative number and positive number is negative
$ak - bk < 0$	by distributive axiom
$ak < bk$	by (2.25)

■   If $a < b$ and $k < 0$, then $ak > bk$ (5.6)

PROOF

$a < b$	given
$a - b < 0$	by (2.25)
$(a-b)k > 0$	since product of two negative numbers is positive
$ak - bk > 0$	by distributive axiom
$ak > bk$	by (2.24)

Inequality

An *inequality* in one variable is an open sentence of the type $f(x) > g(x)$ or $f(x) < g(x)$. If $f(x)$ and $g(x)$ are polynomials of degree 1 in x, or if one of them is a polynomial of degree 1 and the other is a constant, the inequality is *linear*.

For example,

$$2x + 1 > x + 2, \quad 3x - 1 < 2x + 3, \quad \text{and} \quad x + 3 > 4$$

are linear inequalities.

Solution Set

The *solution set* of an inequality is the set of replacements for the variable for which the inequality is a true statement. For example, we can verify that $x - 1 > 3$ is true if x is replaced by a number greater than 4. Hence, the solution set of the inequality is $\{x \mid x > 4\}$ or stated in another way,

$$\{x \mid x - 1 > 3\} = \{x \mid x > 4\}$$

Equivalent Inequalities

■ Two inequalities are *equivalent* if their solution sets are equal

The procedure for solving a linear inequality is very similar to the method for solving a linear equation, and it involves the property of equivalence. We use the following theorems for this purpose:

■ If $f(x)$, $g(x)$, and $h(x)$ are polynomials, then $f(x) > g(x)$ is equivalent to $f(x) + h(x) > g(x) + h(x)$ $\qquad$ (5.7)

and

$f(x) < g(x)$ is equivalent to $f(x) + h(x) < g(x) + h(x)$ $\qquad$ (5.8)

■ If k is a positive constant, then $f(x) > g(x)$ is equivalent to $k \cdot f(x) > k \cdot g(x)$ $\qquad$ (5.9)

and

$f(x) < g(x)$ is equivalent to $k \cdot f(x) < k \cdot g(x)$ $\qquad$ (5.10)

■ If c is a negative constant, the $f(x) > g(x)$ is equivalent to $c \cdot f(x) < c \cdot g(x)$ $\qquad$ (5.11)

and

$f(x) < g(x)$ is equivalent to $c \cdot f(x) > c \cdot g(x)$ $\qquad$ (5.12)

The proofs of Theorems (5.7) to (5.12) are similar to those of Theorems (5.1) and (5.2) and are based on Axioms (2.10) and (2.33) and Theorems (5.4), (5.5), and (5.6). The details will be left as an exercise for the student.

EXAMPLE 1

Find the solution set of the inequality $5x - 9 > 2x + 3$.

SOLUTION

$$5x - 9 > 2x + 3 \qquad \text{given}$$

$$5x - 9 - 2x + 9 > 2x + 3 - 2x + 9 \qquad \text{by (5.7) with } h(x) = -2x + 9$$

$$3x > 12 \qquad \text{combining terms}$$

$$x > 4 \qquad \text{by (5.9) with } k = \tfrac{1}{3}$$

Hence,

$$\{x \,|\, 5x - 9 > 2x + 3\} = \{x \,|\, x > 4\}$$

EXAMPLE 2

Find the solution set of $\frac{1}{6}x - \frac{3}{4} < \frac{3}{8}x + \frac{1}{2}$.

SOLUTION

$$\frac{1}{6}x - \frac{3}{4} < \frac{3}{8}x + \frac{1}{2} \qquad \text{given}$$

$$24(\tfrac{1}{6}x - \tfrac{3}{4}) < 24(\tfrac{3}{8}x + \tfrac{1}{2}) \qquad \text{by (5.10) with } k = 24$$

$$4x - 18 < 9x + 12 \qquad \text{by left-hand distributive axiom}$$

$$4x - 18 - 9x + 18 < 9x + 12 - 9x + 18 \qquad \text{by (5.8) with } h(x) = -9x + 18$$

$$-5x < 30 \qquad \text{combining terms}$$

$$x > -6 \qquad \text{by (5.11) with } c = -\tfrac{1}{5} \text{ (note that since } -\tfrac{1}{5} \text{ is negative, we reverse the inequality sign)}$$

Hence,

$$\{x \,|\, \tfrac{1}{6}x - \tfrac{3}{4} < \tfrac{3}{8}x + \tfrac{1}{2}\} = \{x \,|\, x > -6\}$$

EXERCISE 5.4
Linear Inequalities

Find the solution set of each of the following inequalities.

1 $3x + 5 > 8$ **2** $2x - 7 > 3$ **3** $5x - 1 < 9$

4 $4x + 5 < -7$ **5** $-2x - 4 > -8$ **6** $-5x + 2 < -3$

7 $-4x + 12 > -8$ **8** $-7x + 12 < 5$ **9** $4x - 7 > 2x + 1$

10 $9x + 2 < 6x - 4$ **11** $5x - 3 > 2x + 6$

12 $7x + 8 < 2x - 2$ **13** $2x + 7 > 5x - 2$

14 $3x - 5 < 5x + 1$ **15** $-x - 8 > 5x + 4$

16 $-7x + 4 < 5x - 8$ **17** $\dfrac{2x}{3} - \dfrac{1}{2} > \dfrac{2}{3} + \dfrac{3}{2}$

18 $\dfrac{3x}{5} + \dfrac{3}{10} < \dfrac{1}{5} - \dfrac{1}{2}$

19 $\dfrac{5x}{6} - \dfrac{5}{8} < \dfrac{1}{12} + \dfrac{3}{4}$

20 $\dfrac{5x}{4} + \dfrac{2}{3} < \dfrac{5}{6} - \dfrac{7}{12}$

21 $\dfrac{x}{2} + \dfrac{2}{3} > \dfrac{x}{6} + \dfrac{1}{4}$

22 $\dfrac{2x}{3} - \dfrac{1}{2} > \dfrac{x}{4} + \dfrac{1}{3}$

23 $\dfrac{5x}{8} - \dfrac{1}{4} < \dfrac{x}{2} - \dfrac{1}{16}$

24 $\dfrac{5x}{6} + \dfrac{2}{3} < \dfrac{5x}{9} - \dfrac{1}{6}$

25 $\dfrac{x}{4} + \dfrac{3}{2} < \dfrac{3x}{8} + \dfrac{5}{4}$

26 $\dfrac{x}{7} + \dfrac{2}{3} > \dfrac{x}{6} - \dfrac{5}{14}$

27 $\dfrac{3x}{4} - \dfrac{2}{9} > \dfrac{5x}{6} - \dfrac{1}{3}$

28 $\dfrac{3x}{4} + \dfrac{2}{3} < \dfrac{7x}{8} + \dfrac{1}{2}$

29 If $a > b$, prove that $\frac{1}{2}(a + b) < a$.

30 If $a < 3b$, prove that $a - b < \frac{1}{2}(a + b)$.

31 If $a > b$, prove that $3a + 2b > 2a + 3b$.

32 If $3b < a$, prove that $a + 2b < 2a - b$.

5.7
FRACTIONAL INEQUALITIES

Fractional Inequality

A *fractional inequality* is an inequality in which at least one fraction with the variable in the denominator appears.

In this section we shall consider only inequalities that are equivalent to

$$\frac{ax + b}{cx + d} > 0 \qquad \text{or} \qquad \frac{ax + b}{cx + d} < 0$$

We use the law of signs for division to solve such inequalities. For example, if $(ax + b)/(cx + d)$ is greater than zero, or positive, each member of the fraction must be positive or each member must be negative. Similarly, if the fraction is less than zero, one member must be positive and the other negative.

EXAMPLE 1 Find the solution set of

$$\frac{-13x - 34}{3x + 6} > -5$$

SOLUTION

$$\frac{-13x - 34}{3x + 6} > -5 \qquad \text{given}$$

$$\frac{-13x - 34}{3x + 6} + 5 > 0 \qquad \text{adding 5 to each member}$$

$$\frac{-13x - 34 + 15x + 30}{3x + 6} > 0 \qquad \text{performing indicated addition}$$

$$\frac{2x - 4}{3x + 6} > 0 \qquad \text{combining terms}$$

The last fraction is positive if

$$2x - 4 > 0 \qquad\qquad\qquad\qquad\qquad\qquad\qquad\qquad (1)$$

and

$$3x + 6 > 0 \qquad\qquad\qquad\qquad\qquad\qquad\qquad\qquad (2)$$

or if

$$2x - 4 < 0 \qquad\qquad\qquad\qquad\qquad\qquad\qquad\qquad (3)$$

and

$$3x + 6 < 0 \qquad\qquad\qquad\qquad\qquad\qquad\qquad\qquad (4)$$

Note The required replacement for x must satisfy *both* (1) and (2) *or both* (3) and (4). We designate the solution sets of these inequalities by S_1, S_2, S_3, and S_4, respectively.

Now, any replacement for x that satisfies *both* (1) and (2) is an element of S_1 and S_2 and therefore belongs to $S_1 \cap S_2$. We determine S_1 and S_2 as follows:

$$2x - 4 > 0 \qquad \text{inequality (1)}$$
$$2x > 4 \qquad \text{adding 4 to each member}$$
$$x > 2 \qquad \text{multiplying by } \tfrac{1}{2}$$

Therefore,

$$S_1 = \{x \,|\, x > 2\}$$

Similarly, by solving inequality (2) we get

$$S_2 = \{x \,|\, x > -2\}$$

Now, by referring to Fig 5.1, we see that

$$S_1 \cap S_2 = \{x \,|\, x > 2\} \cap \{x \,|\, x > -2\}$$
$$= \{x \,|\, x > 2\}$$

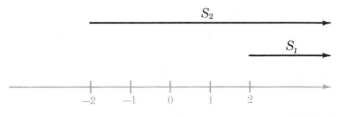

FIGURE 5.1

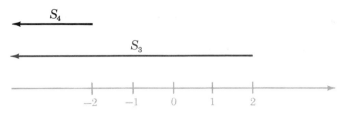

FIGURE 5.2

Likewise, any replacement for x that satisfies both inequalities (3) *and* (4) is an element of $S_3 \cap S_4$. By solving (3) and (4), we have

$$S_3 = \{x \,|\, x < 2\} \qquad \text{and} \qquad S_4 = \{x \,|\, x < -2\}$$

Then by referring to Fig. 5.2, we see that

$$S_3 \cap S_4 = \{x \,|\, x < 2\} \cap \{x \,|\, x < -2\}$$
$$= \{x \,|\, x < -2\}$$

Therefore, since a replacement for x that satisfies both (1) and (2) *or* both (3) and (4) is an element of the *union* of $S_1 \cap S_2$ and $S_3 \cap S_4$, **we have**

$$\left\{ x \,\middle|\, \frac{-13x - 34}{3x + 6} > -5 \right\} = (S_1 \cap S_2) \cup (S_3 \cap S_4)$$
$$= \{x \,|\, x > 2\} \cup \{x \,|\, x < -2\}$$

EXAMPLE 2 Find the solution set of

$$\frac{x - 1}{2x + 4} < 0$$

SOLUTION A fraction is negative if one member is positive and the other is negative. Hence, we seek a replacement for x such that

$$x - 1 > 0 \tag{5}$$

and

$$2x + 4 < 0 \tag{6}$$

or

$$x - 1 < 0 \tag{7}$$

and

$$2x + 4 > 0 \tag{8}$$

FIGURE 5.3

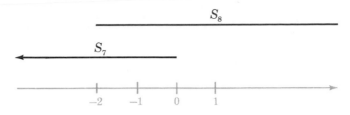

FIGURE 5.4

We designate the solution sets of (5), (6), (7), and (8) by S_5, S_6, S_7, and S_8, respectively. By solving these inequalities, we get

$S_5 = \{x \mid x > 1\}$

$S_6 = \{x \mid x < -2\}$

$S_7 = \{x \mid x < 1\}$

$S_8 = \{x \mid x > -2\}$

A replacement for x that satisfies (5) *and* (6) belongs to $S_5 \cap S_6$. Also a replacement that satisfies (7) *and* (8) belongs to $S_7 \cap S_8$. Furthermore, a replacement that belongs to $S_5 \cap S_6$ *or* to $S_6 \cap S_7$ is an element of $(S_5 \cap S_6) \cup (S_7 \cap S_8)$.

From Fig. 5.3 we see that no element of S_5 belongs to S_6. Hence,

$S_5 \cap S_6 = \varnothing$

or the empty set. Furthermore, from Fig. 5.4, it is evident that

$S_7 \cap S_8 = \{x \mid -2 < x < 1\}$

Therefore,

$$(S_5 \cap S_6) \cup (S_7 \cap S_8) = \varnothing \cup \{x \mid -2 < x1\}$$
$$= \{x \mid -2 < x < 1\}$$

Consequently,

$$\left\{x \,\middle|\, \frac{x-1}{2x+4} < 0\right\} = \{x \mid -2 < x < 1\}$$

5.8
INEQUALITIES INVOLVING ABSOLUTE VALUES

In this section we discuss methods for solving inequalities of the type of $|ax + b| < c$ and $|rs + t| > c$. In Sec. 1.4 we stated that the absolute value of a number z is the undirected distance from the point (0) on the number line to the point (z). Hence, if $|z| < c$ and $c > 0$, then the point (z) is between points ($-c$) and (c), and therefore $-c < z < c$. Now from Fig. 5.5 it is evident that the set

$\{z \mid -c < z < c\}$

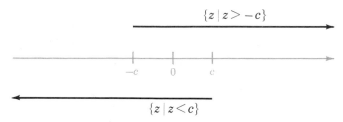

FIGURE 5.5

is the intersection of the sets

$\{z \,|\, z > -c\}$ and $\{z \,|\, z < c\}$

then

$\{z \,|\, |z| < c\} = \{z \,|\, z > -c\} \cap \{z \,|\, z < c\}$

On the other hand, if $|z| > c$, then the point (z) is to right of (c) *or* to the left of $(-c)$, as indicated in Fig. 5.6. Hence, an element of $\{z \,|\, |z| > c\}$ is an element of

$\{z \,|\, z > c\}$ or of $\{z \,|\, z < -c\}$

Therefore, $\{z \,|\, |z| > c\}$ is the *union* of $\{z \,|\, z < -c\}$ and $\{z \,|\, z > c\}$. That is,

$\{z \,|\, |z| > c\} = \{z \,|\, z > c\} \cup \{z \,|\, z < -c\}$

Similarly, if $|ax + b| < c$, then the replacement for x must be such that $-c < ax + b < c$. Hence, x must satisfy both of the inequalities

$ax + b < c$ (1)

and

$ax + b > -c$ (2)

Therefore, the solution set of $|ax + b| < c$ is the *intersection* of the solution sets of (1) and (2).

Likewise, if $|rx + t| > c$, the replacement for x must satisfy

$rs + t > c$ (3)

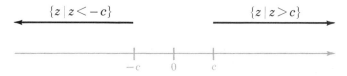

FIGURE 5.6

118

or

$$rs + t < -c \tag{4}$$

Hence, in this case the solution set of $|rs + t| > c$ is the *union* of the solution sets of (3) and (4).

EXAMPLE 1 Solve the inequality

$$|3x - 5| < 13 \tag{5}$$

SOLUTION The replacement for x that satisfies inequality (5) must satisfy both

$$3x - 5 < 13 \tag{6}$$

and

$$3x - 5 > -13 \tag{7}$$

We let S_5, S_6, and S_7 stand for the solution sets of (5), (6), and (7), respectively. Now, since an element of S_5 must be an element of S_6 *and* of S_7, then S_5 is the *intersection* of S_6 and S_7. Consequently, $S_5 = S_6 \cap S_7$.

We now determine S_6 and S_7 by solving (6) and (7).

$$3x - 5 < 13 \qquad \text{inequality (6)}$$
$$3x < 18 \qquad \text{adding 5 to each member}$$
$$x < 6 \qquad \text{multiplying by } \tfrac{1}{3}$$

Hence,

$$S_6 = \{ x \mid x < 6 \}$$

and

$$3x - 5 < -13 \qquad \text{inequality (7)}$$
$$3x > -8 \qquad \text{adding 5 to each member}$$
$$x > -\tfrac{8}{3} \qquad \text{multiplying by } \tfrac{1}{3}$$

Therefore,

$$S_7 = \{ x \mid x > -\tfrac{8}{3} \}$$

Consequently,

$$S_5 = S_6 \cap S_7 = \{ x \mid x < 6 \} \cap \{ x \mid x > -\tfrac{8}{3} \}$$
$$= \{ x \mid -\tfrac{8}{3} < x < 6 \}$$

EXAMPLE 2 Find the solution set of the inequality

$$|4x + 7| > 15 \tag{8}$$

SOLUTION Now, $|4x + 7| > 15$ if

$$4x + 7 > 15 \tag{9}$$

or

$$4x + 7 < -15 \tag{10}$$

We let S_8, S_9, and S_{10} stand for the solution sets of (8), (9), and (10), respectively. Then, an element of S_8 must belong to S_9 *or* to S_{10} and hence to the *union* of S_9 and S_{10}. In order to determine S_9 and S_{10}, we solve (9) and (10) as follows:

$$4x + 7 > 15 \qquad \text{inequality (9)}$$
$$4x > 8 \qquad \text{adding } -7 \text{ to each member}$$
$$x > 2 \qquad \text{multiplying by } \tfrac{1}{4}$$

Hence,

$$S_9 = \{x \mid x > 2\}$$

and

$$4x + 7 < -15 \qquad \text{inequality (10)}$$
$$4x < -22 \qquad \text{adding } -7 \text{ to each member}$$
$$x < -\tfrac{11}{2} \qquad \text{multiplying by } \tfrac{1}{4} \text{ and reducing to lowest terms}$$

Therefore,

$$S_{10} = \{x \mid x < -\tfrac{11}{2}\}$$

Thus,

$$S_8 = S_9 \cup S_{10} = \{x \mid x > 2\} \cup \{x \mid x < -\tfrac{11}{2}\}$$

EXERCISE 5.5
Inequalities

Find the solution set of the inequality in each of Probs. 1 to 20.

1 $\dfrac{x + 3}{x - 2} > 0$ \qquad\qquad **2** $\dfrac{x - 1}{x + 5} > 0$

3 $\dfrac{x+6}{x-3} < 0$

4 $\dfrac{x-3}{x+2} < 0$

5 $\dfrac{x}{(x+2)(x-1)} < 0$

6 $\dfrac{x+2}{(x+3)(x-2)} < 0$

7 $\dfrac{x-5}{(x+1)(x+4)} > 0$

8 $\dfrac{x+3}{(x-4)(x-5)} > 0$

9 $\dfrac{x^2 - 2x - 2}{x-2} > x+1$

10 $\dfrac{x^2 + 3x + 15}{x+3} > x+5$

11 $\dfrac{6 + 2x - x^2}{x-3} < 2 - x$

12 $\dfrac{x^2 + x + 2}{x+1} < 2 + x$

13 $|x-4| < 5$

14 $|3x+4| < 2$

15 $|2x-3| < 7$

16 $|-5x+2| < 3$

17 $|x+4| > 3$

18 $|3x-2| > 4$

19 $|-4x+1| > 7$

20 $|-2x-3| > 5$

5.9
SOLVING STATED PROBLEMS

A stated problem is a word description of a situation that involves both known and unknown quantities and certain relations between these quantities. If the problem is solvable by means of one equation, it must be possible to find two combinations of the quantities in the problem that are equal so that an equation can be written. Furthermore, at least one of the combinations must involve the unknown.

The procedure for solving a stated problem by means of an equation is not always simple, and considerable practice is necessary before one becomes adept at it. The following approach is suggested:

1   Read the problem carefully, and study it until the situation is thoroughly understood.
2   Identify both the known and the unknown quantities that are involved in the problem.
3   Select one of the unknowns, and represent it by a symbol, usually x, and then express the other unknowns in terms of this symbol.
4   Search the problem for information that tells which quantities or combinations are equal.
5   When the desired combinations are found, set them equal to each other and thus obtain an equation.
6   Solve the equation thus obtained, and check the solution set in the original problem.

It is usually helpful to tabulate the data given in the problem, as is done in the following illustrative examples, which show the methods for solving several types of stated problems.

PROBLEMS INVOLVING MOTION AT A UNIFORM VELOCITY   Problems that involve motion usually state a relation between the velocities (or speeds), between the distances traveled, or between the periods of time involved. The fundamental formulas used in such problems are

$$d = vt, \quad v = \frac{d}{t}, \quad \text{or} \quad t = \frac{d}{v}$$

where d represents distance, v represents velocity (or speed), and t represents a period of time. When one or more of these formulas is used in the same problem, d and v must be expressed in terms of the same linear unit and v and t must be expressed in the same unit of time.

EXAMPLE 1

A party of hunters made a trip of 380 miles to a hunting lodge in 7 hours. They traveled 4 hours on a paved highway and the remainder of time on a pasture road. If the average speed through the pasture was 25 miles per hour less than that on the highway, find the average speed and the distance traveled on each part of the trip.

SOLUTION   The unknown quantities here are the two speeds and the distances traveled on each part of the trip. We shall let

$x =$ speed on highway in miles per hour

Then

$x - 25 =$ speed through pasture

We now tabulate the data of the problem with the unknown quantities expressed in terms of x and printed in color. Note that the formula $d = vt$ is used to obtain the unknown distances.

	Time t, hours	Velocity v, miles per hour	Distance $d = vt$, miles
On paved highway	4	x	$4x$
On pasture road	$7 - 4 = 3$	$x - 25$	$3(x - 25)$
Total	7		380

The last column gives us two quantities that are equal:

Distance on highway + distance through pasture = total distance

or

$4x + 3(x - 25) = 380$

Hence, the desired equation is

$$4x + 3(x - 25) = 380$$

and we solve it as follows:

$4x + 3x - 75 = 380$	left-hand distributive axiom
$4x + 3x = 380 + 75$	adding 75 to each member
$7x = 455$	combining terms
$x = 65$	multiplying each member by $\frac{1}{7}$

Thus 65 miles per hour is the velocity on the highway,

$$65 - 25 = 40$$

so that 40 miles per hour is the velocity through the pasture;

$$4(65) = 260 \text{ miles traveled on highway}$$

$$3(40) = 120 \text{ miles traveled through the pasture}$$

Check

$$260 + 120 = 380$$

EXAMPLE 2

Airports A, B, and C are located on a north-south line. B is 645 miles north of A, and C is 540 miles north of B. A pilot flew from A to B, delayed 2 hours, and continued to C. The wind blew from the south at 15 miles per hour during the first part of the trip, but during the delay it changed to the north, with a velocity of 20 miles per hour. If each flight required the same period of time, find the airspeed, that is, the speed delivered by the propeller, of the plane.

SOLUTION   If we let

$x =$ airspeed of plane

then

$x + 15 =$ groundspeed of plane while wind was from the south

and

$x - 20 =$ groundspeed of plane while wind was from the north

We now use the formula $t = d/v$ to obtain the time required for each flight and tabulate the data:

	Distance d, miles	Airspeed of Plane, miles per hour	Speed v of Wind, miles per hour	Groundspeed of Plane, miles per hour	Period of Time $t = d/v$, hours
From A to B	645	x	15, from south	$x + 15$	$\dfrac{645}{x + 15}$
From B to C	540	x	20, from north	$x - 20$	$\dfrac{540}{x - 20}$

According to the statement of the problem, the following quantities are equal:

Time in hours from A to B = time in hours from B to C

$$\frac{645}{x + 15} = \frac{540}{x - 20}$$

Hence, the required equation is

$$\frac{645}{x + 15} = \frac{540}{x - 20}$$

and we solve it as follows:

$$(x - 20)(x + 15)\frac{645}{x + 15} = (x - 20)(x + 15)\frac{540}{x - 20} \qquad \text{multiplying each member by LCM of denominators}$$

$$(x - 20)645 = (x + 15)540 \qquad \text{performing indicated multiplication}$$

$$645x - 12,900 = 540x + 8,100 \qquad \text{right-hand distributive axiom}$$

$$105x = 21,000 \qquad \text{adding } -540x + 12,900 \text{ to each member and combining terms}$$

$$x = 200 \qquad \text{multiplying each member by } \tfrac{1}{105}$$

Check The airspeed is 200 miles per hour, the groundspeed from A to B is $200 + 15 = 215$ miles per hour, the groundspeed from B to C is $200 - 20 = 180$ miles per hour, and $\frac{645}{215} = \frac{540}{180} = 3$.

WORK PROBLEMS Problems that involve the rate of performance can often be solved by first finding the fractional part of the task done by each individual or agent in one unit of time and then finding a relation between several fractional parts. In this method the unit 1 used represents the entire job.

EXAMPLE 3

A farmer can plow a field in 4 days using a tractor. His hired hand can plow the same field in 6 days using a smaller tractor. How many days will be required for the plowing if they work together?

SOLUTION We let

x = number of days required to plow the field if they work together

Now we tabulate the data and complete the solution:

	Farmer	Helper	Together
Days required to plow field	4	6	x
Part plowed in 1 day	$\dfrac{1}{4}$	$\dfrac{1}{6}$	$\dfrac{1}{x}$

Quantities that are equal:

Part done in 1 day by farmer + part done by helper = part done by both

$$\frac{1}{4} + \frac{1}{6} = \frac{1}{x} \qquad \text{desired equation}$$

SOLUTION

$$3x + 2x = 12 \qquad \text{multiplying each member by } 12x$$
$$5x = 12$$
$$x = 2\frac{2}{5}\text{ days}$$

Check In $2\frac{2}{5}$ days the farmer plows $\frac{12}{5}\left(\frac{1}{4}\right) = \frac{3}{5}$ of the field. The helper plows $\frac{12}{5}\left(\frac{1}{6}\right) = \frac{2}{5}$ of the field, and $\frac{3}{5} + \frac{2}{5} = 1$, or the entire field.

EXAMPLE 4

If, in Example 3, the hired hand worked 1 day with the smaller machine and then was joined by the employer, how many days were required to finish the plowing?

SOLUTION Since the hired hand plowed $\frac{1}{6}$ of the field in 1 day, $\frac{5}{6}$ of it remained unplowed. We shall let

x = number of days required to finish the plowing

Then $x/4$ is the part plowed by the farmer, and $x/6$ is the part plowed by the helper. Hence,

$$\frac{x}{4} + \frac{x}{6} = \frac{5}{6}$$

$3x + 2x = 10$ multiplying each member by 12

$5x = 10$

$x = 2$

MIXTURE PROBLEMS   Many problems involve the combination of certain substances of known strengths, usually expressed in percentages, into a mixture of required strength in one of the substances. Others involve the mixing of certain commodities of specified prices. In such problems, it should be remembered that the total amount of any given element in a mixture is equal to the sum of the amounts of that element in the substances combined and that the monetary value of any mixture is the sum of the values of the substances that are put together.

EXAMPLE 5

How many gallons of a liquid that is 74 percent alcohol must be combined with 5 gallons of one that is 90 percent alcohol in order to obtain a mixture that is 84 percent alcohol?

SOLUTION   If we let x represent the number of gallons needed of the first liquid and remember that 74 percent of x is $0.74x$, then the following table showing the data in the problem is self-explanatory.

	Number of Gallons	Percentage of Alcohol	Number of Gallons of Alcohol
First liquid	x	74	$0.74x$
Second liquid	5	90	$0.90(5) = 4.5$
Mixture	$x + 5$	84	$0.84(x + 5)$

Quantities that are equal:

Number of gallons of alcohol in first liquid
    +number of gallons of alcohol in second liquid
        = number of gallons of alcohol in mixture

Hence, the equation is

$0.74x + 4.5 = 0.84(x + 5)$

and the solution, obtained by the usual method, is $x = 3$.

There is a wide variety of problems that can be solved by means of equations. The fundamental approach to all of them is the same and consits in finding two quantities, one or both of which include the unknown, that are equal.

EXERCISE 5.6
Problem Solving with Equations

1 Find three consecutive integers whose sum is 72.

2 A salesman called on 20 customers in three days. If the second day he called on 1 more customer than he did the first day, and the third day he called on 3 more than the second day, how many calls did he make each day?

3 Two rare coins were worth $90. If the value of one was $1\frac{1}{2}$ that of the other, what was the value of each?

4 A farmer used 1960 ft of fencing to enclose a rectangular plot of ground. If the width of the plot was three-fourths of the length, what were its dimensions?

5 Fred and Mary have a total of $559 in their bank accounts. If Fred has $213 more in his account than Mary has in hers, how much is in each account?

6 David found that when he completed the 14 hr he was taking that semester, he would have half the hours credit he needed for his degree. If at the beginning of the semester he had had 0.4 of the hours he needed, how many hours were required for graduation?

7 Dean and Leon caught 21 lb of fish. If Dean's catch weighed 3.4 lb more than Leon's, how many pounds of fish did each one catch?

8 A company safety officer found that of the 62 accidents with company vehicles the previous year, 8 more occurred while the vehicle was backing than all other types of accidents combined. How many backing accidents were there?

9 Bill, Joe, and Jack worked a total of 21 hr overtime. If Bill and Joe together worked 15 hr overtime and Jack worked 2 hr more than Bill, how much overtime did each one work?

10 Mr. Duff got a ticket for speeding. He was driving 17 miles per hour faster than the speed limit allowed. If he had been driving 18 miles per hour faster, he would have been driving twice as fast as the law allowed. What was the speed limit?

11 Three of four poker players lost money. If Mr. Green lost one-fourth of the money, Mr. Brown lost $10 more than Mr. Green, and Mr. White lost $6 more than Mr. Brown, how much did the fourth player, Mr. Black, win?

12 An amusement park charged $1.80 per person for a ticket, but had discount tickets available for $1.55. If $635.90 was collected from the sale of 363 tickets, how many of the tickets were purchased at the discount price?

13 Mrs. Cox counted a total of 27 honor points in her bridge hand from aces, kings, and queens by using the system that allows 4 points per ace, 3 points per king, and 2 points per queen. If she had one more ace than kings and 5 of her 13 cards were not honor cards, how many aces did she have?

14 A man bought a suit at a sale at a 10 percent reduction, and his wife bought a dress at a 15 percent reduction. The original price total for both garments was $160.00, and they paid $141.00. What was the original price of the suit?

15 One month Madge and Anne earned $30.80 by each working 30 hr as a baby sitter. Except for one job on which Anne earned 60 cents an hour, each was paid 50 cents an hour. How many hours did Anne work at the higher rate?

16 A bookstore received $628.75 from the sale of 470 copies of a particular book. If the book was available in both a hardback edition costing $4.00 and a paperback edition costing 75 cents, how many copies of each type of edition were sold?

17 In the annual candy sale, a campfire girl had twice as many boxes of mint sticks to sell as she had boxes of assorted chocolates. After she sold two boxes of each kind of candy, she had three times as many boxes of mint sticks as assorted chocolates. How many boxes of candy remained to be sold?

18 During the first 6 weeks of the summer session of a state college, there were three times as many out-of-state students as there were in-state students enrolled. Of this number 6000 out-of-state and 1000 in-state students did not enroll for the second summer session, but 9000 new in-state students enrolled. If there were then twice as many in-state students as out-of-state students, what was the enrollment for the first session?

19 In a certain community, there are 400 more registered Republicans than Democrats. If the voting age were dropped from 21 to 18, both the Republicans and Democrats would gain 100 eligible voters and there would then be twice as many Republicans as Democrats. How many registered voters over 21 are there in the community?

20 An overweight man weighed twice as much as his wife. After he lost 60 lb, he weighed $1\frac{1}{2}$ times his wife's weight. How much did his wife weigh?

21 The second number of the combination to unlock a locker was twice the first, and the third number was one-fourth the second. If the sum of the numbers in the combination was 42, what was the combination?

22 The snack bar at a Little League ball game took in $43.20 during one game. If there were 20 one-dollar bills, 20 pennies, 15 more dimes than quarters, and 15 more nickels than dimes, how many nickels, dimes, and quarters were there?

23 If 20 gal of water at a temperature of 65°F is mixed with 20 gal of water at a temperature of 81°F, what is the temperature of the mixture?

24 A 5-liter flask designated for 15 percent hydrochloric acid solution was found to be filled with 25 percent hydrochloric acid by mistake. How much should be drained off and replaced with distilled water to provide the desired 15 percent solution?

25 A tour group rode a bus at an average rate of 50 miles per hour to visit a city where they took a walking tour. If they walked at a rate of $\frac{3}{4}$ mile per hour, their day's trip took $7\frac{1}{2}$ hr, and they traveled 178 miles, how long was the walking tour?

26 The tour group of Prob. 25 spent their next day taking a scenic drive along a river to a spot 100 miles from the city. They returned to the city by boat. If the average speed of the boat was five-eighths that of the bus and the boat trip took $1\frac{1}{2}$ hr longer than the bus trip, how long did the group travel that day?

27 A delivery-truck driver drove 30 miles on a freeway from a downtown warehouse to the suburbs where he delivered parcels along a 24-mile route. He spent $2\frac{2}{3}$ as much time delivering parcels as he did driving to the suburbs. Find his average speed on each part of his trip if he drove 35 miles per hour faster on the freeway than he did in the suburbs.

28 Airport B is 90 miles due north of airport A. One morning a pilot in a small plane flew from A to B, had lunch, and started back to A, but when he had flown the same length of time he had in the morning, he was only 60 miles of the way back. If there was a wind blowing from the south at a uniform rate of 20 miles per hour all day, find the airspeed of the plane.

29 Two neighbors whose lots are equal in width spent 12 and 16 hr, respectively, building fences across the backs of their yards. How long will it take them to build a fence of equal length between their yards if they work together?

30 A volunteer worker required 2 hr to address a group of envelopes for a fund drive, while a second worker required 3 hr to address a similar group. How long would it take the two volunteers working together to address a third similar group of envelopes?

31 A maintenance man needed 8 hr to wash the windows in a certain building. The next month his assistant took 10 hr to wash the windows. If the two men worked together, how much time would be needed to wash the windows?

32 Jean could read the proof for the school newspaper in 1 hr, while Lois needed $1\frac{1}{4}$ hr to read the same amount of proof. One day the girls worked together for $\frac{1}{3}$ hr, and then Jean had to leave and Lois finished the proofreading. How long did it take her?

33 Matt and Mary worked together on a jigsaw puzzle for 3 hr. Then Matt left and Mary finished the puzzle in 30 min. Matt had worked similar puzzles alone in 6 hr. How long would Mary have taken to work the puzzle alone?

34 A political worker could fold enough campaign literature for a mailing list in 10 hr. His wife could do the job in 12 hr, and his daughter in 15 hr. If they all worked together, how long would the job require?

35 Todd, Mark, and Brad had 900 handbills to distribute. Each boy could distribute an average of 120 handbills per hour. Todd started at 8:00 a.m., Mark at 8:30, and Brad at 9:00. What time did they finish?

36 A farmer has a storage tank for irrigation that can be filled by the intake pipe in 10 hr and drained by the outlet in 8 hr. If, at the start of an irrigation job, the tank is full and both pipes are open, how long will it take to drain the tank?

37 Three maids could clean the rooms in a certain motel in 8 hr, but two maids would need 12 hr to clean all the rooms. One day two of the maids started work on time, but the third was 3 hr late. How many hours did it take the three maids to finish cleaning the motel?

38 A father and his two sons decided to install a sprinkler system in their backyard. The father could have done the work in 8 hr, the older boy in 12 hr, and the younger in 16 hr. They started work together, but after 2 hr the younger boy left to play baseball and an hour later the older boy went to band practice. How much longer did the father work alone to finish the sprinkler system?

39 New members of a labor crew were paid $22.50 a day, and experienced members of the crew were paid $25.80 a day. If the average daily wage paid to members of a certain crew was $24 and the total daily wages paid to experienced members of the crew was $129.00, how many new members were on the crew?

40 How many gallons of a 25 percent salt solution must be mixed with 10 gal of a 15 percent solution to produce a 20 percent salt solution?

41 A 1 percent solution of insecticide was needed to spray a tree, but because of a mixing error, the 8-gal sprayer was filled with a solution that was 0.6 percent insecticide. How much of the solution must be drawn off and replaced with 8.6 percent solution to provide the proper concentration?

42 A chemist mixed 35 ml of 4 percent nitric acid solution with 65 ml of 12 percent nitric acid solution. He used a portion of the mixture, and replaced it with distilled water. The new solution tested 6.9 percent nitric acid. How much of the original solution was used?

43 On its last trip of the day, a dump truck carrying a load of dirt traveled 8 miles from an excavation to a construction site, dumped its dirt, and then drove 33 miles to a storage yard. Its loaded speed was two-thirds its empty speed, and the entire traveling time was 1 hr. Find the average speeds of the truck when it was loaded and when it was empty.

44 A commuter drove his car at a rate of 30 miles per hour from his home to a train station where he waited 10 min and then boarded a train that

averaged 45 miles an hour to the city. If his entire trip was 35 miles and took 1 hr, how far did he live from the train station?

45 Airfield *A* is 325 miles due north of airfield *B*, and airfield *C* is 416 miles due east of airfield *B*. The pilot of a small plane flew from field *A* to field *B* on one day and on to field *C* on the next day. The wind was blowing from the south at a rate of 12 miles per hour the first day and from the west at a rate of 16 miles per hour the second day. Find the airspeed of the plane if both flights took the same length of time.

46 A truck carrying a group of construction workers left the construction camp for the work site. Six minutes later the resident engineer left the camp in his jeep and passed the truck 5 miles from the camp. If traveling time from the camp to the site was 24 min for the truck and 12 min for the jeep, find the average speed of the truck.

47 Mr. Peterson borrowed $1000 from a loan company and agreed to repay it in equal monthly installments plus interest of 1 percent per month on the outstanding principal. At the end of 5 months, he had paid a total of $45 in interest. Find the amount of his monthly installment.

48 A man invested $1440 in the common stock of one company and $2160 in the stock of another. The price per share of the second stock was three-fourths that of the first. The next day the price of the more expensive stock declined by $0.75 a share, while the price of the other advanced $1.50 a share, and as a result, the value of his investment increased $45. Find the price per share of the more expensive stock.

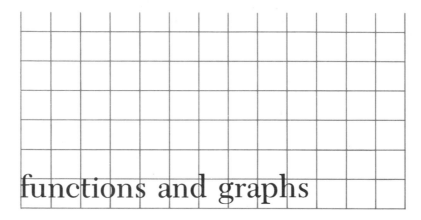

functions and graphs

In Sec. 1.3 we discussed the concept of one-to-one correspondence between the elements of two sets. In this chapter we shall discuss a specific type of correspondence leading to the concept of a function. Functions are very important in mathematics, and we shall use them extensively in the remainder of this book. In our discussion we shall frequently refer to *variables* and *constants*. The reader should review the definitions of these terms in Sec. 1.2.

6.1
ORDERED PAIRS OF NUMBERS

The fundamental ideas to be discussed in this chapter are illustrated by the following situations.

Two weeks before a concert 500 tickets were printed, to be sold at $2 each. If x stands for the number of tickets sold up to a certain time and y stands for the value of these tickets in dollars, then

$$y = 2x \tag{1}$$

The replacement set D for x in (1) is $D = \{x \mid x$ is an integer and $0 \le x \le 500\}$. The replacement set R for y is $R = \{y \mid y = 2x$ and x belongs to $D\}$. If at the end of each of the first 4 days the total number of tickets sold was 100, 175, 200, and 250, respectively, we have the following set of corresponding values of x and y:

x	100	175	200	250
y	200	350	400	500

We can express this correspondence as the ordered pairs (100, 200), (175, 350), (200, 400), (250, 500). The first number in each pair is the replacement for x, and the second is the replacement for y. This illustrates the following definition.

Ordered Pair of Numbers

An *ordered pair of numbers*, written (x, y), is a pair of numbers in which the first number is a member of some set A, and the second is a member of another set B.

As a second illustration, consider the formula

$$s = 16.1t^2 + v_0 t \qquad (2)$$

In this formula, s is the vertical distance in feet that a compact body will fall in t seconds if it is given an initial vertical velocity of v_0 feet per second. If a steel bearing is thrown downward with a velocity of 3.5 feet per second from the top of a tower 420 feet high, it will strike the ground at the end of 5 seconds, since by (2)

$$s = (16.1)5^2 + (3.5)5 = 402.5 + 17.5 = 420$$

In (2), v_0 is a constant since its replacement set contains only one number, in this case, 3.5; t is a variable whose replacement set is $D = \{t \mid 0 \le t \le 5\}$; s is a variable whose replacement set is $R = \{s \mid 420 \ge s \ge 0\}$. Furthermore, (2) establishes a one-to-one correspondence between the elements of D and the elements of R. Six of the pairs of corresponding numbers are

t	0	1	2	3	4	5
s	0	19.6	71.4	155.4	271.6	420

Consequently Formula (2) establishes a set of ordered pairs of numbers that includes (0, 0), (1, 19.6), (2, 71.4), (3, 155.4), (4, 271.6), and (5, 420), where the first number in each pair is a replacement for t, and the second is the corresponding replacement for s. Of course the number of ordered pairs established by Formula (2) is unlimited since we may assign any real number in the set D to t and calculate the corresponding replacement in the set R for s.

6.2
FUNCTIONS

In each of the illustrations in Sec. 6.1, a formula establishes a set of ordered pairs of corresponding numbers. The first number in each pair is an element of a set D, and the second is an element of a set R. Such a set of ordered pairs is called a function, defined more precisely as follows:

Function, Domain, Range

If (1) D is a set of numbers, (2) there exists a rule such that for each element x in D, one and only one number y is determined, and (3) R is the set of all numbers y, then the set of ordered pairs $\{(x, y)\}$ is called a *function* with *domain D* and *range R*.

For example, in the first illustration of Sec. 6.1, the equation $y = 2x$ is the rule that defines a function where $D = \{x \mid x$ is an integer and $0 \le x \le 500\}$, $R = \{y \mid y = 2x\}$, and the function is $\{(x, y) \mid x$ belongs to D and $y = 2x$ belongs to $R\}$.

Independent and Dependent Variables

The *independent variable* associated with a function is the variable whose replacement set is the set of first numbers in the ordered pairs of the function, and the *dependent variable* is the variable whose replacement set is the set of second numbers in the ordered pairs. For example, in the function $\{(x, y) \mid y = f(x)\}$, x is the independent variable and y is the dependent variable, and in $\{(t, s) \mid s = 16.1t^2 + v_0 t\}$, the independent and dependent variables are t and s, respectively.

Since the definition of a function requires that each element in the domain correspond to one and only one element in the range, no two pairs in the function will have the same first element. Consequently, the two pairs $(3, 1)$ and $(3, -2)$ could not occur in a function since then we should have the element 3 in the domain associated with two elements in the range.

The rule that establishes the correspondence between the elements of the domain and range is not necessarily an equation. The rule can be, and frequently is, an agreement that is not expressed in algebraic symbols. For example, a set of ordered pairs of numbers is established by a table of statistics composed of two columns of numbers if we agree that the first number in each pair is a number in the first column and the second number in each pair is the corresponding number in the second column.

The following agreement also establishes a function: Let

$$D = \{1, 2, 3, \ldots r, \ldots, n - 2, n - 1, n\}$$

and

$$R = \{n + 1, n + 2, n + 3, \ldots, n + r, \ldots, 2n - 2, 2n - 1, 2n\}$$

Furthermore, let the rth term of D be paired with the $[n - (r - 1)]$st term of R. Therefore, if $r = 1$, then $n - (r - 1) = n$, and if $r = 2$, then

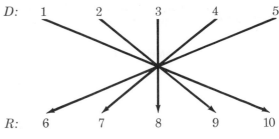

FIGURE 6.1

$n - (r - 1) = n - 1$, etc. Therefore, the first term of D is paired with the nth term $2n$ of R, the second term of D is paired with the $(n - 1)$st term $2n - 1$ of R, and so on, until the nth term of D is paired with the first term $n + 1$ of R. The set of ordered pairs thus determined is the function. If $n = 5$, then D and R are the sets indicated in Fig. 6.1, and the numbers connected by the arrows are the elements in each of the ordered pairs that constitute the function. Thus the function is $\{(1, 10), (2, 9), (3, 8), (4, 7), (5, 6)\}$.

6.3
FUNCTIONAL NOTATION

It has been customary for some time to designate by the letter f the function whose rule is $y = f(x)$, and either of the following notations may be used for this purpose:

$$f = \{(x, y) \mid y = f(x)\}$$
$$f = \{(x, f(x))\}$$

The second notation illustrates the fact that it is not necessary to use the letter y for the dependent variable, since $f(x)$ stands for the second number in the ordered pair with first number x. Thus, in the function $f = \{(x, f(x))\}$, f designates the function and $f(x)$ is the value of the function for a specified replacement for x. For example, if $y = 3x^2 - 2x + 4$ and both D and R are sets of real numbers, we may designate the function f in any one of the following ways:

$$f = \{(x, y) \mid y = 3x^2 - 2x + 4\}$$
$$f = \{(x, f(x)) \mid f(x) = 3x^2 - 2x + 4\}$$
$$f = \{(x, 3x^2 - 2x + 4)\}$$

Furthermore, the value of f for $x = 5$ is $f(5) = 3(5^2) - 2(5) + 4 = 75 - 10 + 4 = 69$.

If more than one function is involved in a particular discussion, it is customary to designate one of them by f and the others by letters other than f.

Thus the functions defined by $y = h(x)$, $y = g(x)$, and $y = k(x)$ are $h = \{(x, h(x))\}$, $g = \{(x, g(x))\}$, and $k = \{(x, k(x))\}$, respectively.

EXAMPLE 1   If

$$f(x) = \frac{x - 2}{x + 1}$$

find the function values $f(2)$, $f(\frac{1}{2})$, and $f(-\frac{3}{4})$.

SOLUTION

$$f(2) = \frac{2 - 2}{2 + 1} = \frac{0}{3} = 0$$

$$f(\tfrac{1}{2}) = \frac{\frac{1}{2} - 2}{\frac{1}{2} + 1}$$

$$= \frac{1 - 4}{1 + 2} \qquad \text{multiplying each member of the fraction by 2}$$

$$= -\tfrac{3}{3} = -1$$

$$f(-\tfrac{3}{4}) = \frac{-\frac{3}{4} - 2}{-\frac{3}{4} + 1} = \frac{-3 - 8}{-3 + 4} = -11$$

EXAMPLE 2   If

$$f(x) = x^2 - x - 3, \qquad g(x) = \frac{x^2 - 1}{x + 2}, \qquad \text{and} \qquad h(x) = f(x) + g(x)$$

find $h(2)$.

SOLUTION

$$h(2) = f(2) + g(2)$$
$$f(2) = 2^2 - 2 - 3 = -1$$
$$g(2) = \frac{4 - 1}{2 + 2} = \frac{3}{4}$$

Hence,

$$h(2) = -1 + \tfrac{3}{4}$$
$$= -\tfrac{1}{4}$$

EXAMPLE 3

Find the set of ordered pairs $\{(x, y)\}$ if $y = x^2 - 2x - 3$ and $D = \{x \mid x \text{ is an integer and } 1 \le x \le 4\}$.

SOLUTION   We first note that $D = \{1, 2, 3, 4\}$. Then we have

$$y = \begin{cases} 1^2 - 2(1) - 3 = -4 & \text{for } x = 1 \\ 2^2 - 2(2) - 3 = -3 & \text{for } x = 2 \end{cases}$$

Similarly, if $x = 3$, then $y = 0$, and if $x = 4$, then $y = 5$. Thus, $\{(x, y)\} = \{(1, -4), (2, -3), (3, 0), (4, 5)\}$.

EXAMPLE 4

If $D = \{x \mid x \text{ is an integer and } -2 \le x \le 1\}$, find the function

$$\{(x, f(x)) \mid f(x) = x^3 - 3 \text{ and } x \text{ belongs to } D\}$$

SOLUTION

$$D = \{-2, -1, 0, 1\}$$

Thus,

$$f(-2) = -8 - 3 = -11$$
$$f(-1) = -1 - 3 = -4$$
$$f(0) = -3$$

and

$$f(1) = 1 - 3 = -2$$

Hence,

$$f = \{(x, f(x)) \mid f(x) = x^3 - 3 \text{ and } x \text{ belongs to } D\}$$
$$= \{(-2, -11), (-1, -4), (0, -3), (1, -2)\}$$

EXAMPLE 5

If $f(x) = 3x + 4$ and $D = \{x \mid -1 \le x \le 3\}$, find the range of $f(x)$.

SOLUTION   We first prove that the value of $3x + 4$ increases with x. If $X > x$, then

$$3X > 3x \qquad \text{by (2.33)}$$

Hence,

$$3X + 4 > 3x + 4 \qquad \text{by (2.10)}$$

Hence, if x belongs to D, the function value $f(x) = 3x + 4$ is least when $x = -1$ and greatest when $x = 3$. Consequently, since

$$f(-1) = -3 + 4 = 1 \qquad \text{and} \qquad f(3) = 9 + 4 = 13$$

then

$$R = \{y \mid 1 \le y \le 13\}$$

Occasionally the independent variable x in a function is expressed in terms of another variable, as in the statement $y = f(x)$, where $x = g(t)$. In such cases y can be expressed in terms of t by replacing x by $g(t)$ in $f(x)$.

EXAMPLE 6

If $y = f(x) = (x^2 - 2)/(x^2 + 4)$ and $x = t + 1$, express y as a function of t.

SOLUTION

$$y = f(x) = f(t + 1) = \frac{(t+1)^2 - 2}{(t+1)^2 + 4} = \frac{t^2 + 2t - 1}{t^2 + 2t + 5}$$

6.4
RELATIONS

An essential requirement in the definition of a function is that the rule in the definition must establish a correspondence between the elements of the domain D and the range R such that each element of D corresponds to one and only one element of R. As a consequence, no two pairs in the set of ordered pairs defined by the rule can have the same first element and different second elements.

We now consider the equation

$$y^2 = x \tag{1}$$

where $D = \{0, 1, 4, 9\}$. If we replace x by each element in D and solve for y, we obtain the following set of ordered pairs: $\{(0, 0), (1, 1), (1, -1), (4, 2), (4, -2), (9, 3), (9, -3)\}$. Notice that each of the two pairs $(1, 1)$ and $(1, -1)$, $(4, 2)$ and $(4, -2)$, $(9, 3)$ and $(9, -3)$ have the same first element. Hence, according to our definition, $y^2 = x$ does not define a function. In classical literature, the set of ordered pairs defined by $y^2 = x$ is called a *double-valued function*. The tendency among modern writers, however, is to call such a set of pairs a *relation*.

Relation

The definition of a relation differs from the definition of a function only in the fact that it is not necessary for the first elements of pairs that constitute a relation to be different numbers. Although in our example each value of x except zero was paired with two values of y, a relation may contain more than two ordered pairs with the same first element but with different second elements. For example, $(0,0)$, $(0, 1)$, and $(0, -1)$ are ordered pairs in the relation defined by $y^3 - y = x$.

EXAMPLE 1

Find the relation defined by $y^2 = 25 - x^2$, where x belongs to $D = \{0, 3, 4, 5\}$.

SOLUTION If we assign each of the numbers in D to x and solve for y, we obtain the following corresponding values:

x	0	3	4	5
y	± 5	± 4	± 3	0

Hence,

$\{(x, y) \mid y^2 = 25 - x^2$ and x belongs to $\{0, 3, 4, 5\}\} = \{(0, -5), (3, -4), (4, -3),$
$(5, 0), (4, 3), (3, 4), (0, 5)\}$

EXAMPLE 2

Find the set of ordered pairs defined by $y = \sqrt{x + 1}$ and $D = \{-1, 0, 3, 8\}$. Is the set a function or a relation?

SOLUTION Since y is the principal square root of $x + 1$, it stands for a non-negative number. Hence, we have the following set of corresponding values of x and y:

x	-1	0	3	8
y	0	1	2	3

Consequently,

$\{(x, y) \mid y = \sqrt{x + 1}$ and x belongs to $D = \{-1, 0, 3, 8\}\}$ is $\{(-1, 0), (0, 1),$
$(3, 2), (8, 3)\}$

Since no two first elements in this set of pairs are the same, the set is a function.

EXERCISE 6.1
Function Values

1. If $f(x) = 2x + 6$, find $f(0)$, $f(2)$, $f(\frac{1}{2})$.
2. If $h(x) = 3x - 1$, find $h(3)$, $h(-2)$, $h(-\frac{2}{3})$.
3. If $F(x) = \dfrac{3x}{4} - \dfrac{5}{8}$, find $F(4)$, $F(3)$, $F(\frac{5}{6})$.
4. If $g(x) = \dfrac{x}{3} - \dfrac{7}{9}$, find $g(3)$, $g(5)$, $g(\frac{2}{3})$.
5. If $F(x) = 3x^2 - 2x + 2$, find $F(-2)$, $F(4)$, $F(\frac{1}{3})$.
6. If $G(y) = 6y^2 - 3y - 1$, find $G(2)$, $G(-3)$, $G(\frac{2}{3})$.

7   If $H(z) = 3z^3 - z + 2$, find $H(-1)$, $H(-2)$, $H(\frac{1}{3})$.

8   If $f(w) = 2w^3 - 3w^2 + 2w - 1$, find $f(-2)$, $f(3)$, $f(\frac{1}{2})$.

9   If $f(x) = 2x - 5$, $g(x) = x + 3$, and $F(x) = f(x) + g(x)$, find $F(1)$, $F(-3)$, $F(2)$.

10 If $f(x) = 3x - 1$, $g(x) = 3x + 1$, and $F(x) = f(x) \cdot g(x)$, find $F(3)$, $F(\frac{1}{3})$, $F(\frac{1}{6})$.

11 If $f(x) = x^2 - 3x - 4$, $g(x) = x + 1$, and $F(x) = \dfrac{f(x)}{g(x)}$, find $F(8)$, $F(-3)$, $F(\frac{1}{4})$.

12 If $f(x) = x^3 + 4x - 5$, $g(x) = x - 1$, and $F(x) = \dfrac{f(x)}{[g(x)]^2}$, find $F(3)$, $F(-2)$, $F(\frac{3}{2})$.

13 If $f(x) = 3x + 5$, find $f(x - 3) + f(x)$.

14 If $F(t) = t^3 - 1$, find $F(t - 1) - F(t)$.

15 If $G(s) = \dfrac{s + 2}{s}$, find $G(s - 3) - G(s)$.

16 If $f(x) = x^2 + x - 1$, find $\dfrac{f(x + h) - f(x)}{h}$.

17 If $y = 2x + 5$ and $x = t^2 - 3$, express y in terms of t.

18 If $y = x^2 + 3x - 2$, and $x = t - 1$, express y in terms of t.

19 If $y = 3x^2 - x + 2$ and $x = \dfrac{2}{t - 2}$, express y in terms of t.

20 If $y = \dfrac{x - 1}{x + 2}$ and $x = \dfrac{t + 2}{t - 1}$, express y in terms of t.

In each of Probs. 21 to 24, find the set of ordered pairs (x, y) if x and y are related by the given equation and the domain D is the indicated set of numbers.

21 $y = 2x - 5$, $D = \{-2, 0, 2, 4, 6\}$

22 $y = x^2 + 3$, $D = \{-3, -2, 0, 1, 2\}$

23 $y = \dfrac{x - 1}{x}$, $D = \{x \mid x$ is a positive integer and $x \le 5\}$

24 $y = \dfrac{x^2 - 4}{x + 3}$, $D = \{x \mid x$ is an integer and $-2 \le x \le 2\}$

25 If $D = \{x \mid x$ is an integer and $-2 \le x \le 3\}$, find the set of ordered pairs $\{(x, f(x)) \mid f(x) = x^2 - 3\}$.

26 If $D = \{x \mid x$ is an integer and $-1 \le x \le 2\}$, find the set of ordered pairs $\left\{(x, f(x)) \mid f(x) = \dfrac{x^2 - 1}{x^2 + 2}\right\}$

27 If $D = \{x \mid x$ is an integer and $0 \le x \le 4\}$, find the set of ordered pairs $\left\{(x, f(x)) \mid f(x) = \dfrac{x^3 - 3}{x + 1}\right\}$

28 If $D = \{x \mid x$ is an integer and $-3 \le x \le 1\}$, find the set of ordered pairs $\left\{(x, f(x)) \mid f(x) = \dfrac{\sqrt{x^2 + 2}}{x + 4}\right\}$

In each of Probs. 29 to 32, find the range of the function indicated in the problem for the specified domain.

29 $y = 2x - 5, D = \{x \mid -3 \le x \le 7\}$
30 $y = 6 - 5x, D = \{x \mid 1 \le x \le 6\}$
31 $y = 3x^2 + 2, D = \{x \mid 0 \le x \le 7\}$
32 $y = 5 - 2x^2, D = \{x \mid -5 \le x \le 0\}$

Determine the set of ordered pairs $\{(x, y)\}$ in each of Probs. 33 to 36. State whether each set is a function or a relation, and give your reason.

33 $\{(x, y) \mid x^2 + y^2 = 36\}, D = \{x \mid x \text{ is an integer and } 1 \le x \le 6\}$
34 $\{(x, y) \mid y^2 = 2x - 3\}, D = \{x \mid x \text{ is an integer and } 2 \le x \le 6\}$
35 $\{(x, y) \mid y = 2x^2 - 1\}, D = \{x \mid x \text{ is an integer and } 0 \le x \le 4\}$
36 $\{(x, y) \mid y^2 = 3x^2 - 2\}, D = \{x \mid x \text{ is an integer and } 3 \le x \le 6\}$

6.5
THE RECTANGULAR COORDINATE SYSTEM

In this section we introduce a device for associating an ordered pair of numbers with a point in a plane. Invented by the French mathematician and philosopher Rene Descartes (1596–1650), it is called the *rectangular* or *Cartesian coordinate system*.

Coordinate Axes

In order to set up this system, we construct two perpendicular number lines in the plane and choose a suitable scale on each. For convenience these lines are horizontal and vertical, and the unit length on each is the same (see Fig. 6.2), although neither restriction is necessary. The two lines are called the *coordinate axes*, the horizontal line being the X axis and the vertical line the Y axis. The intersection of the two lines is the *origin*, designated by the letter O. The coordinate axes separate the plane into four sec-

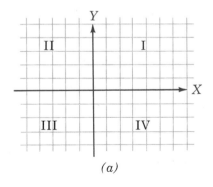

(a)

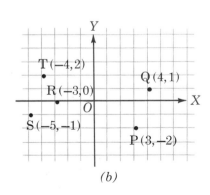

(b)

FIGURE 6.2

tions called *quadrants*. These quadrants are numbered I, II, III, and IV counterclockwise, as indicated in Fig. 6.2*a*.

Directed Distance

Next, it is agreed that horizontal distances measured to the right from the *Y* axis are positive, and horizontal distances measured to the left are negative. Similarly, vertical distances measured upward from the *X* axis are positive, and vertical distances measured downward are negative. These distances, because of their signs, are called *directed distances*. Finally, we agree that the first number in an ordered pair of numbers represents the directed distance from the *Y* axis to a point and the second number in the pair represents the directed distance from the *X* axis to the point. It follows then that an ordered pair of numbers uniquely determines the position of a point in the plane. For example, (4, 1) determines the point *Q* in Fig. 6.2*b* that is 4 units to the right of the *Y* axis and 1 unit above the *X* axis. Similarly, the ordered pair $(-5, -1)$ determines the point *S* in Fig. 6.2*b* that is 5 units to the left of the *Y* axis and 1 unit below the *X* axis. Conversely, each point in the plane determines a unique ordered pair of numbers. For example, the point *P* in Fig. 6.2*b* is 3 units to the right of the *Y* axis and 2 units below the *X* axis, and so *P* determines the ordered pair $(3, -2)$.

Cartesian Plane

A plane in which the coordinate axes have been constructed is called a *Cartesian plane*.

Coordinates; Abscissa and Ordinate

The two numbers in an ordered pair that is associated with a point in the Cartesian plane are called the *coordinates* of the point. The first number is called the *abscissa* of the point, and it is the directed distance from the *Y* axis to the point. The second number in the pair is the *ordinate* of the point, and it represents the directed distance from the *X* axis to the point.

Plotting a Point

The procedure for locating a point in the plane by means of its coordinates is called *plotting* the point. The notation $P(a, b)$ means that *P* is the point whose coordinates are (a, b). In order to plot the point $T(-4, 2)$ we count 4 units to the left of the origin on the *X* axis and then upward 2 units and thus arrive at the point. Similarly, the point $R(-3, 0)$ is 3 units to the left of the origin and on the *X* axis. The general point and its coordinates are written $P(x, y)$.

6.6
THE GRAPH OF A FUNCTION

By use of the rectangular coordinate system, we can obtain a geometric representation, or a geometric "picture" of a function. For this purpose, we

require that each ordered pair of numbers (x, y) of a function be the coordinates of a point in the Cartesian plane with x as the abscissa and y as the ordinate. We then define the graph of a function as follows.

Graph of a Function

The *graph of a function* is the totality of points (x, y) whose coordinates
constitute the set of ordered pairs of the function with x a number in the
domain D and y the corresponding number in the range R.

The graphs of most functions we shall discuss in this chapter are
smooth† continuous curves. When we say that the graph of a function is a
curve, we mean that the point determined by each ordered pair of numbers
in the function is on the curve and that the coordinates of each point on the
curve are an ordered pair of numbers in the function.

We shall illustrate the procedure for obtaining the graph of a function by
explaining the steps in the construction of the graph of the function defined
by $y = x^2 - 3x - 1$ for $-2 \leq x \leq 5$. Note that this function is

$$\{(x, y) \mid y = x^2 - 3x - 1 \text{ and } x \text{ belongs to } D = \{x \mid -2 \leq x \leq 5\}\}$$

The first step is to assign each of the integers in D to x and then
calculate each corresponding value of y by using the defining equation
$y = x^2 - 3x - 1$. Before doing this, however, it is advisable to make a table
like the one below in which to record the corresponding values.

Now we shall calculate the corresponding value of y when x is assigned the
numbers $-2, -1, 0, 1, 2, 3, 4,$ and 5.

$$y = \begin{cases} (-2)^2 - 3(-2) - 1 = 9 & \text{for } x = -2 \\ (-1)^2 - 3(-1) - 1 = 3 & \text{for } x = -1 \\ 0^2 - 3(0) - 1 = -1 & \text{for } x = 0 \end{cases}$$

Continuing this process, we obtain the additional ordered pairs $(2, -3)$, $(3,$
$-1)$, $(4, 3)$, and $(5, 9)$ and enter the results in the table:

x	-2	-1	0	1	2	3	4	5
y	9	3	-1	-3	-3	-1	3	9

Now we plot the points (x, y) thus determined, as shown in Fig. 6.3. Because
there is an ordered pair in the function for every intermediate real value of x,

†At present we are not in a position to give a rigorous definition of a "smooth continuous curve." For our purposes, however, the following *description* will suffice. A
smooth continuous curve contains no breaks or gaps, and there are no sudden or
abrupt changes in its direction.

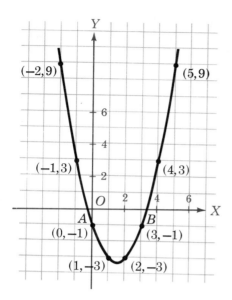

FIGURE 6.3

we connect the plotted points with a smooth curve. This is the portion of the graph defined by $y = x^2 - 3x - 1$ over the specified domain.

We now investigate the nature of the graph as x increases through values greater than 5 and decreases through values less than -2. For this purpose we first replace x in $y = x^2 - 3x - 1$ by $5 + h$ and get $y = h^2 + 7h + 9$. From the latter equation we see that as h increases from zero, y increases from 9. Hence the graph extends upward and to the right from the point $(5, 9)$. Similarly, if we replace x by $-2 - k$, we get $y = k^2 + 7k + 9$, and we therefore conclude that as k increases from zero, y increases from 9. Hence the graph extends upward and to the left from the point $(-2, 9)$.

Zero of a Function

A *zero* of a function defined by $y = f(x)$ is a value of the independent variable x for which $y = 0$. Hence the zeros of a function are the abscissas of the points where the graph crosses the X axis. The zeros of many classes of functions can be obtained by algebraic methods, but we must depend upon graphical methods for others. The zeros of the function $\{(x, y) | y = x^2 - 3x - 1\}$ are the abscissas of the points A and B in Fig. 6.3 and are approximately -0.3 and 3.3.

If the domain of a function is not specified, it is assumed to be the real-number system. In such cases, to get a set of ordered pairs for constructing the graph, it is usually advisable to start by assigning consecutive small integers to x and continue the process until a sufficient number of points are obtained to determine the nature of the graph. At times the points obtained by assigning consecutive integers to x are too far apart to enable one to sketch

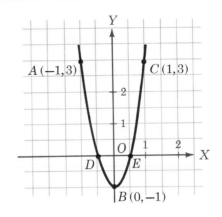

FIGURE 6.4

the curve. For example, in the function defined by $y = 4x^2 - 1$, if we assign $-1, 0,$ and 1 to x, we get the pairs $(-1, 3), (0, -1),$ and $(1, 3)$. These pairs determine the points $A, B,$ and C in Fig. 6.4, and it is evident that these points alone do not show the nature of the graph. We can get two more points by assigning $-\frac{1}{2}$ and $\frac{1}{2}$ to x and obtaining the pairs $(-\frac{1}{2}, 0)$ and $(\frac{1}{2}, 0)$. When these points are plotted, we get the points D and $E,$ and then the curve in Fig. 6.4 can be sketched.

6.7
LINEAR FUNCTIONS

It is proved in analytic geometry that if the domain is the set of real numbers, the graph of the function $\{(x, y)\,|\,y = ax + b\}$ is a straight line. Such a function is called a *linear function*, and its graph is completely determined by two points. If the graph does not pass through the origin or is not parallel to either axis, the points where the graph crosses the axes are readily determined by assigning 0 to x in the equation $y = ax + b$ to obtain the point $(0, b)$ and then by assigning 0 to y and solving for x to obtain $(-b/a, 0)$. These points determine the graph, but it is advisable to calculate a third point as a check.

The X Intercept and Y Intercept

The abscissa of the point where the line crosses the X axis is called the X *intercept*, and the ordinate of the point where it crosses the Y axis is the Y *intercept*. The graph of $\{(x, y)\,|\,y = ax\}$ passes through the origin since $(0, 0)$ satisfies the equation $y = ax$. Hence, we must assign a number other than zero to x to get a second point to determine the graph and a third number as a check.

The graph of $\{(x, y)\,|\,x = a\}$ is the line parallel to the Y axis at the directed distance of a units from it. The graph of $\{(x, y)\,|\,y = b\}$ is the line

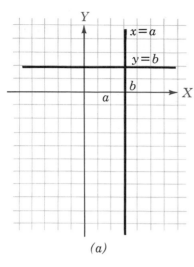

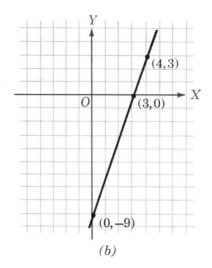

(a) (b)

FIGURE 6.5

parallel to the X axis and at the directed distance of b units from it. These graphs are shown in Fig. 6.5a.

EXAMPLE Construct the graph of the function defined by $y = 3x - 9$.

SOLUTION We find the intercepts by assigning 0 to x and solving for y and by assigning 0 to y and solving for x. We find a third point by assigning 4 to x and solving for y. Thus we get the following table of corresponding numbers:

x	0	3	4
y	-9	0	3

We now plot the points determined by these pairs of numbers, draw a straight line through them, and obtain the graph in Fig. 6.5b.

EXERCISE 6.2
Linear Function

1 On one set of axes, plot the points determined by the following ordered pairs of numbers: $(2, 6)$, $(7, 3)$, $(5, -1)$, $(3, -4)$, $(-5, 6)$, $(-1, 4)$, $(-6, -4)$, $(-2, -1)$, $(0, 7)$, $(0, -5)$, $(-4, 0)$, $(7, 0)$.

2 Describe the line on which all points determined by each of the following requirements are located: (a) the abscissa of each point is zero; (b) the ordinate of each point is zero; (c) the two coordinates of each point are numerically equal but opposite in sign; (d) the two coordinates of each point are equal.

3 Describe the line on which all points determined by each of the following requirements are located: (*a*) the abscissa of each point is 3; (*b*) the
ordinate of each point is 6; (*c*) the abscissa of each point is -5; (*d*) the
ordinate of each point is -1.

4 If n is a positive real number, state the quadrant in which each of the
following points is located: $(3, n)$, $(n, 5)$, $(n, -7)$, $(-2, -n)$, (n, n), $(n, -n)$,
$(-n, n)$, $(-n, -n)$.

Construct the graph of the function defined by the equation in each of Probs. 5 to 20,
and estimate the zero or zeros of each.

5 $y = x - 2$ **6** $y = x + 7$

7 $y = 2x + 5$ **8** $y = 3x - 8$

9 $y = -2x + 10$ **10** $y = -3x - 9$

11 $y = -2x + 13$ **12** $y = -4x - 1$

13 $y = x^2$ **14** $y = x^2 - 9$

15 $y = x^2 + 2$ **16** $y = -x^2 - 5$

17 $y = x^2 - 3x - 4$ **18** $y = -x^2 - 3x + 10$

19 $y = -x^2 + 4x - 4$ **20** $y = x^2 - 5x + 6$

In Probs. 21 to 24, find the intercepts of the graph of the indicated function; then construct the graph for the indicated domain.

21 $\{(x, y) \mid y = 3x - 12\}$, $D = \{x \mid 0 \le x \le 10\}$

22 $\{(x, f(x)) \mid f(x) = -2x + 16\}$, $D = \{x \mid -3 \le x \le 9\}$

23 $\{(x, y) \mid y = 4x + 8\}$, $D = \{x \mid -4 \le x \le 5\}$

24 $\{(x, f(x)) \mid f(x) = -5x - 15\}$, $D = \{x \mid -6 \le x \le 2\}$

6.8
SOME SPECIAL FUNCTIONS

In this section we shall discuss examples of functions whose graphs are not
continuous curves. As a first example we shall consider the function

$$f: \begin{cases} \{(x, y) \mid y = x + 2 \text{ and } x < 3\} \\ \{(x, y) \mid y = -x + 4 \text{ and } x \ge 3\} \end{cases}$$

If $x < 3$, the graph of $y = x + 2$ extends to the left and downward from the
point $(3, 5)$†, and if $x \ge 3$, the graph of $y = -x + 4$ extends to the right and
downward from the point $(3, 1)$. Consequently, the graph of the function f
consists of the two rays shown in Fig. 6.6.

---

† The point $(3, 5)$ is not on the graph since $x < 3$. However, if $P(a, b)$ is on the
graph and a and b increase, P moves toward $(3, 5)$ and the distance from P to $(3, 5)$
becomes less than any preassigned number.

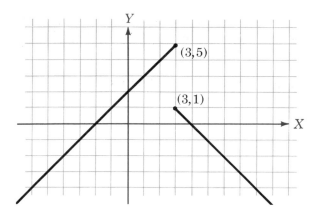

FIGURE 6.6

An important function in the theory of numbers is the *bracket function* $\{(x, y)\,|\,y = [x]\}$, where the notation $[x]$ means the greatest integer that is less than or equal to x. Hence, in the equation $y = [x]$, if $x = \frac{1}{2}$, then $y = 0$, since 0 is the greatest integer less than $\frac{1}{2}$. Similarly, if $x = 2\frac{1}{2}$, then $y = [2\frac{1}{2}] = 2$. Furthermore, if n is an integer, $y = [n] = n$. Hence, by the above and similar arguments, we have the following corresponding values of x and y for the bracket function:

$$y = \begin{cases} 0 & 0 \le x < 1 \\ 1 & 1 \le x < 2 \\ 2 & 2 \le x < 3 \\ \cdots & \cdots\cdots\cdots \\ n & n \le x < n+1 \quad \text{for } n \text{ an integer} \end{cases}$$

Consequently, the graph of the bracket function is a set of horizontal line segments, four of which are shown in Fig. 6.7.

6.9
THE INVERSE OF A FUNCTION

In order to illustrate the meaning of the inverse of a function, we shall compare the functions determined by

$$y = f(x) = 3x - 2 \tag{1}$$

and

$$y = g(x) = \frac{x + 2}{3} \tag{2}$$

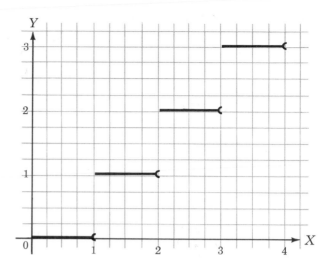

FIGURE 6.7

Equation (2) is obtained by interchanging x and y in (1) and solving the resulting equation for y, as follows:

$x = 3y - 2$ **interchanging x and y in Eq. (1)**

$y = \dfrac{x + 2}{3}$ **solving for y**

We first obtain the function f determined by (1) with the domain $D = \{0, 1, 2, 3, 4, 5\}$. Replacing x in (1) by each number in D and computing the corresponding value of $y = f(x)$, we get the following table:

x	0	1	2	3	4	5
y	-2	1	4	7	10	13

Hence, the function f is

$f = \{(x, y) \mid y = f(x) = 3x - 2 \text{ and } x \in D\}$

$\quad = \{(0, -2), (1, 1), (2, 4), (3, 7), (4, 10), (5, 13)\}$ \hfill (3)

Since the range R of f is the set whose elements are the second numbers in the pairs in Eq. (3),

$R = \{-2, 1, 4, 7, 10, 13\}$

Next we obtain the function g determined by Eq. (2) with the domain equal to the range of f. Hence, for this function, $D = \{-2, 1, 4, 7, 10, 13\}$. Replacing x in (2) by each number in D and computing the corresponding value of y, we have the following table:

x	-2	1	4	7	10	13
y	0	1	2	3	4	5

Consequently,

$$g = \{(x, y) \mid y = g(x) = \frac{x + 2}{3} \text{ and } x \in D\}$$

$$= \{(-2, 0), (1, 1), (4, 2), (7, 3), (10, 4), (13, 5)\}$$

Now if we compare the ordered pairs in functions f and g, we see that each pair in g is a pair in f with the first and second elements interchanged. Furthermore, the domain of f is the range of g, and the range of f is the domain of g.

We next show that the above statement is true if the domain of f and of g is the set of all real numbers. Let (b, a) be an ordered pair of g. Then

$$a = g(b) = \frac{b + 2}{3}$$

Now,

$$f(a) = 3a - 2 \qquad \text{since } f(x) = 3x - 2$$

$$= 3\left(\frac{b + 2}{3}\right) - 2 \qquad \text{since } a = \frac{b + 2}{3}$$

$$= b + 2 - 2$$

$$= b$$

Therefore, $(a, f(a)) = (a, b)$, and thus (a, b) is an ordered pair in f if (b, a) is an ordered pair in g.

We call the function g the *inverse* of the function f. We shall define the inverse of a function more precisely later.

The graphs of the functions f and g are shown in Fig. 6.8, and from the figure we see that the position of the graph of f with respect to the Y axis is the same as the position of g with respect to the X axis.

The definition of a function requires that each value of the independent variable correspond to one, and only one, value of the dependent variable. Hence no two of the ordered pairs of a function can have the same first element and different second elements. We now consider the function $f = \{(x, y) \mid y = f(x)\}$ in which no two of the pairs have the same second element. If we interchange x and y in $y = f(x)$ and solve for y, we get an equation of the type of $y = g(x)$, and this determines the set of ordered pairs $g = \{(x, y) \mid y = g(x)\}$ each of which is a pair of f with the elements interchanged. Now since no two pairs of f have the same *second* element, then no two pairs

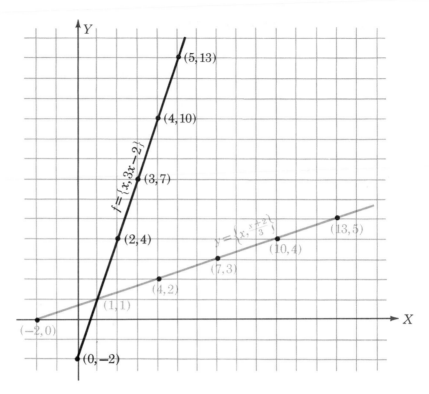

FIGURE 6.8

of g will have the same *first* element. Therefore, the set of ordered pairs g is a function.

Inverse of a Function Defined

We now define the inverse of a function as follows:

■ If f is a function such that no two of its ordered pairs have the same second element and different first elements, the function g obtained by interchanging the elements in each pair of f is called the *inverse* of f.

The customary notation for the inverse of f is f^{-1}, and we caution the reader that the number -1 is not to be interpreted as an exponent. In terms of this notation, the function g is the inverse f^{-1} of f. Also, f is the inverse g^{-1} of g.

EXAMPLE 1

If $D = \{-2, -1, 0, 1, 2\}$ and $f = \{(x, y) \mid y = f(x) = 2x + 5 \text{ and } x \in D\}$, then first, write the function f; second, obtain the equation that defines the inverse f^{-1} of f; third, obtain the domain of f^{-1}; fourth, write the function f^{-1}; fifth, show that $f(f^{-1}(x)) = x$ and that $f^{-1}(f(x)) = x$.

SOLUTION   First, we replace x by each element in D in $y = 2x + 5$ and obtain the following table of values:

x	-2	-1	0	1	2
y	1	3	5	7	9

Hence,

$$f = \{(-2, 1), (-1, 3), (0, 5), (1, 7), (2, 9)\}$$

Second, we interchange x and y in $y = 2x + 5$ and solve for y; thus,

$$x = 2y + 5 \qquad \text{interchanging } x \text{ and } y \text{ in } y = 2x + 5$$

$$y = \frac{x - 5}{2}$$

Hence,

$$f^{-1}(x) = \frac{x - 5}{2}$$

Third, the domain of f^{-1} is the range of f and is therefore

$$D = \{1, 3, 5, 7, 9\}$$

Fourth, replacing x in $y = (x - 5)/2$ by each number in D and computing the corresponding value of y, we get

$$f^{-1} = \{(1, -2), (3, -1), (5, 0), (7, 1), (9, 2)\}$$

Fifth, to get $f[f^{-1}(x)]$, we replace x in $2x + 5$ by $f^{-1}(x) = (x - 5)/2$ and get

$$f[f^{-1}(x)] = 2\left(\frac{x - 5}{2}\right) + 5 = x - 5 + 5 = x$$

Similarly,

$$f^{-1}(f(x)) = \frac{2x + 5 - 5}{2} \qquad \text{replacing } x \text{ in } (x - 5)/2 \text{ by } 2x + 5$$

$$= \frac{2x}{2} = x$$

EXAMPLE 2   Given

$$f = \{(x, y) \mid y = F(x) = \sqrt{x^2 + 16} \text{ and } x \geq 0\}$$

find the range of F^{-1}, and then obtain the function F^{-1}.

SOLUTION   The defining equation of F is

$$y = \sqrt{x^2 + 16}$$

Since $\sqrt{x^2 + 16}$ increases with x and is least for $x = 0$, the range of F is $\{x \mid x \geq 4\}$. Furthermore, the range and domain of F are the domain and range of F^{-1} respectively. Consequently, the domain of F^{-1} is $\{x \mid x \geq 4\}$ and the range is $\{y \mid y \geq 0\}$. To get the defining equation of F^{-1}, we proceed as follows:

$x = \sqrt{y^2 + 16}$ interchanging x and y in Eq. (4)

$x^2 = y^2 + 16$ equating the squares of the members of Eq. (5)

$y = \sqrt{x^2 - 16}$ solving for y

Since the range of F^{-1} is $\{y \mid y \geq 0\}$, we use only the positive values of the radical. Consequently,

$$F^{-1} = \{(x, y) \mid y = \sqrt{x^2 - 16} \text{ and } x \geq 4\}$$

6.10
GRAPHICAL SOLUTION OF INEQUALITIES

In order to find the solution set of an inequality of the type $f(x) > 0$, we first set y equal to $f(x)$ and obtain the equation

$$y = f(x) \tag{1}$$

Next, we construct the graph of Eq. (1). Now, if $P(x, y)$ is on the portion of the graph that is above the X axis, then $y > 0$ and, hence, $f(x) > 0$. Therefore the solution set of $f(x) > 0$ is the set of abscissas of the points on the graph of (1) that are above the X axis.

Similarly, the solution set of $f(x, y) < 0$ is the set of abscissas of the points on the graph of $y = f(x)$ that are below the X axis.

EXAMPLE 1   Find the solution sets of both

$3x - 6 > 0$ and $3x - 6 < 0$

by the graphical method.

SOLUTION   We let

$$y = 3x - 6 \tag{2}$$

By referring to Sec. 6.7 we see that the graph of Eq. (2) is a straight line, and we construct the line by use of the following table:

x	0	2	3
y	−6	0	3

The graph is shown in Fig. 6.9, and from it we see that the graph is above the X axis at the right of the point $P(2, 0)$. Hence the ordinate $y = 3x - 2$ of each point on this portion of the graph is positive and the solution set of $3x - 6 > 0$ is

$$\{x \mid x > 2\}$$

Likewise, the solution set of $3x - 6 < 0$ is

$$\{x \mid x < 2\}$$

since the graph of (2) is below the X axis at the left of $P(2, 0)$.

EXAMPLE 2 Determine the solution sets of both $x^2 - 3x - 4 > 0$ and $x^2 - 3x - 4 < 0$

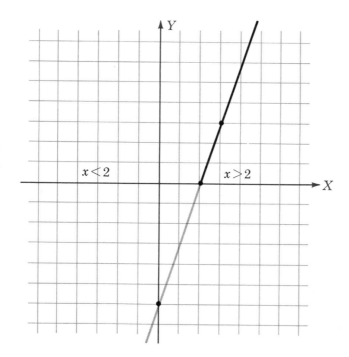

FIGURE 6.9

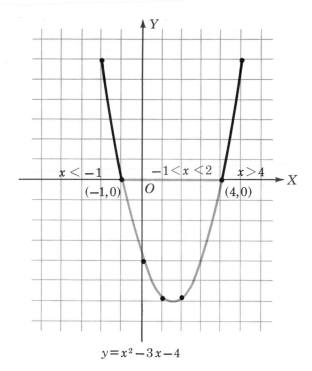

$$y = x^2 - 3x - 4$$

FIGURE 6.10

SOLUTION   We let

$$y = x^2 - 3x - 4 \tag{3}$$

and construct the graph of Eq. (2) by using the following table of values:

x	-2	-1	0	1	2	3	4	5
y	6	0	-4	-6	-6	-4	0	6

and obtain the curve shown in Fig. 6.10. The graph is above the X axis if $x > 4$ *or* if $x < -1$. Hence, the solution set of $x^2 - 3x - 4 > 0$ is

$$\{x \,|\, x > 4\} \cup \{x \,|\, x < -1\}$$

Furthermore, the graph is below the X axis if $-1 < x < 4$. Therefore, the solution set of $x^2 - 3x - 4 < 0$ is

$$\{x \,|\, -1 < x < 4\}$$

EXERCISE 6.3

Graphs, Inverses, and Inequalities

Construct the graph of the function in each of Probs. 1 to 4.

1 $f = \begin{cases} f(x) = x & \text{if } x \geq 0 \\ f(x) = -x & \text{if } x < 0 \end{cases}$

2 $f = \begin{cases} f(x) = x - 2 & \text{if } x \geq 2 \\ f(x) = x + 2 & \text{if } x < 2 \end{cases}$

3 $f = \begin{cases} f(x) = x^2 & \text{if } x \geq 0 \\ f(x(= -x & \text{if } x < 0 \end{cases}$

4 $f = \begin{cases} f(x) = \sqrt{9 - x^2} & \text{if } 0 \leq x \leq 3 \\ f(x) = 2x - 6 & \text{if } x > 3 \end{cases}$

Construct the graph of the function defined by the equation in each of Probs. 5 to 8.

5 $y = [x] - 1$ **6** $y = 3[x]$ **7** $y = [x] + x$ **8** $y = |x|$

Note   If $y = |x|$, then

$$y = \begin{cases} x & \text{if } x > 0 \\ 0 & \text{if } x = 0 \\ -x & \text{if } x < 0 \end{cases}$$

Find the inverse f^{-1} of the function in each of Probs. 9 to 16.

9 $f = \{(x, x + 3) \mid x \geq 0\}$

10 $f = \{(x, 3x - 1) \mid x \geq \frac{1}{3}\}$

11 $f = \left\{ \left(x, \frac{x}{3} + 3\right) \middle| x \leq -3 \right\}$

12 $f = \left\{ \left(x, \frac{x - 4}{3}\right) \middle| x \leq 1 \right\}$

13 $f = \{(x, x^2) \mid x \geq 0\}$ Why is the restriction $x \geq 0$ necessary if f is to have an inverse?

14 $f = \{(x, x^2 + 2) \mid x \geq 0\}$ Why is the restriction $x \geq 0$ needed?

15 $f = \{(x, x^2 + 5) \mid x \geq 0\}$ Why is the restriction $x \geq 0$ needed?

16 $f = \{(x, \sqrt{36 - x^2}) \mid 0 \leq x \leq 6\}$ Why is the restriction on x needed?

Find the equation that defines the inverse of the function defined by the equation in each of Probs. 17 to 20.

17 $y = \dfrac{x + 3}{x - 2}$ $x \neq 2$

18 $y = \dfrac{x^2 + 4}{2}$ $x \geq 1$

19 $y - 3 = \sqrt{x + 4}$ $x \geq -4$

20 (a) $y = x^2 + 6$ $x \geq 0$
   (b) $y = x^2 + 6$ $x \leq 0$

Find the solution set of the following inequalities by the graphical method.

21 $3x - 2 > 0$ **22** $4x + 1 > 0$

23 $2 - 5x < 0$ **24** $4 + 3x > 0$

25 $5x - 2 < 3x + 4$ **26** $7x + 5 > 9x - 3$

27 $4x + 3 < 2x - 5$ **28** $2x + 3 > 5x - 9$

29 $x^2 + x - 2 > 0$ **30** $x^2 + 2x - 3 < 0$

31 $x^2 + 2x - 8 > 0$ **32** $-x^2 + 3x + 4 > 0$

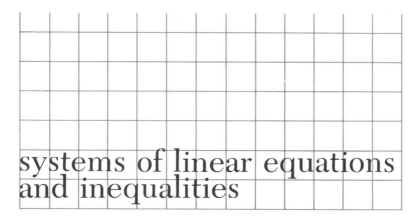

systems of linear equations and inequalities

Chapter

7

The statement "3 times a number less 2 is equal to 7" can be expressed in algebraic language as the equation $3x - 2 = 7$, and we know by the methods of Chap. 5 that the solution set of this equation is $\{3\}$. In order to express the more general statement "the sum of two numbers is 9" as an equation, we must have two variables, x and y, and then the equation is $x + y = 9$. It is evident that we can obtain as many *pairs* of numbers as we please that will satisfy the equation $x + y = 9$, because if we assign any value to y and then subtract that value from 9, we obtain the corresponding value of x. For example, if y is 1, x is 8; if y is 5, x is 4. If, however, we also require that the difference of the two numbers be 1, we have the two equations $x + y = 9$ and $x - y = 1$ that must be satisfied *by the same pair of numbers.*

Solving Equations Simultaneously

In this chapter we shall explain methods that will show that there is only one pair of numbers, $x = 5$ and $y = 4$, that satisfies both $x + y = 9$ and $x - y = 1$. The process of finding this pair of numbers is called *solving the equations simultaneously.* The problem of solving two equations in two variables simultaneously occurs frequently in all fields in which algebra is used.

It is the purpose of this chapter to present the methods most often used when the two equations are linear.

7.1
DEFINITIONS

Linear Equation in Two Variables

■ An equation of the type $ax + by = c$, where a, b, and c are arbitrary constants and neither a nor b is zero, is a *linear equation in two variables*.

Solution Pair

■ An ordered pair of numbers, the first a replacement for x and the second a replacement for y, that satisfies an equation in two variables is a *solution pair* of the equation.

For example, each of the ordered pairs $(1, 9)$, $(3, 3)$, and $(4, 0)$ is a solution pair of $3x + y = 12$.

Simultaneous Solution Set

■ An ordered pair of numbers that satisfies each of two equations in two variables is a *simultaneous solution pair* of the two equations, and the set of all such pairs is the *simultaneous solution set*.

For example, $(5, -3)$ is a simultaneous solution pair of the equations $3x + y = 12$ and $x - 3y = 14$.

Equivalent Equations

■ Two equations in two variables are *equivalent* if every solution pair of each is a solution pair of the other.

Note   In the above definitions we have used the letters x and y as the variables. This is the usual custom, but frequently other pairs of letters such as u and v or z and w are used.

7.2
GRAPHICAL SOLUTION OF TWO LINEAR
EQUATIONS IN TWO VARIABLES

If we solve

$$ax + by = c \qquad (1)$$

for y, we obtain

$$y = \frac{-ax}{b} + \frac{c}{b} \qquad b \neq 0 \qquad (2)$$

Consequently, (1) and (2) are equivalent equations. As pointed out in Sec. 6.7, the graph of (2) is a straight line; hence the graph of (1) is a straight line. At this point we remind the reader that, by the definition of the graph, every solution pair of the equation is a pair of numbers that are the coordinates of a point on the graph. Conversely, the coordinates of every point on the graph are the elements of a solution pair of the equation.

If the graph of a second equation

$$Ax + By = C \tag{3}$$

intersects the graph of (1), the coordinates of the point of intersection are the elements of the simultaneous solution pair of (1) and (3), since that point is on the graph of each equation. Consequently, we can use the graphical method to obtain the simultaneous solution pair of two linear equations in two variables. The possibility that the two equations may not have a simultaneous solution is dealt with in the next section.

Since the graph of a linear equation in two variables is a straight line, it is completely determined by two points, but it is advisable to obtain a third point as a check. We can obtain the coordinates of a point on the graph either by assigning a value to x and solving for y or by assigning a value to y and solving for x. If $c \neq 0$, two points are easily obtained by first setting $x = 0$ and then setting $y = 0$. A third point is obtained by setting x equal to some value not zero.

EXAMPLE   Solve the equations

$$3x - 4y = 8 \tag{1}$$

$$2x + 5y = 10 \tag{2}$$

simultaneously by the graphical method.

SOLUTION   We shall assign 0 and 4 to x and 0 to y in each of the equations and in each case solve for the other unknown and thus get the following tables of corresponding values:

(1)

x	0	$\frac{8}{3}$	4
y	-2	0	1

(2)

x	0	5	4
y	2	0	$\frac{2}{5}$

We next plot the points determined by the pairs of corresponding values in each table and draw a straight line through each set and get the lines in Fig. 7.1. By observation we see that these lines intersect at a point whose coordinates to the nearest tenth are (3.5, .6). Therefore we say that the graphical simultaneous solution set of Eqs. (1) and (2) is $\{(3.5, .6)\}$. We check this solution by substituting these values in the left members of Eqs. (1) and (2) and get

$$3(3.5) - 4(.6) = 10.5 - 2.4 = 8.1 \quad \textbf{from (1)}$$

$$2(3.5) + 5(.6) = 7 + 3 = 10 \qquad \textbf{from (2)}$$

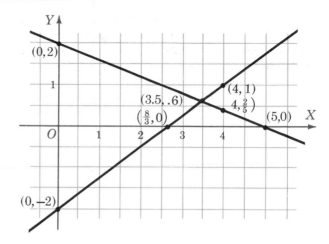

FIGURE 7.1

Since the right members of Eqs. (1) and (2) are 8 and 10, respectively, we are fairly safe in assuming that our solution set is correct to one decimal place.

7.3
INDEPENDENT, INCONSISTENT, AND DEPENDENT EQUATIONS

Obviously, if the graphs of two linear equations are parallel, the equations have no simultaneous solution set. Furthermore, if the graphs are coincident, every solution pair of one is a solution pair of the other. If the graphs are neither parallel nor coincident, they intersect in one and only one point. This illustrates the following definition:

■ Two linear equations in two variables are *independent* if their simultaneous solution set contains only one ordered pair of numbers, *inconsistent* if their simultaneous solution set is the empty set ∅, and *dependent* if every solution pair of one is a solution pair of the other.

We shall next derive a simple criterion that will enable us to decide whether two linear equations in two variables are independent, inconsistent, or dependent. We shall consider the equations from Sec. 7.2:

$$ax + by = c \tag{1}$$

$$Ax + By = C \tag{3}$$

The graphs of (1) and (3) intersect the X axis at $R(c/a, 0)$ and $S(C/A, 0)$, respectively, and the Y axis at the points $T(0, c/b)$ and $U(0, C/B)$, respectively,

161

as illustrated in Fig. 7.2. We can see that the graphs of (1) and (3) are parallel if the segments RT and SU are parallel, and these two segments are parallel if and only if the triangles ORT and OSU are similar. The triangles are similar if and only if $OR/OS = OT/OU$.

Now $OR/OS = c/a \div C/A = Ac/aC$, and $OT/OU = c/b \div C/B = Bc/bC$. Consequently, $OR/OS = OT/OU$ if and only if $Ac/aC = Bc/bC$. If we multiply each member of the last equation by C/c, we obtain $A/a = B/b$. Therefore, the graphs of (1) and (3) are parallel if and only if $A/a = B/b$. If in addition to $A/a = B/b$ we have $A/a = B/b = C/c = k$, then $A = ak$ and $C = ck$. Hence, $C/A = ck/ak = c/a$ and the points R and S coincide. Therefore, since the graphs are parallel and have one point in common, they coincide. Hence, we have the following theorem.

Independent, Inconsistent, and Dependent Systems

■ Two linear equations $ax + by = c$ and $Ax + By = C$ are independent if and only if $A/a \neq B/b$, inconsistent if and only if $A/a = B/b \neq C/c$, and dependent if and only if $A/a = B/b = C/c$.

The following three examples illustrate the application of the above theorem.

The equations

$$2x - 3y = 4$$

$$5x + 2y = 8$$

are independent, since $\frac{2}{5} \neq -\frac{3}{2}$.

The equations

$$3x - 9y = 1$$

$$2x - 6y = 2$$

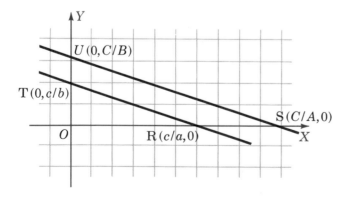

FIGURE 7.2

are inconsistent, since $3/2 = -9/-6 \neq 1/2$.
The equations

$$2x - 4y = 12$$

$$3x - 6y = 18$$

are dependent, since $2/3 = -4/-6 = 12/18$.

EXERCISE 7.1
Graphical Solution of Systems of Equations

Construct the graphs of the equations in each of the following problems. Then find to one decimal place the simultaneous solution set of each pair of equations by estimating the coordinates of the intersection of the graphs. *any 2 points for X*

1 $2x + 3y = 12$
    $x - 3y = -3$

2 $2x + y = 6$
    $2x - y = 2$

3 $2x - y = -8$
    $2x + y = -4$

4 $3x + y = -3$
    $x - y = -5$

5 $2x + 3y = 12$
    $2x - 3y = -6$

6 $3x - 2y = 8$
    $6x - 4y = 2$

7 $x + 2y = 6$
    $3x - 2y = 6$

8 $4x - y = 12$
    $4x + y = 16$

9 $2x - 4y = 7$
    $3x - 6y = 2$

10 $x + 2y = 4$
    $3x + 4y = 9$

11 $2x - 3y = 7$
    $4x + 3y = -4$

12 $3x + 4y = 2$
    $2x - y = 5$

13 $2x + 4y = -7$
    $x - 3y = 9$

14 $4x - 5y = 20$
    $8x - 10y = 10$

15 $2x + 3y = 3$
    $x - 3y = 3$

16 $x + y = 4$
    $3x - y = 6$

17 $2x - y = 2$
    $2x + y = 4$

18 $x + 5y = 5$
    $3x - 10y = 0$

19 $8x + 12y = 16$
    $6x + 9y = 12$

20 $5x + 4y = 16$
    $5x - 2y = -2$

21 $2x - 3y = 15$
    $8x + y = 8$

22 $5x - 3y = 10$
    $10x + 7y = 7$

23 $4x - 5y = -5$
    $3x - 10y = -20$

24 $5x + 10y = 5$
    $3x + 6y = 3$

25 $3x + 4y = 12$
    $2x - 3y = 6$

26 $5x + y = 10$
    $x - 3y = 6$

27 $9x - 12y = 3$
    $6x - 8y = 2$

28 $3x - 4y = 12$
    $2x + y = 4$

29 $4x - 5y = 10$
$5x + 2y = 5$

30 $3x + 6y = 2$
$5x + 10y = 7$

31 $6x - 5y = 15$
$5x + 8y = 20$

32 $5x + 8y = 20$
$4x - 3y = -18$

33 $10x - 5y = 7.5$
$6x - 3y = 4.5$

34 $4x + 3y = 9$
$7x - 8y = 28$

35 $5x - 2y = 10$
$3x + 4y = 9$

36 $7x - 5y = 14$
$3x + 8y = 8$

7.4
ELIMINATION BY ADDITION OR SUBTRACTION

Eliminating a Variable

To solve two independent linear equations in two variables algebraically, we combine the two equations in such a way as to obtain one equation in one variable whose root is one of the numbers in the simultaneous solution pair of the two given equations. This process is called *eliminating a variable*. After one of the numbers in the solution pair is found, the other is obtained by substituting the first number in one of the given equations and solving the resulting equation for the remaining variable.

One of the methods for eliminating a variable is a process called *elimination by addition or subtraction*. We shall explain the method by means of two examples.

EXAMPLE 1 Find the simultaneous solution set of the equations

$$2x + 3y = 8 \tag{1}$$

$$x - 3y = -5 \tag{2}$$

by using the method of elimination by addition or subtraction.

SOLUTION In order to solve (1) and (2) simultaneously by the method of elimination by addition or subtraction, we reason in this way: If the right and left members of each of (1) and (2) are equal for some pair of values of x and y, the equation obtained by adding the corresponding members will be equal for the same pair of values. If we add the corresponding members of (1) and (2), we get $3x = 3$, and then if we divide each member of the last equation by 3, we get $x = 1$. Therefore, $x = 1$ is one number in the solution pair sought. We now substitute 1 for x in Eq. (1) and solve for y as follows:

$2 + 3y = 8$ substituting 1 for x in Eq. (1)

$3y = 6$ adding -2 to each member

$y = 2$ dividing each member by 3

Consequently, the simultaneous solution set of Eqs. (1) and (2) is $\{(1, 2)\}$. We check this solution by substituting 1 for x and 2 for y in the left member of Eq. (2) and obtain $1 - 6 = -5$. Therefore, since the right member of Eq. (2) is also -5, then $x = 1$, $y = 2$ satisfies the equation.

EXAMPLE 2   Solve the equations

$$3x + 4y = -6 \tag{1}$$

$$5x + 6y = -8 \tag{2}$$

by the method of elimination by addition or subtraction.

SOLUTION   In Eqs. (1) and (2) neither variable is eliminated if we add or subtract the corresponding members. We notice, however, that if we multiply the members of Eq. (1) by 3 and the members of Eq. (2) by 2, we get Eqs. (3) and (4), in which the coefficients of y are equal,

$9x + 12y = -18$	multiplying Eq. (1) by 3	(3)
$10x + 12y = -16$	multiplying Eq. (2) by 2	(4)
$-x \qquad = -2$	subtracting Eq. (4) from Eq. (3)	
$x = 2$	dividing each member by -1	

By substituting 2 for x in Eq. (1) and solving for y, we get

$6 + 4y = -6$	substituting 2 for x in Eq. (1)
$4y = -12$	adding -6 to each member
$y = -3$	dividing each member by 4

Therefore, the simultaneous solution set of Eqs. (1) and (2) is $\{(2, -3)\}$. To check the solution, we substitute 2 for x and -3 for y in the left member of Eq. (2) and obtain

$$5(2) + 6(-3) = 10 - 18 = -8$$

Hence, since the right member of Eq. (2) is also -8,

$$x = 2, \qquad y = -3$$

satisfies the equation.

The reader should note that an alternative procedure is to multiply Eq. (1) by 5 and Eq. (2) by 3 and then solve the resulting equations for y.

7.5
ELIMINATION BY SUBSTITUTION

If one of the equations in a pair of linear equations in two variables is easily solvable for one of the variables in terms of the other, a very efficient method for eliminating one of the variables is the *method of substitution*. We shall first list the steps in the process and then illustrate the procedure.

Steps in Elimination by Substitution

1. Solve one of the equations for one variable in terms of the other.
2. Substitute the solution obtained in step 1 for that variable in the second equation and thus obtain an equation in one variable.
3. Solve the equation obtained in step 2.
4. Substitute the value found in step 3 in the solution obtained in step 1 and solve for the other variable.
5. Write the solution set in the form of $\{(x, y)\}$ replacing x and y by the appropriate values found in steps 3 and 4.
6. Check the solution by substituting the values for x and for y, step 5, in the given equation not used in step 1.

EXAMPLE Solve the equations

$$3x + 5y = 5 \tag{1}$$

$$x + 4y = 11 \tag{2}$$

simultaneously by the method of elimination by substitution.

SOLUTION To solve Eqs. (1) and (2) simultaneously by the substitution method, we first notice that (2) is readily solvable for x in terms of y, and the solution is

$$x = 11 - 4y \qquad \text{step 1, adding } -4y \text{ to each member of (2)} \tag{3}$$

We now proceed as directed in steps 2 and 3 and obtain

$$3(11 - 4y) + 5y = 5 \qquad \text{step 2, substituting } 11 - 4y \text{ for x in Eq. (1)}$$

$$33 - 12y + 5y = 5 \qquad \text{by distributive axiom}$$

$$-12y + 5y = 5 - 33 \qquad \text{adding } -33 \text{ to each member}$$

$$-7y = -28 \qquad \text{combining terms}$$

$$y = 4 \qquad \text{step 3, solve for y by dividing each member by } -7$$

We now substitute 4 for y in Eq. (3), step 4, and get

$$x = 11 - 16$$

$$x = -5$$

Therefore, the solution set, step 5, is $\{(-5, 4)\}$. As in step 6, we check the solution by substituting the values for x and y in the left member of Eq. (1). Thus, we get $-15 + 20 = 5$, and since the right member of Eq. (1) is also 5, $x = -5$, $y = 4$ satisfies the equation.

Solution of Systems of Equations by Algebraic Methods

By use of elimination by addition or subtraction, find the simultaneous solution set of the following pairs of equations.

1 $2x + 5y = 1$
$3x - 5y = 14$

2 $5x - 2y = 20$
$3x + 2y = -4$

3 $2x + 7y = 13$
$2x - 5y = -23$

4 $5x + 2y = -18$
$5x - 3y = 2$

5 $3x - 4y = 7$
$4x + 2y = 13$

6 $5x + 9y = 19$
$2x + 3y = 8$

7 $10x - 5y = 1$
$5x + 2y = 5$

8 $8x + 3y = 15$
$4x - 2y = -3$

9 $3x + 2y = 2$
$2x + 3y = 1$

10 $4x + 8y = 3$
$3x + 7y = 2$

11 $5x - 12y = 2$
$7x - 8y = 5$

12 $16x - 6y = 3$
$12x + 5y = 7$

Find the simultaneous solution set of each of the following pairs of equations by the method of substitution.

13 $3x + y = 3$
$5x + 2y = 4$

14 $6x - 5y = 2$
$2x + y = 6$

15 $x + 5y = 8$
$4x - 2y = -1$

16 $7x + 5y = 1$
$x - 3y = 2$

17 $2x - 5y = -1$
$3x - 4y = 2$

18 $5x - 6y = 1$
$2x + 4y = -2$

19 $3x + 7y = 1$
$5x + 9y = -1$

20 $5x - 4y = 1$
$3x + 6y = -3$

21 $4x - 3y = 18$
$3x + 2y = 5$

22 $2x + 5y = 10$
$5x - 3y = -37$

23 $3x + 8y = 26$
$4x + 7y = 31$

24 $5x - 2y = 29$
$9x - 5y = 62$

25 $8x + 3y = -4$
$12x - 11y = 25$

26 $6x - 15y = 21$
$7x + 10y = 19$

27 $8x + 3y = 21$
$12x + 5y = 33$

28 $12x - 5y = -20$
$18x - 2y = -19$

29 $9x + 8y = 23$
$6x + 10y = 27$

30 $8x - 15y = -7$
$12x - 10y = -3$

31 $12x - 6y = -19$
$18x + 8y = -3$

32 $14x - 9y = 13$
$21x + 6y = -13$

Find the simultaneous solution set of each of the following pairs of equations by any algebraic method.

33 $6x + 5y = 3$
$4x + 3y = 5$

34 $7x + 12y = 3$
$5x + 9y = 2$

35 $3x - 10y = 2$
$4x - 15y = 7$

36 $35x - 6y = 3$
$42x - 7y = 4$

37 $10x - 3y = 6$
$16x + 5y = 4$

38 $7x + 6y = 5$
$8x + 9y = 10$

39 $18x - 2y = 3$
$15x + 7y = 9$

40 $6x + 8y = 3$
$4x + 12y = 5$

Solve each of the following pairs of equations simultaneously for 1/x and 1/y, and then find the replacements for x and y.

41 $\dfrac{6}{x} - \dfrac{2}{y} = 1$

$\dfrac{9}{x} + \dfrac{8}{y} = 7$

42 $\dfrac{2}{x} + \dfrac{3}{y} = 2$

$\dfrac{6}{x} - \dfrac{5}{y} = -1$

43 $\dfrac{3}{x} - \dfrac{2}{y} = 1$

$\dfrac{12}{x} + \dfrac{4}{y} = 1$

44 $\dfrac{6}{x} - \dfrac{12}{y} = 2$

$\dfrac{12}{x} + \dfrac{16}{y} = -1$

7.6
THREE LINEAR EQUATIONS IN THREE VARIABLES

An equation that is of the form $ax + by + cz = d$, where a, b, c, and d are arbitrary nonzero constants, is a linear equation in three variables. A set of three values, one for x, one for y, and one for z, in the order (x, y, z), that satisfies the equation is called a *solution triple* of the equation.

If a simultaneous solution set of a system of three linear equations in three variables exists, we obtain it by means of the following steps.

Method for Solving a System of Three Linear Equations in Three Variables

1 Eliminate the same variable from two pairs of the given equations, and thereby get two equations in two variables.
2 Solve the two equations obtained in step 1 simultaneously for the other two variables.
3 Substitute the values obtained in step 2 in one of the given equations, and solve for the third variable.

Example 1 illustrates the method.

EXAMPLE 1   Solve the following system of equations simultaneously:

$$2x + 3y + 4z = 4 \tag{1}$$

$$3x - 2y - 6z = 7 \tag{2}$$

$$5x + 7y + 8z = 9 \tag{3}$$

SOLUTION   In order to solve the given system simultaneously, we shall arbitrarily select z as the first variable to be eliminated. In Eqs. (1) and (2) the LCM of the coefficients of z is 12. Therefore, as step 1, we multiply Eq. (1) by 3 and Eq. (2) by 2 and equate the sums of the corresponding members of the resulting equations.

$6x + 9y + 12z = 12$	Eq. (1) times 3	(4)
$6x - 4y - 12z = 14$	Eq. (2) times 2	(5)
$12x + 5y \qquad = 26$	adding corresponding members of (4) and (5)	(6)

Since the coefficient of z in Eq. (3) is twice the coefficient of z in Eq. (1), we multiply the members of Eq. (1) by 2 and subtract the members of the resulting equation from Eq. (3) as follows. This is the second part of step 1.

$5x + 7y + 8z = 9$	Eq. (3) copied	(3)
$4x + 6y + 8z = 8$	Eq. (1) times 2	(7)
$x + y \qquad = 1$	subtracting each member of Eq. (7) from the corresponding member of Eq. (3)	(8)

As step 2, we next solve Eqs. (6) and (8) simultaneously for x and y.

$12x + 5y = 26$	Eq. (6) copied	(6)
$5x + 5y = 5$	Eq. (8) times 5	(9)
$7x \qquad = 21$	subtracting each member of Eq. (9) from the	
$x \qquad = 3$	corresponding member of Eq. (6)	

We can now obtain the value of y by substituting $x = 3$ in either Eq. (6) or (8). We shall use the latter and get

$$3 + y = 1$$

$$y = -2$$

Finally, as step 3, we can get the value of z by substituting $x = 3$ and $y = -2$ in any one of the three given equations. Using Eq. (1), we get

$$2(3) + 3(-2) + 4z = 4 \qquad \text{substituting } x = 3 \text{ and } y = -2 \text{ in Eq. (1)}$$

$$6 - 6 + 4z = 4$$

$$4z = 4$$

$$z = 1$$

Therefore, the solution set is $\{(3, -2, 1)\}$.

We check the solution by substituting these values in either Eq. (2) or Eq. (3). Using Eq. (3), we have

$$5(3) + 7(-2) + 8(1) = 15 - 14 + 8 = 9$$

and since the right member of Eq. (3) is also 9,

$$x = 3 \qquad y = -2 \qquad z = 1$$

satisfies the equation.

Frequently, in a system of three linear equations in three variables, not all three variables appear in every equation. We shall illustrate the method for dealing with such a system in Example 2.

EXAMPLE 2 Solve the following system of equations simultaneously:

$$2x - 3y = 4 \tag{1}$$

$$6x + 2z = 9 \tag{2}$$

$$5y + 4z = 7 \tag{3}$$

SOLUTION Since Eq. (1) involves x and y, (2) involves x and z, and (3) involves y and z, we shall, as step 1, eliminate z from (2) and (3) and solve the resulting equation simultaneously with (1). The procedure in detail is as follows:

$12x \qquad + 4z = 18$	Eq. (2) times 2	(4)
$\underline{\qquad 5y + 4z = \ \ 7}$	Eq. (3) copied	(3)
$12x - \ \ 5y \qquad = 11$	subtracting Eq. (3) from Eq. (4)	(5)
$\underline{12x - 18y \qquad = 24}$	Eq. (1) times 6	
$13y \qquad = -13$	subtracting Eq. (6) from Eq. (5)	(6)
$y = -1$		

As steps 2 and 3, we now substitute $y = -1$ in Eq. (1), solve for x, then substitute $y = -1$ in Eq. (3), solve for z and obtain

$$2x + 3 = 4 \qquad \text{substituting } y = -1 \text{ in Eq. (1)}$$

$$2x = 1$$

$$x = \tfrac{1}{2}$$

$$-5 + 4z = 7 \qquad \text{substituting } y = -1 \text{ in Eq. (3)}$$

$$4z = 12$$

$$z = 3$$

Hence, the solution set is $\{\tfrac{1}{2}, -1, 3\}$.

Since Eqs. (1) and (3) were used in obtaining $x = \tfrac{1}{2}$ and $z = 3$, we check the solution by substituting $x = \tfrac{1}{2}$ and $z = 3$ in Eq. (2). For these values, the left member of Eq. (2) becomes $6(\frac{1}{2}) + 2(3) = 3 + 6 = 9$. Therefore, since the right member of Eq. (2) is also 9, $x = \frac{1}{2}, y = -1, z = 3$, satisfies the equation.

EXERCISE 7.3
Systems of Three Linear Equations

Find the simultaneous solution set of the following systems of equations.

1 $x - 2y + z = 5$
$2x + y - 3z = 0$
$3x + 4y + z = 3$

2 $2x - y + z = 7$
$4x + y - z = -1$
$x - 2y - 4z = -7$

3 $x + y + z = 1$
$x - 2y - z = 9$
$2x + y + 3z = 3$

4 $4x - y - z = 2$
$2x + y + 2z = 9$
$x + 2y - z = -10$

5 $3x + 2y - 4z = 4$
$5x + 6y - 3z = 11$
$2x + 4y + 6z = 2$

6 $2x + 4y - 3z = -4$
$x - y + 2z = 8$
$4x + 2y - z = 8$

7 $3x + y - 2z = -7$
$x + 3y + 4z = 11$
$4x - 2y - 6z = -16$

8 $x + 3y - 2z = -8$
$2x - 4y + 5z = -1$
$4x + 5y - 3z = -28$

9 $2x + y - z = 6$
$4x + 2y + z = 3$
$6x - 3y + 2z = -9$

10 $2x - 3y + 5z = 11$
$4x + 6y + 3z = 7$
$3x + 9y - 2z = -2$

11 $x + 2y - 4z = 5$
$3x + 5y + 8z = 1$
$2x + y + 12z = -2$

12 $4x - 2y + 5z = 4$
$3x + y - 15z = 8$
$2x - 3y + 10z = -5$

13 $6x + 4y + z = 6$
$4x - 2y - 3z = -5$
$2x + 3y - 2z = -13$

14 $2x - 3y + 5z = 3$
$3x - 6y - 2z = -3$
$5x - 12y + 4z = 5$

15 $3x + 5y - 2z = 8$
$x - 10y + 3z = -5$
$2x - 15y + z = 7$

16 $3x + 2y + 4z = 1$
$2x + y - 8z = 7$
$5x + 3y + 12z = -4$

17 $3x + y - 4z = 2$
$2x + y + 2z = 3$
$x + 2y + 2z = 4$

18 $3x + 2y + 2z = 2$
$4x + 2y - 2z = 3$
$3x + 4y - 2z = 4$

19 $x + 2y - 3z = 1$
$2x - y + 4z = 2$
$3x + 4y - z = -3$

20 $2x + 5y + 3z = 4$
$4x - 10y + 2z = 2$
$6x - 15y + z = 2$

21 $x + 3y + 2z = -1$
$4x - 2y + z = 3$
$3x + y - 6z = -3$

22 $x + y - z = 1$
$3x + 6y - 4z = 4$
$3x - 3y + 5z = -2$

23 $2x - 3y - 4z = 4$
$x - 6y + 2z = 6$
$x - 3y - 2z = 3$

24 $2x + y + 6z = 4$
$4x - y + 2z = 2$
$10x + 4y + 8z = 11$

7.7

PROBLEMS LEADING TO SYSTEMS OF LINEAR EQUATIONS

Many stated problems contain more than one unknown quantity, and often the equation for solving such a problem can be more easily obtained if more than one unknown letter is introduced. However, before the problem can be completely solved, the number of equations formed must be equal to the number of unknown letters used. The general procedure for obtaining the equations is the same as that in Sec. 5.9, and the student is advised to reread that section before he studies the following examples or attempts the problems in Exercise 7.4.

EXAMPLE 1

A real estate dealer received $1200 in rents on two dwellings in 1 year; one of them brought $10 per month more than the other. How much did he receive per month for each if the more expensive house was vacant for 2 months?

SOLUTION If we let

$x =$ the monthly rental on the more expensive house

and

$y =$ the monthly rental on the other

then

$$x - y = 10 \tag{1}$$

since one rented for $10 more per month than the other. Furthermore, since the first of the above houses was rented for 10 months and the other was rented for 12 months, we know that $10x + 12y$ is the total amount received in rentals. Hence,

$$10x + 12y = 1200 \tag{2}$$

We now have the two equations (1) and (2) in the unknowns x and y, and we shall solve them simultaneously by eliminating y. The solution follows:

$$12x - 12y = 120 \qquad \text{Eq. (1) times 12} \tag{3}$$
$$\underline{10x + 12y = 1200}$$
$$22x = 1320 \qquad \text{Eq. (3) plus Eq. (2)} \tag{4}$$

Hence,

$$x = 60$$

Substituting 60 for x in Eq. (1), we get

$$60 - y = 10$$

Hence,

$$-y = 10 - 60 = -50$$

and

$$y = 50$$

Therefore, the montly rentals were $60 and $50, respectively.

EXAMPLE 2

A tobacco dealer mixed one grade of tobacco worth $1.40 per pound with another worth $1.80 per pound in order to obtain 50 pounds of a blend that sold for $1.56 per pound. How much of each grade did he use?

SOLUTION   We shall let

x = the number of pounds of the $1.40 grade used

and

y = the number of pounds of the $1.80 grade used

Then

$$x + y = 50 \tag{1}$$

since there were 50 pounds in the mixture. Furthermore, $1.40x$ is the value in dollars of the first grade, $1.80y$ is the value in dollars of the second, and $(1.56)50 = 78$ is the value in dollars of the mixture. Therefore,

$$1.40x + 1.80y = 78 \tag{2}$$

Hence (1) and (2) are the two required equations, and we shall solve them by eliminating x.

$$
\begin{array}{lll}
1.40x + 1.40y = 70 & \text{Eq. (1) times 1.40} & (3) \\
1.40x + 1.80y = 78 & & (2) \\
\hline
-0.40y = -8 & \text{Eq. (3) minus Eq. (2)} &
\end{array}
$$

Therefore,

$$y = \frac{-8}{-0.40} = 20$$

Substituting 20 for y in Eq. (1), we have

$$x + 20 = 50$$
$$x = 30$$

Hence, the dealer used 30 pounds of the $1.40 grade and 20 pounds of the $1.80 grade in the mixture.

EXAMPLE 3

Two airfields A and B are 400 miles apart, and B is due east of A. A plane flew from A to B in 2 hours and then returned to A in $2\frac{1}{2}$ hours. If the wind blew with a constant velocity from the west during the entire trip, find the speed of the plane in still air and the speed of the wind.

SOLUTION   Let

$x = $ the speed of the plane in still air

and

$y = $ the speed of the wind

Then, since the wind was blowing from the west,

$x + y = $ the speed of the plane from A to B

and

$x - y = $ the speed of the plane on the return trip

Hence,

$$\frac{400}{x + y} = \text{the time required for the first half of the trip}$$

and

$$\frac{400}{x - y} = \text{the time required to return}$$

Therefore,

$$\frac{400}{x + y} = 2 \tag{1}$$

$$\frac{400}{x - y} = 2\frac{1}{2} \tag{2}$$

Now we multiply both members of Eq. (1) by $x + y$ and of Eq. (2) by $2(x - y)$ and get

$$400 = 2x + 2y \tag{3}$$

and

$$800 = 5x - 5y \qquad (4)$$

We shall solve Eqs. (3) and (4) simultaneously by first eliminating y.

$2000 = 10x + 10y$	Eq. (3) times 5	(5)
$1600 = 10x - 10y$	Eq. (4) times 2	(6)
$3600 = 20x$	Eq. (5) plus Eq. (6)	

Hence,

$$x = 180$$

Substituting 180 for x in Eq. (3), we have

$$400 = 2(180) + 2y$$

$$400 = 360 + 2y$$

$$2y = 40$$

$$y = 20$$

Hence, the speed of the plane in still air was 180 miles per hour, and the speed of the wind was 20 miles per hour.

EXAMPLE 4

A cash drawer contains $50 in nickels, dimes, and quarters. There are 802 coins in all, and 10 times as many nickels as dimes. How many coins of each denomination are in the drawer?

SOLUTION   Let

q = the number of quarters

d = the number of dimes

n = the number of nickels

We now form the following three linear equations in q, d, and n:

$25q + 10d + 5n = 5000$	since $50 = 5000 cents	(1)
$q + d + n = 802$	since there are 802 coins in all	(2)
$n = 10d$	since there are 10 times as many nickels as dimes	(3)

If we substitute $10d$ for n [given by Eq. (3)] in Eqs. (1) and (2), we obtain two linear equations in q and d. From Eq. (1) we get

$$25q + 10d + 5(10d) = 5000$$

which reduces to

$$25q + 60d = 5000 \tag{4}$$

Furthermore, from Eq. (2) we have

$$q + d + 10d = 802$$

or

$$q + 11d = 802 \tag{5}$$

We may eliminate q from Eqs. (4) and (5) as follows:

$$
\begin{array}{llr}
25q + 60d = & 5,000 & \text{(4)} \\
\underline{25q + 275d = } & \underline{20,050} & \text{Eq. (5) times 25} \quad \text{(6)} \\
 -215d = & -15,050 & \text{Eq. (4) minus Eq. (6)} \\
 d = & 70 &
\end{array}
$$

Now, substituting 70 for d in Eq. (3), we get

$$n = 10(70) = 700$$

Finally, substituting $d = 70$ in Eq. (5), we have

$$q + 11(70) = 802$$

Hence,

$$q = 802 - 770 = 32$$

Consequently, there are 32 quarters, 70 dimes, and 700 nickels in the cash drawer.

EXERCISE 7.4
Problems Leading to Systems of Linear Equations

1 The sum of two numbers is three times the smaller, and their difference exceeds one-half the smaller by 12. Find the numbers.

2 The sum of two numbers is twice their difference, and the larger exceeds twice the smaller by 6. Find the numbers.

3 The quotient of the sum and the difference of the same two numbers is 5, and three times the larger number exceeds twice the smaller by 60. Find the numbers.

4 The sum of two numbers is four times the smaller. If the smaller number is increased by 15 and the larger is decreased by 13, the results are equal. Find the numbers.

5   A man invested \$10,900 for stock in two companies at \$55 and \$36 per share, respectively. The more expensive stock yielded an annual dividend of \$2.20 per share and the other an annual dividend of \$1.20. If the total income from the two was \$400 per year, find the number of shares of each stock that was bought.

6   During a certain year some apartments in a building rented for \$125 per month, the remainder for \$160 per month, and the total monthly rental was \$4900. The next year the monthly rent on the cheaper apartments was increased by \$5, and on the others by \$6. If the monthly income was thereby increased by \$190, how many apartments of each type are in the building?

7   The dwellings in a new housing tract were priced at \$30,000 and \$35,000, respectively, and the value of the tract was \$3,200,000. At the end of 6 months one-half of the more expensive houses and two-thirds of the others had been sold. If the amount received from the sales was \$1,900,000, how many dwellings of each type are in the tract?

8   The monthly rental for a beach cottage is higher during the 3 summer months than during the remainder of the year. If it is occupied during the entire year, the rental amounts to \$2700. In a certain year, however, because of fire damage, the cottage was occupied for 2 summer months and 5 off-season months, and the rental amounted to \$1600. Find the monthly rental for each portion of the year.

9   Two boys paddled a canoe 6 miles downstream in 1 hr. On the return trip they were joined by a companion and, with the three paddling, the rate of the canoe in still water was increased by 1 mile per hour. If the return trip required 2 hr, find the rate of the current and the rate the first two boys could paddle in still water.

10   Airfield B is 960 miles north of field A. A pilot left A to fly to B, and 30 min later a pilot left B to fly to A. The two met in 1 hr and 50 min after the departure of the first plane, and the first pilot reached B 1 hr and 10 min later. If the airspeeds of the two planes were the same, and the wind was blowing from the south at a constant velocity during the flight, find the airspeed of the two planes and the velocity of the wind.

11   Airfield B is 990 miles due north of field A. A pilot flew from A to B and back, and there was a north wind with a constant velocity during the entire flight. The northward trip required $5\frac{1}{2}$ hr and the return trip required $4\frac{1}{2}$ hr. Find the airspeed of the plane and the velocity of the wind.

12   Airfield B is due west of field A, and field C is due south of B. A pilot left A at 8 a.m. and flew to B, delayed 3 hr, and then flew to C, arriving at 3 p.m. The wind blew from the west at 20 miles per hour during the westward flight, but changed to the north at 30 miles per hour during the delay. If the airspeed of the plane was 270 miles per hour and the total distance flown was 1080 miles, find the distances from A to B and from B to C.

13   A student traveled from the campus to an airport on a bus that averaged 45 miles per hour and then flew to his home on a plane at 400 miles per hour. The time required for the trip, including a 24-min wait at the airport, was 3 hr. The bus fare was 5 cents per mile, the plane fare was 8 cents per mile, and the cost of the trip was $65.35, find the distance traveled by each method.

14 A rancher rode a horse at 6 miles per hour to his ranch headquarters. He then drove his car at 50 miles per hour to an airport, where he boarded a plane and flew at 300 miles per hour to a cattlemen's convention. The total distance traveled was 340 miles, and the traveling time was 4 hr. If he was on the plane twice as long as in the car, find the time required for each part of the trip.

15 One morning a salesman left his home and called on customers in towns A and B. In the afternoon he returned to his home from B on a modern highway. His average speed in the morning was 30 miles per hour, and in the afternoon it was 65 miles per hour. If he traveled a total of 7 hr and was on the road 3 hr longer in the morning than in the afternoon, find the distance traveled on each part of the trip.

16 A tour bus and a car left a resort hotel at the same time, but traveled in opposite directions around a scenic loop. When they met, the bus had traveled 32 miles and the car 48 miles. Find the average speed of each if the car reached the hotel 1 hr and 20 min ahead of the bus.

17 A man fenced a rectangular plot with one of the shorter sides along a highway. At the same time, he divided it into two parts with a fence parallel to the highway. The cost of the fence along the highway was 60 cents per ft, and elsewhere it was 50 cents per ft. The total cost of the fencing was $620, and the fence along the highway cost $380 less than the remainder. Find the dimensions of the plot.

18 A building that is 80 ft wide with the longer side fronting a street is divided into three parts by partitions perpendicular to the front wall. The sum of the areas of the two smaller parts is equal to the area of the larger, and the perimeter of the building is 400 ft. If the areas of the two smaller parts are equal, find the dimensions of the two smaller parts and of the larger part.

19 The owners of a shopping center in which 70 percent of the area was used for parking bought an adjacent tract, set aside 85 percent for parking, and used the remainder for buildings. The enlarged area contained 30,000 sq yd with 75 percent used for parking. Find the original area and the area of the land purchased.

20 A rectangular tract of land with the southern boundary running east and west and the eastern boundary 600 yd long is bordered on the north by another rectangular tract whose northern boundary is 500 yd long, and the eastern boundaries of the two tracts form a straight line. The combined areas of the two tracts is 680,000 sq yd. They are enclosed with 3600 yd of fencing with no fence along the common boundary. Find the unknown dimension of each tract.

21 An automobile dealer has 45 cars in stock made up of sedans, sports cars, and station wagons. There are twice as many sports cars as station

wagons. The sedans are priced at \$4000, the sports cars at \$3500, and the station wagons at \$4200. If the retail value of the cars is \$172,000, how many cars of each type are in stock?

22 A cotton buyer declined an offer of \$175 per bale for his consignment of baled cotton. Two months later, when the price had increased by \$5 per bale, he sold the consignment for \$140 more than he would have received on the first offer. If in the meantime, 2 bales had been destroyed by fire, find the number of bales in the original consignment and the amount he would have received if he had accepted the first offer.

23 Mr. Smith had 52 shares of stock A and 32 shares of stock B. On a certain day his holdings were worth \$2810. The next day, the price of stock B was 1 percent lower, the price of stock A was 2 percent higher, and his holdings were worth \$31.96 more. Find the price of each stock on the first day.

24 A family on vacation traveled an average of 220 miles a day at a cost of 5 cents per mile. Their meals averaged \$14 per day, and their motel costs averaged \$15 per night. The total cost of the vacation was \$505, and the motel costs were \$37 more than the mileage costs. Find the number of miles traveled and the number of nights spent in a motel.

25 On a certain day when his irrigation reservoir was full, a farmer opened the inlet pipe and the outlet pipe, and the reservoir was empty after 24 hr. On another day, when the reservoir was empty, he opened the inlet pipe and allowed it to run 3 hr. Then he opened the outlet, and the reservoir was drained in 9 hr. How long does it take the inlet to fill the reservoir if the outlet is closed, and how long will it take the outlet to drain the reservoir if it is full and the inlet is closed?

26 Two brothers mowed a lawn together in $2\frac{2}{9}$ hr. The next week the older boy worked alone for 3 hr and the younger boy finished the job in $1\frac{1}{4}$ hr. How long will it take each boy to mow the lawn alone?

27 A tobacco dealer mixed two grades of tobacco worth \$3 and \$3.50 per lb, respectively, and obtained 20 pounds of a mixture worth \$3.20 per lb. How many pounds of each grade were used?

28 A traveler left his home with his 5 gal radiator filled with a solution that was 20 percent antifreeze. When he stopped for gasoline, he discovered that the radiator was leaking and the attendant added water until it was full. He then went to the nearest repair shop where the solution was tested. After repairs $\frac{1}{2}$ gal of a solution that was 56 percent antifreeze was added to bring the radiator solution up to the original strength. Find the number of gallons lost before the first stop and the percent of antifreeze revealed by the test.

29 On Monday, Joe, Tom, and Bill working together polished Joe's car in $1\frac{1}{2}$ hr. On Tuesday, Joe helped Bill polish his car in 2 hr, and on Wednesday, Bill and Tom polished Tom's car in $2\frac{1}{4}$ hr. If all cars were the same make and model with the same surface conditions, how long would it take each boy to polish his car alone?

30 There are 52 desks in 3 offices, and the number in the second office is one-half the number in the first. The first office has 48 sq ft of floor space per desk, the second has 46 sq ft per desk, and the third has 45 sq ft. If there are 2424 sq ft in the three offices, how many desks are in each?

31 A salesman was allowed 10 cents per mile for the use of his car, $10 per day for meals, and $15 per night for a hotel room. On a certain trip his expense bill was $160, he averaged 120 miles per day, and his hotel bill was $10 more than he spent for meals. Find the number of days and the number of miles traveled, and the number of nights he stayed in a hotel.

32 A man had a total of $3600 in three savings banks. The interest rates were $5\frac{1}{2}, 5$, and $4\frac{3}{4}$ percent, respectively. His total yearly income from the three accounts was $186. He withdrew his money from the third bank and deposited half of it in each of the others and increased his yearly income by $4. Find the amount of each of his original deposits.

7.8
SYSTEMS OF LINEAR INEQUALITIES

In this section we discuss the graphical method for obtaining the solution sets of inequalities of the type $y \geq ax + b$ and $y \leq ax + b$. We first consider $y > ax + b$ and prove that the solution set of this inequality is the set of coordinates of all points that are above the graph of $y = ux + b$.

Suppose the graph of $y = ax + b$ is the line in Fig. 7.3 and that $P_1(x_1, y_1)$ is above the line. We construct a line parallel to the Y axis through P_1 intersecting the graph at $P(x, y)$. We now prove that $y_1 > ax_1 + b$ as follows:

$x = x_1$	since P_1P is parallel to the Y axis
$y = ax + b$	since $P(x, y)$ is on the graph
$y = ax_1 + b$	since $x = x_1$
$y_1 > y$	since P_1 is above P
$y_1 > ax_1 + b$	replacing y by $ax_1 + b$

Therefore, (x_1, y_1) is a solution pair of $y > ax + b$. Furthermore, if S_1 is the set of solution pairs $\{(x, y)\}$ of $y \geq ax + b$, then

$$S_1 = \{(x, y) \,|\, P(x, y) \text{ is a point on or } above \text{ the graph of } y = ax + b\}$$

Similarly, we can prove that if S_2 is the set of solution pairs of $y \leq ax + b$, then

$$S_2 = \{(x, y) \,|\, P(x, y) \text{ is } on \text{ or } below \text{ the graph of } y = ax + b\}$$

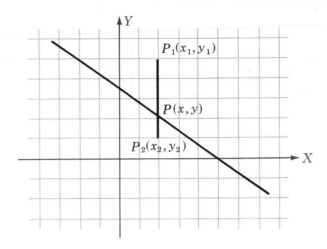

FIGURE 7.3

We now consider the system of inequalities

$$y \leq -x + 4 \tag{1}$$

$$y \leq 2x + 10 \tag{2}$$

$$y \geq -\tfrac{1}{3}x + \tfrac{2}{3} \tag{3}$$

and shall obtain the region of the plane such that the coordinates of each point in or on the boundary of the region are a simultaneous solution pair of Inequalities (1), (2), and (3). For this purpose we construct the graphs of

$$y = -x + 4 \tag{4}$$

$$y = 2x + 10 \tag{5}$$

$$y = -\tfrac{1}{3}x + \tfrac{2}{3} \tag{6}$$

and obtain lines L_4, L_5, and L_6 in Fig. 7.4. Then the solution sets of (1), (2), and (3) are, respectively,

$S_1 = \{(x, y) \,|\, P(x, y)$ is on or below $L_4\}$

$S_2 = \{(x, y) \,|\, P(x, y)$ is on or below $L_5\}$

$S_3 = \{(x, y) \,|\, P(x, y)$ is on or above $L_6\}$

Hence, the simultaneous solution set of Inequalities (1), (2), and (3) is the in-

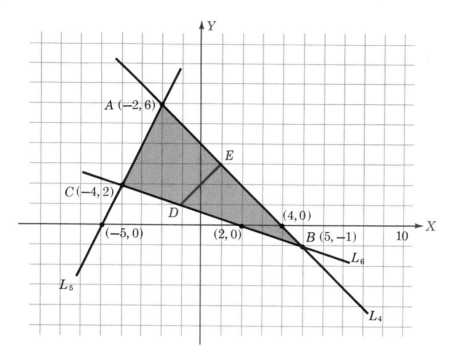

FIGURE 7.4

tersection of S_1, S_2, and S_3 or $S_1 \cap S_2 \cap S_3$. Furthermore,

$$S_1 \cap S_2 \cap S_3 = \{(x, y) \mid P(x, y) \text{ is in the triangle } ABC \text{ or on the boundary}\}$$

Therefore, $P(x, y)$ is in or on the boundary of the shaded region in Fig. 7.4.

If we replace x by 2 and y by 4 in the binomial $3x - 5y$, we get $6 - 20 = -14$, and we say that -14 is the *function value* of $3x - 5y$ at the point $(2, 4)$. The set of all function values of $3x - 5y$ at all points in a given set is called the *set of function values* over the given set.

EXAMPLE 1

Describe the solution set S of the system of inequalities (1), (2), and (3) and the equation $x + y = 2$. Construct a figure that shows this set of points.

SOLUTION

$S = \{P(x, y) \mid P$ is within or on the perimeter of the triangle bounded by the graphs of Eqs. (4), (5), and (6) and on the graph of $x + y = 2\}$.

This set of points is the line segment DE in Fig. 7.4.

The problem of finding the maximum and minimum values of a binomial of the type $ax + by$ over the set of points in a region of the plane bounded by straight lines is very important in linear programming. In the discussion

182

of this problem we shall use the following statement that can be proved by the methods of analytic geometry:

■ The maximum and minimum function values of the binomial $ax + by$ over a convex region bounded by straight lines occur at a vertex.

Consequently, the maximum function value of $3x - 5y$ over the region in Fig. 7.4 occurs at one vertex of the triangle, and the minimum function value occurs at another. If we solve each of the pairs of Eqs. (4) and (5), (4) and (6), and (5) and (6) simultaneously, we find that the coordinates of the vertices of the triangle are

$$A(-2, 6), \quad B(5, -1), \quad \text{and} \quad C(-4, 2)$$

At $(-2, 6)$

$$3x - 5y = -6 - 30 = -36$$

at $(5, -1)$

$$3x - 5y = 15 + 5 = 20$$

and at $(-4, 2)$

$$3x - 5y = -12 - 10 = -22$$

Hence, the maximum and minimum function values of $3x - 5y$ over the region are 20 and -36, respectively.

EXAMPLE 2
Find the maximum and minimum function values of $1.5x - 0.3y$ over the region determined by the following inequalities:

$$x \geq 0 \tag{1}$$

$$y \geq 0 \tag{2}$$

$$y \leq \tfrac{1}{2}x + 4 \tag{3}$$

$$y \geq 2x - 6 \tag{4}$$

SOLUTION We construct the graphs of

$$x = 0 \tag{5}$$

$$y = 0 \tag{6}$$

$$y = \tfrac{1}{2}x + 4 \tag{7}$$

$$y = 2x - 6 \tag{8}$$

and obtain the quadrilateral $OPQR$ in Fig. 7.5. By solving each of the pairs of Eqs. (5) and (6), (6) and (8), (8) and (7), and (5) and (7) simultaneously, we obtain the coordinates of the vertices indicated on the figure. The function value of $1.5x - 0.3y$ at each of these points is tabulated below.

Point	Function Value
$(0, 0)$	0
$(3, 0)$	$1.5(3) - 0.3(0) = 4.5$
$(\frac{20}{3}, \frac{22}{3})$	$1.5(\frac{20}{3}) - 0.3(\frac{22}{3}) = 7.8$
$(0, 4)$	$1.5(0) - 0.3(4) = -1.2$

Hence, the maximum function value is 7.8, and the minimum function value is -1.2.

EXERCISE 7.5

Describe the simultaneous solution set S of the pairs of inequalities in each of Probs. 1 to 4.

1 $y > 2x - 12$
    $y < 3x + 6$

2 $y > 3x + 15$
    $y < 2x - 8$

3 $y > 4x + 7$
    $y > 2x - 5$

4 $y < 5x + 3$
    $y < 3x - 1$

Construct a figure and show the solution set of the systems in Probs. 5 to 8.

5 $x + y \le 4$
    $2x - y \ge 6$
      $x = 2$

6 $3x + y \ge 6$
    $x - y \ge 1$
      $y = 3$

7 $2x - y \le -2$
    $x - 2y \ge -2$
    $3x + 2y = 6$

8 $2x - 3y \ge 6$
    $5x + y \ge 5$
    $x + y = 2$

Construct a figure and shade the region in which $P(x, y)$ lies if $\{(x, y)\}$ is the solution set of the system of inequalities in each of Probs. 9 to 20.

9 $x \ge 0$
     $y \ge 0$
   $x + y \le 4$

10 $x \ge 0$
      $y \le 0$
   $x + y \ge 3$

11 $x \le 0$
       $y \ge 0$
   $x - 2y \le -4$

12 $x \le 0$
       $y \le 0$
  $3x + 4y \ge -12$

13 $x \ge 2$
    $y \le 5$
    $y \ge x$

14 $x \le 3$
       $y \ge 2$
  $2x - 3y \le -6$

15 $x \le -1$
     $y \ge 2$
  $x - y \le -5$

16 $x \le 1$
      $y \ge -3$
  $3x + 2y \le 3$

17 $2x - 3y \le -3$
   $5x - 2y \ge 9$
   $x + 4y \ge -7$

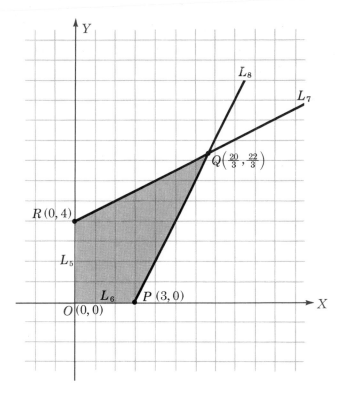

FIGURE 7.5

18 $x + \ y \le 2$
      $x - \ y \le -6$
      $x + 3y \ge -2$

19 $2x - 7y \le -13$
      $6x - 5y \ge 9$
      $2x + \ y \ge -5$

20 $3x + 8y \le 28$
      $5x - 6y \ge 8$
      $4x + \ y \ge -11$

Find the maximum and minimum function values of the indicated linear function over the region described in the following problems.

21 $2x + 3y$, the region determined by the inequalities in Prob. 17

22 $x + 0.5y$, the region determined by the inequalities in Prob. 18

23 $0.2x - 1.4y$, the region determined by the inequalities in Prob. 19

24 $1.5x + 0.6y$, the region determined by the inequalities in Prob. 20

25 $x - 2y$, the region determined by $3x - 5y \le -15$, $6x - y \ge 24$, $x \ge 0$, $y \ge 0$

26 $x + 4y$, the region determined by $x + 5y \le 10$, $4x - 3y \ge 17$, $x \ge 0$, $y \ge -3$

27 $3x - 10y$, the region determined by $3x + y \le 11$, $x - 3y \ge 7$, $x \ge -2$, $y \le 2$

28 $2x + 5y$, the region determined by $x - 2y \ge 11$, $7x - 2y \le -19$, $x \le 3$, $y \le -1$

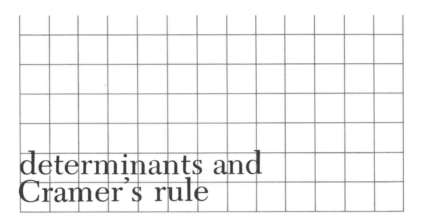

determinants and Cramer's rule

Chapter

8

A notation called a *determinant* was discovered independently by the Japanese mathematician Kiowa in 1683 and by Leibnitz in 1693. In 1750, Cramer devised the method for using determinants to obtain the solution sets of systems of linear equations. In this chapter we shall discuss determinants of the second and third orders, and shall explain Cramer's rule for using them to solve systems of two and three linear equations.

8.1
DETERMINANTS OF THE SECOND ORDER

The square array

$$\begin{vmatrix} a & b \\ c & d \end{vmatrix}$$

is a *determinant* of the second order. It stands for the binomial $ad - bc$, and this binomial is called the *value* or the *expansion* of the determinant. The letters a, b, c, and d are the *elements* of the determinant. The terms in the binomial $ad - bc$ are the products of the elements connected by the arrows in the following diagram, and each product is preceded by the sign at the point of the arrow.

Expansion of a Determinant

$$\begin{vmatrix} a & b \\ c & d \end{vmatrix} = ad - bc \tag{8.1}$$

If in Eq. (8.1) we replace a, b, c, and d by 3, 2, 5, and 7, respectively, we have

$$\begin{vmatrix} 3 & 2 \\ 5 & 7 \end{vmatrix} = 21 - 10 = 11$$

Similarly,

$$\begin{vmatrix} -3 & 6 \\ -4 & 7 \end{vmatrix} = (-3)(7) - (6)(-4) = -21 + 24 = 3$$

We now consider the system of linear equations

$$ax + by = e \tag{1}$$

$$cx + dy = f \tag{2}$$

We obtain the solution set as follows:

$adx + bdy = ed$	multiplying Eq. (1) by d	(3)
$bcx + bdy = bf$	multiplying Eq. (2) by b	(4)
$adx - bcx = ed - bf$	subtracting	
$x(ad - bc) = ed - bf$	by the distributive axiom	

$$x = \frac{ed - bf}{ad - bc} \qquad ad - bc \neq 0 \tag{5}$$

Note that if $ad - bc = 0$, the equations are not independent.
    Similarly, solving for y we get

$$y = \frac{af - ec}{ad - bc} \tag{6}$$

We next show that each of the values of x and y in Eqs. (5) and (6) is the quotient of two determinants. For this purpose, we write the determinant

$$D = \begin{vmatrix} a & b \\ c & d \end{vmatrix} = ad - bc$$

whose elements are the coefficients of x and y in Eqs. (1) and (2). The expansion of D is the denominator in each of Eqs. (5) and (6).

Now we replace the column of coefficients of x in D by the column of constant terms e and f and get

$$D_x = \begin{vmatrix} e & b \\ f & d \end{vmatrix} = ed - bf$$

Finally, we replace the column of coefficients of y in D by e and f and obtain

$$D_y = \begin{vmatrix} a & e \\ c & f \end{vmatrix} = af - ec$$

Now referring to Eqs. (5) and (6), we see that D_x is the numerator of the fraction in (5) and D_y is the numerator in (6). Consequently,

$$x = \frac{D_x}{D} \quad \text{and} \quad y = \frac{D_y}{D}$$

Cramer's Rule

The procedure just given for obtaining the simultaneous solution set of a system of linear equations is known as *Cramer's rule*, and it consists of the following steps:

1 Arrange the terms in the equations so that the x terms are first, the y terms are second, and the constant terms are at the right of the equality sign.
2 Write the determinant D, whose elements are the coefficients of x and y, in the order in which they occur in the equations.
3 Replace the column of coefficients of x in D by the column of constant terms, and obtain the determinant D_x.
4 Replace the column of coefficients of y in D by the column of constant terms, and get determinant D_y.
5 Then the elements in the solution set are

$$x = \frac{D_x}{D} \quad \text{and} \quad y = \frac{D_y}{D}$$

EXAMPLE 1

Use Cramer's rule to get the simultaneous solution set of

$$3x - 5y = -7$$

$$2x + 3y = 8$$

1 $D = \begin{vmatrix} 3 & -5 \\ 2 & 3 \end{vmatrix} = 9 + 10 = 19$

2 $D_x = \begin{vmatrix} -7 & -5 \\ 8 & 3 \end{vmatrix} = (-7)(3) - (-5)(8) = -21 + 40 = 19$

3 $D_y = \begin{vmatrix} 3 & -7 \\ 2 & 8 \end{vmatrix} = 3(8) - (-7)(2) = 24 + 14 = 38$

4 $x = \dfrac{D_x}{D} = \dfrac{19}{19} = 1 \qquad y = \dfrac{D_y}{D} = \dfrac{38}{19} = 2$

Hence the solution set is $\{(1, 2)\}$.

EXERCISE 8.1
Determinants of Order 2, Cramer's Rule

Expand the determinants in Probs. 1 to 16.

1 $\begin{vmatrix} 3 & 2 \\ 1 & 4 \end{vmatrix}$ **2** $\begin{vmatrix} 5 & 2 \\ 3 & 6 \end{vmatrix}$ **3** $\begin{vmatrix} 7 & 2 \\ 3 & 1 \end{vmatrix}$ **4** $\begin{vmatrix} 8 & 6 \\ 4 & 2 \end{vmatrix}$

5 $\begin{vmatrix} 5 & -3 \\ 2 & 1 \end{vmatrix}$ **6** $\begin{vmatrix} 4 & 6 \\ -2 & 3 \end{vmatrix}$ **7** $\begin{vmatrix} 9 & -2 \\ -4 & 3 \end{vmatrix}$ **8** $\begin{vmatrix} -7 & 4 \\ 2 & -3 \end{vmatrix}$

9 $\begin{vmatrix} 3 & 6 \\ 6 & 12 \end{vmatrix}$ **10** $\begin{vmatrix} 2 & -1 \\ 8 & -4 \end{vmatrix}$ **11** $\begin{vmatrix} 9 & -3 \\ -12 & 4 \end{vmatrix}$ **12** $\begin{vmatrix} -18 & 12 \\ 12 & -8 \end{vmatrix}$

13 $\begin{vmatrix} 8 & -3 \\ -4 & -2 \end{vmatrix}$ **14** $\begin{vmatrix} -6 & 4 \\ -2 & -3 \end{vmatrix}$ **15** $\begin{vmatrix} -3 & -2 \\ -5 & -7 \end{vmatrix}$ **16** $\begin{vmatrix} -8 & -4 \\ -5 & -3 \end{vmatrix}$

By expanding the determinants, prove that the statements in Probs. 17 to 20 are true.

17 $\begin{vmatrix} a+1 & c \\ b+2 & d \end{vmatrix} = \begin{vmatrix} a & c \\ b & d \end{vmatrix} + \begin{vmatrix} 1 & c \\ 2 & d \end{vmatrix}$

18 $\begin{vmatrix} a+2d & d \\ b+2c & c \end{vmatrix} = \begin{vmatrix} a & d \\ b & c \end{vmatrix}$

19 $\begin{vmatrix} 3a & c \\ 3b & d \end{vmatrix} = 3 \begin{vmatrix} a & c \\ b & d \end{vmatrix}$

20 $\begin{vmatrix} 2a & 4c \\ 2b & 4d \end{vmatrix} = 8 \begin{vmatrix} a & c \\ b & d \end{vmatrix}$

By use of Cramer's rule, find the simultaneous solution set of the pair of equations in each of the following problems.

21 $2x + 3y = 4$
      $x + 2y = 1$

22 $3x - 2y = 6$
      $4x - 3y = 2$

23 $2x + 4y = 5$
      $3x + 6y = 2$

24 $5x + 2y = 1$
      $8x + 3y = 2$

25 $5x + 3y = 2$
      $4x + 2y = 3$

26 $6x - 5y = 3$
      $3x - 2y = 1$

27 $6x + 5y = 4$
 $3x + 2y = 2$

28 $4x + 8y = 1$
 $3x + 6y = 3$

29 $3x - 9y = 1$
 $2x + 4y = -1$

30 $4x - 2y = 3$
 $2x + 5y = 6$

31 $7x - 14y = -3$
 $3x + 2y = 1$

32 $5x + 2y = 3$
 $7x - 5y = -1$

33 $3x - 5y = -8$
 $4x - 2y = 1$

34 $5x - 2y = 3$
 $4x - 7y = 9$

35 $15x - 9y = 2$
 $10x - 6y = 1$

36 $13x - 3y = -1$
 $3x - 5y = 3$

8.2
DETERMINANTS OF THE THIRD ORDER

The square array

$$■ \quad D = \begin{vmatrix} a_1 & b_1 & c_1 \\ a_2 & b_2 & c_2 \\ a_3 & b_3 & c_3 \end{vmatrix}$$

is a determinant of order 3.

Expansion of a Determinant

The expansion is defined to be the polynomial whose terms are of the type $a_i b_j c_k$ as follows:

$$■ \quad D = a_1 b_2 c_3 + a_2 b_3 c_1 + a_3 b_1 c_2 - a_3 b_2 c_1 - a_2 b_1 c_3 - a_1 b_3 c_2 \qquad (1)$$

The expansion in this form is difficult to remember and cumbersome to apply. Fortunately we can express it in terms of second-order determinants by a method that requires little memorization. The method involves the *minor* and the *cofactor* of an element. We define these terms now.

Minor of an Element

The *minor* of a specified element of a determinant D is the determinant whose elements are all elements of D that are not in the same row or column as the specified element.

Cofactor of an Element

If an element of D appears in the ith row and the jth column, the cofactor of the element is the minor of the element if $i + j$ is an even number, and it is the negative of the minor if $i + j$ is an odd number.

We shall designate the minors of a_i, b_i, and c_i by $m(a_i)$, $m(b_i)$, and $m(c_i)$, respectively, and the cofactors by A_i, B_i, and C_i, where $i = 1, 2,$ or 3.

According to the above definition

$$m(c_2) = \begin{vmatrix} a_1 & b_1 & c_1 \\ a_2 & b_2 & c_2 \\ a_3 & b_3 & c_3 \end{vmatrix} = \begin{vmatrix} a_1 & b_1 \\ a_3 & b_3 \end{vmatrix} = a_1 b_3 - a_3 b_1$$

Furthermore, since c_2 is in the second row and third column of D, it follows that $i + j = 2 + 3 = 5$ is an odd number; hence, the cofactor of c_2 is $C_2 = -m(c_2)$. Similarly, since a_3 is in the third row and first column, and $3 + 1 = 4$ is an even number,

$$A_3 = m(a_3) = \begin{vmatrix} b_1 & c_1 \\ b_2 & c_2 \end{vmatrix} = b_1 c_2 - b_2 c_1$$

We now show that the expansion of D in (1) can be expressed as the algebraic sum of the products of each element in a given row (or column) and its cofactor. We first consider the second row of D and group the terms of (1) and factor each group as follows:

$$D = -(b_1 c_3 - b_3 c_1)a_2 + (a_1 c_3 - a_3 c_1)b_2 - (a_1 b_3 - a_3 b_1)c_2$$

Now we note that

$$b_1 c_3 - b_3 c_1 = \begin{vmatrix} b_1 & c_1 \\ b_3 & c_3 \end{vmatrix} = m(a_2)$$

Likewise,

$$a_1 c_3 - a_3 c_1 = m(b_2) \quad \text{and} \quad a_1 b_3 - a_3 b_1 = m(c_2)$$

Consequently,

$$D = -m(a_2) \cdot a_2 + m(b_2) \cdot b_2 - m(c_2) \cdot c_2$$

Furthermore, since a_2 is in the second row and first column of D and $2 + 1 = 3$ is odd, $A_2 = -m(a_2)$. Similarly, $B_2 = m(b_2)$ and $C_2 = -m(c_2)$. Therefore, in terms of the cofactors of the elements of the second row, we have

$$D = A_2 a_2 + B_2 b_2 + C_2 c_2$$

Now we consider the third column of D and group and factor the terms of (1) as follows:

$$D = (a_2 b_3 - a_3 b_2)c_1 - (a_1 b_3 - a_3 b_1)c_2 + (a_1 b_2 - a_2 b_1)c_3$$
$$= m(c_1) \cdot c_1 - m(c_2) \cdot c_2 + m(c_3) \cdot c_3$$
$$= C_1 c_1 + C_2 c_2 + C_3 c_3$$

since

$$C_1 = m(c_1), \quad C_2 = -m(c_2), \quad \text{and} \quad C_3 = m(c_3)$$

By repeated application of this procedure, we can prove that the expansion of D can be expressed in terms of the cofactors of the elements of the rows in any one of the following ways:

$$A_1 a_1 + B_1 b_1 + C_1 c_1 = m(a_1)[a_1] - m(b_1)[b_1] + m(c_1)[c_1] \tag{2}$$

$$A_2 a_2 + B_2 b_2 + C_2 c_2 = -m(a_2)[a_2] + m(b_2)[b_2] - m(c_2)[c_2] \tag{3}$$

$$A_3 a_3 + B_3 b_3 + C_3 c_3 = m(a_3)[a_3] - m(b_3)[b_3] + m(c_3)[c_3] \tag{4}$$

Likewise, we can prove that the expansion in terms of the cofactors of the elements of the columns has one of the forms:

$$A_1 a_1 + A_2 a_2 + A_3 a_3 = m(a_1)[a_1] - m(a_2)[a_2] + m(a_3)[a_3] \tag{5}$$

$$B_1 b_1 + B_2 b_2 + B_3 b_3 = -m(b_1)[b_1] + m(b_2)[b_2] - m(b_3)[b_3] \tag{6}$$

$$C_1 c_1 + C_2 c_2 + C_3 c_3 = m(c_1)[c_1] - m(c_2)[c_2] + m(c_3)[c_3] \tag{7}$$

It is convenient to use the following diagram to determine the sign that must be placed before a minor of an element to get the cofactor.

$$+ \quad - \quad +$$

$$- \quad + \quad -$$

$$+ \quad - \quad +$$

For example, the cofactors of a_2, b_2, and c_2 in the second row are $-m(a_2)$, $+m(b_2)$, and $-m(c_2)$, and the signs preceding the respective minors are the signs in the second row of the diagram. Likewise, using the signs in the third column of the diagram, we have $+m(c_1)$, $-m(c_2)$, and $+m(c_3)$ respectively, as the cofactors of the elements of the third column of D. We shall illustrate the application of this method with three examples.

EXAMPLE 1   Expand the determinant

$$D_1 = \begin{vmatrix} 3 & 2 & 4 \\ 1 & 5 & 2 \\ 4 & 7 & 6 \end{vmatrix}$$

in terms of the cofactors of the elements of the first row.

SOLUTION   The cofactors of the elements in the first row of D_1 are the minors of the elements preceded, respectively, by the signs $+$, $-$, $+$. Then, since

$a_1 = 3$, $b_1 = 2$, and $c_1 = 4$, we have

$$A_1 = +m(3) = \begin{vmatrix} 5 & 2 \\ 7 & 6 \end{vmatrix} \qquad B_1 = -m(2) = -\begin{vmatrix} 1 & 2 \\ 4 & 6 \end{vmatrix}$$

$$C_1 = +m(4) = \begin{vmatrix} 1 & 5 \\ 4 & 7 \end{vmatrix}$$

Hence,

$$D_1 = A_1(3) + B_1(2) + C_1(4) \qquad \textbf{by Eq. (2)}$$

$$= \begin{vmatrix} 5 & 2 \\ 7 & 6 \end{vmatrix} 3 - \begin{vmatrix} 1 & 2 \\ 4 & 6 \end{vmatrix} 2 + \begin{vmatrix} 1 & 5 \\ 4 & 7 \end{vmatrix} 4$$

$$= (30 - 14)3 - (6 - 8)2 + (7 - 20)4$$

$$= (16)3 - (-2)2 + (-13)4 = 48 + 4 - 52 = 0$$

EXAMPLE 2 Expand the determinant

$$D_2 = \begin{vmatrix} -2 & 4 & 3 \\ 1 & -5 & -6 \\ 3 & 1 & -2 \end{vmatrix}$$

in terms of the cofactors of the elements of the third column, and check the result by expanding in terms of the cofactors of the second row.

SOLUTION Since

$$c_1 = 3, \qquad c_2 = -6, \qquad \text{and} \qquad c_3 = -2$$

and the signs in the third column of the diagram are $+, -, +$, we have

$$D_2 = \begin{vmatrix} 1 & -5 \\ 3 & 1 \end{vmatrix} 3 - \begin{vmatrix} -2 & 4 \\ 3 & 1 \end{vmatrix}(-6) + \begin{vmatrix} -2 & 4 \\ 1 & -5 \end{vmatrix}(-2)$$

$$= (1 + 15)3 - (-2 - 12)(-6) + (10 - 4)(-2)$$

$$= 48 - 84 - 12 = -48$$

The signs in the second row of the diagram are $-, +, -$, and the elements are $1, -5$, and -6. Hence,

$$D_2 = -\begin{vmatrix} 4 & 3 \\ 1 & -2 \end{vmatrix}(1) + \begin{vmatrix} -2 & 3 \\ 3 & -2 \end{vmatrix}(-5) - \begin{vmatrix} -2 & 4 \\ 3 & 1 \end{vmatrix}(-6)$$

$$= -(-8 - 3)(1) + (4 - 9)(-5) - (-2 - 12)(-6)$$

$$= 11 + 25 - 84 = -48$$

If one or more of the elements of a determinant are zero, it is advisable to expand the determinant in terms of the cofactors of the elements of the row or column that contains the greatest number of zeros.

EXAMPLE 3   Expand the determinant

$$D_3 = \begin{vmatrix} 3 & 2 & 4 \\ 0 & 2 & 0 \\ 1 & 3 & 2 \end{vmatrix}$$

SOLUTION   Here, the second row contains two zeros. Hence, we shall use this row to get the expansion

$$D_3 = -\begin{vmatrix} 2 & 4 \\ 3 & 2 \end{vmatrix}(0) + \begin{vmatrix} 3 & 4 \\ 1 & 2 \end{vmatrix}(2) - \begin{vmatrix} 3 & 2 \\ 1 & 3 \end{vmatrix}(0)$$

$$= 0 + (6 - 4)2 - 0 = 4$$

EXERCISE 8.2
Determinants of Order 3

Expand and find the value of each of the following determinants.

1 $\begin{vmatrix} 3 & 0 & 0 \\ 2 & 5 & 1 \\ 4 & 3 & 2 \end{vmatrix}$ **2** $\begin{vmatrix} 5 & 1 & 3 \\ 0 & 4 & 5 \\ 0 & 2 & 6 \end{vmatrix}$ **3** $\begin{vmatrix} 6 & 0 & 1 \\ 4 & 2 & 7 \\ 2 & 0 & 3 \end{vmatrix}$

4 $\begin{vmatrix} 5 & 4 & 0 \\ 2 & 1 & 6 \\ 7 & 3 & 0 \end{vmatrix}$ **5** $\begin{vmatrix} 8 & 2 & -3 \\ 4 & 5 & 6 \\ 0 & 3 & 0 \end{vmatrix}$ **6** $\begin{vmatrix} 0 & -2 & 0 \\ 6 & 8 & -2 \\ 4 & 1 & -3 \end{vmatrix}$

7 $\begin{vmatrix} 3 & 2 & 5 \\ 0 & 4 & 0 \\ 1 & 4 & 8 \end{vmatrix}$ **8** $\begin{vmatrix} -3 & 0 & 2 \\ 4 & 5 & -1 \\ 1 & 0 & 6 \end{vmatrix}$ **9** $\begin{vmatrix} 3 & 0 & -5 \\ 2 & -4 & 6 \\ 0 & 3 & 7 \end{vmatrix}$

10 $\begin{vmatrix} -5 & 2 & 4 \\ 0 & -3 & 6 \\ 4 & 1 & 0 \end{vmatrix}$ **11** $\begin{vmatrix} 8 & -4 & 0 \\ 2 & 0 & -3 \\ 5 & -7 & 6 \end{vmatrix}$ **12** $\begin{vmatrix} -6 & 8 & -2 \\ 0 & 4 & 1 \\ -3 & 0 & 5 \end{vmatrix}$

13 $\begin{vmatrix} 2 & 1 & 4 \\ 1 & 3 & 5 \\ 3 & 1 & 2 \end{vmatrix}$ **14** $\begin{vmatrix} 4 & 2 & 1 \\ 5 & 3 & 2 \\ 6 & 2 & 3 \end{vmatrix}$ **15** $\begin{vmatrix} 3 & -5 & 7 \\ 2 & 6 & -4 \\ -4 & 1 & 5 \end{vmatrix}$

16 $\begin{vmatrix} 5 & 3 & -2 \\ -4 & 1 & -6 \\ 2 & 4 & 1 \end{vmatrix}$ **17** $\begin{vmatrix} 3 & 6 & -3 \\ 1 & 4 & -2 \\ 2 & 8 & 1 \end{vmatrix}$ **18** $\begin{vmatrix} 2 & -3 & 4 \\ 1 & -5 & 2 \\ 6 & 2 & -2 \end{vmatrix}$

19 $\begin{vmatrix} 4 & -2 & 3 \\ 1 & 6 & 1 \\ 2 & 8 & 6 \end{vmatrix}$ **20** $\begin{vmatrix} 5 & -3 & 8 \\ 2 & 1 & -2 \\ -4 & 3 & -3 \end{vmatrix}$ **21** $\begin{vmatrix} 8 & 5 & 2 \\ 6 & 2 & 1 \\ 5 & -1 & -2 \end{vmatrix}$

22 $\begin{vmatrix} 3 & 2 & 1 \\ -7 & -8 & -2 \\ -3 & -4 & 1 \end{vmatrix}$ **23** $\begin{vmatrix} -2 & 6 & -1 \\ -6 & 1 & -5 \\ 6 & 4 & 4 \end{vmatrix}$ **24** $\begin{vmatrix} 4 & 2 & -5 \\ 2 & -2 & 8 \\ -3 & 1 & -2 \end{vmatrix}$

Find the solution set of each of the following equations.

25 $\begin{vmatrix} x-3 & 0 & 2x \\ 0 & 1 & 0 \\ 1 & 2 & 1 \end{vmatrix} = 0$ **26** $\begin{vmatrix} 2 & 1 & 2 \\ x+2 & 0 & 3x \\ 4 & 3 & 4 \end{vmatrix} = 0$

27 $\begin{vmatrix} x & 2 & 0 \\ x+1 & 3 & 1 \\ x-2 & 0 & 1 \end{vmatrix} = 0$ **28** $\begin{vmatrix} 2 & x-2 & 1 \\ 5 & x+1 & 0 \\ 1 & x-3 & 1 \end{vmatrix} = 0$

Prove that the following statements are true.

29 $\begin{vmatrix} a+1 & 0 & 0 \\ b+1 & 2 & 1 \\ c+1 & 1 & 2 \end{vmatrix} = \begin{vmatrix} a & 0 & 0 \\ b & 2 & 1 \\ c & 1 & 2 \end{vmatrix} + \begin{vmatrix} 1 & 0 & 0 \\ 1 & 2 & 1 \\ 1 & 1 & 2 \end{vmatrix}$

30 $\begin{vmatrix} a+3 & 3 & 0 \\ b+2 & 2 & 0 \\ c+1 & 1 & 2 \end{vmatrix} = \begin{vmatrix} a & 3 & 0 \\ b & 2 & 0 \\ c & 1 & 2 \end{vmatrix}$

31 $\begin{vmatrix} a+2 & b+1 & 0 \\ a+3 & b+2 & 0 \\ a-1 & b-3 & 3 \end{vmatrix} = \begin{vmatrix} a & b & 0 \\ a & b & 0 \\ a & b & 3 \end{vmatrix} + \begin{vmatrix} a & 1 & 0 \\ a & 2 & 0 \\ a & -3 & 3 \end{vmatrix}$

$$+ \begin{vmatrix} 2 & b & 0 \\ 3 & b & 0 \\ -1 & b & 3 \end{vmatrix} + \begin{vmatrix} 2 & 1 & 0 \\ 3 & 2 & 0 \\ -1 & -3 & 3 \end{vmatrix}$$

32 $\begin{vmatrix} 0 & b_1 & c_1 \\ 0 & b_2 & c_2 \\ a_3 & b_3 & c_3 \end{vmatrix} = \begin{vmatrix} 0 & b_1+2c_1 & c_1 \\ 0 & b_2+2c_2 & c_2 \\ a_3 & b_3+2c_3 & c_3 \end{vmatrix}$

8.3
CRAMER'S RULE

In Sec. 8.1 we proved that Cramer's rule yields the simultaneous solution set of a system of two linear equations in two variables. In this section we shall explain the procedure for using Cramer's rule to obtain the solution set of three linear equations in three variables and illustrate the method by solving the system

$$2x - 3y + 4z = 19 \tag{1}$$

$$x + 2y - 2z = -6 \tag{2}$$

$$3x + y + z = 8 \tag{3}$$

First, we set up the determinant of the coefficients

$$D = \begin{vmatrix} 2 & -3 & 4 \\ 1 & 2 & -2 \\ 3 & 1 & 1 \end{vmatrix}$$

In terms of the cofactors of the elements of the first column,

$$D = \begin{vmatrix} 2 & -2 \\ 1 & 1 \end{vmatrix} 2 - \begin{vmatrix} -3 & 4 \\ 1 & 1 \end{vmatrix} 1 + \begin{vmatrix} -3 & 4 \\ 2 & -2 \end{vmatrix} 3$$

$$= (2 + 2)2 - (-3 - 4)1 + (6 - 8)3$$

$$= (4)2 - (-7)1 + (-2)3$$

$$= 8 + 7 - 6 = 9$$

Note that the cofactors of the elements of the first column are 4, $-(-7) = 7$, and $+(-2) = -2$.

Second, we replace the column of coefficients of x in D by the column of constant terms and get

$$D_x = \begin{vmatrix} 19 & -3 & 4 \\ -6 & 2 & -2 \\ 8 & 1 & 1 \end{vmatrix}$$

The expansion of D_x in terms of the cofactors of the elements of the first column is

$$D_x = \begin{vmatrix} 2 & -2 \\ 1 & 1 \end{vmatrix} 19 - \begin{vmatrix} -3 & 4 \\ 1 & 1 \end{vmatrix} (-6) + \begin{vmatrix} -3 & 4 \\ 2 & -2 \end{vmatrix} 8$$

$$= (2 + 2)19 - (-3 \quad 4)(-6) + (6 - 8)8$$

$$= 4(19) + 7(-6) - 2(8)$$

$$= 76 - 42 - 16 = 18$$

Notice that here the cofactors of the elements of the first column are 4, 7, and -2 and these are the same as the cofactors that appear in the expansion of D.

Now we shall eliminate y and z from Eqs. (1), (2), and (3) and show that $x = D_x/D$. We multiply Eq. (1) by 4, (2) by 7, and (3) by -2. Notice that these multipliers are the cofactors appearing in the expansion of D and of D_x.

$8x - 12y + 16z = 76$	multiplying Eq. (1) by 4	(4)
$7x + 14y - 14z = -42$	multiplying Eq. (2) by 7	(5)
$-6x - 2y - 2z = -16$	multiplying Eq. (3) by -2	(6)
$(8 + 7 - 6)x + 0 \cdot y + 0 \cdot z = 76 - 42 - 16$	equating sums of corresponding members of Eqs. (4), (5), and (6)	

$$9x = 18$$

$$x = \frac{18}{9} = 2$$

Hence, since $D_x = 18$ and $D = 9$, it follows that $x = D_x/D$.

Next, we replace the column of coefficients of y in D by the column of constant terms and get

$$D_y = \begin{vmatrix} 2 & 19 & 4 \\ 1 & -6 & -2 \\ 3 & 8 & 1 \end{vmatrix}$$

Expanding in terms of the cofactors of the elements of the second column, we have

$$D_y = -\begin{vmatrix} 1 & -2 \\ 3 & 1 \end{vmatrix} 19 + \begin{vmatrix} 2 & 4 \\ 3 & 1 \end{vmatrix}(-6) - \begin{vmatrix} 2 & 4 \\ 1 & -2 \end{vmatrix} 8$$

$$= -(1+6)19 + (2-12)(-6) - (-4-4)8$$

$$= (-7)19 + (-10)(-6) - (-8)8$$

$$= -133 + 60 + 64$$

$$= -9$$

In order to solve for y, we eliminate x and z from Eqs. (1), (2), and (3) by multiplying (1) by -7, (2) by -10, and (3) by 8. Notice that these multipliers are the cofactors of the elements of the second column of D_y, and equate the sums of the corresponding members. Thus, we get

$$-14x + 21y - 28z = -133$$
$$-10x - 20y + 20z = 60$$
$$\underline{24x + 8y + 8z = 64}$$
$$0 \cdot x + (21 - 20 + 8)y + 0 \cdot z = -133 + 60 + 64$$
$$9y = -9$$
$$y = -\tfrac{9}{9} = -1$$

Hence, since $D_y = -9$ and $D = 9$, it follows that $y = D_y/D$.

Finally, we replace the column of coefficients of z in D by the column of constant terms and obtain

$$D_z = \begin{vmatrix} 2 & -3 & 19 \\ 1 & 2 & -6 \\ 3 & 1 & 8 \end{vmatrix}$$

$$= \begin{vmatrix} 1 & 2 \\ 3 & 1 \end{vmatrix} 19 - \begin{vmatrix} 2 & -3 \\ 3 & 1 \end{vmatrix}(-6) + \begin{vmatrix} 2 & -3 \\ 1 & 2 \end{vmatrix} 8 \qquad \text{in terms of the cofactors of the elements of the third column}$$

$$= (1-6)19 - (2+9)(-6) + (4+3)8$$

$$= -5(19) - 11(-6) + 7(8) = 27$$

Next, we multiply Eq. (1) by -5, Eq. (2) by -11, and Eq. (3) by 7, equate the sums of the corresponding members, and have

$$-10x + 15y - 20z = -95$$
$$-11x - 22y + 22z = 66$$
$$\underline{21x + 7y + 7z = 56}$$
$$0 \cdot x + 0 \cdot y + 9z = 27$$
$$9z = 27$$
$$z = \tfrac{27}{9} = 3$$

Consequently,

$$z = \frac{D_z}{D}$$

Therefore, the solution set is $\{(2, -1, 3)\}$.

Verification

$2(2) - 3(-1) + 4(3) = 4 + 3 + 12 = 19$ from Eq. (1)

$2 + 2(-1) - 2(3) = 2 - 2 - 6 = -6$ from Eq. (2)

$3(2) + (-1) + 3 = 6 - 1 + 3 = 8$ from Eq. (3)

EXAMPLE 1   Use Cramer's rule to get the solution set of

$2x + 5y + 3z = 3$

$x + y + z = 0$

$3x - y + 2z = -5$

SOLUTION   The determinant of the coefficients is

$$D = \begin{vmatrix} 2 & 5 & 3 \\ 1 & 1 & 1 \\ 3 & -1 & 2 \end{vmatrix} = \begin{vmatrix} 1 & 1 \\ -1 & 2 \end{vmatrix} 2 - \begin{vmatrix} 1 & 1 \\ 3 & 2 \end{vmatrix} 5 + \begin{vmatrix} 1 & 1 \\ 3 & -1 \end{vmatrix} 3$$

$$= (2 + 1)2 - (2 - 3)5 + (-1 - 3)3$$

$$= 6 + 5 - 12 = -1$$

The expansion of D was obtained in terms of the cofactors of the elements of the first row.

By replacing the appropriate column of D by the column of constant terms, we get D_x, D_y, and D_z. Since the column of constant terms contains one zero, we expand each of these determinants in terms of the cofactors of this column, and hence each expansion will contain only two terms. Each of these determinants and its expansion follows:

$$D_x = \begin{vmatrix} 3 & 5 & 3 \\ 0 & 1 & 1 \\ -5 & -1 & 2 \end{vmatrix} = \begin{vmatrix} 1 & 1 \\ -1 & 2 \end{vmatrix} 3 + \begin{vmatrix} 5 & 3 \\ 1 & 1 \end{vmatrix} (-5)$$

$$= (2 + 1)3 + (5 - 3)(-5) = 9 - 10 = -1$$

$$D_y = \begin{vmatrix} 2 & 3 & 3 \\ 1 & 0 & 1 \\ 3 & -5 & 2 \end{vmatrix} = -\begin{vmatrix} 1 & 1 \\ 3 & 2 \end{vmatrix} 3 - \begin{vmatrix} 2 & 3 \\ 1 & 1 \end{vmatrix} (-5)$$

$$= -(2 - 3)3 - (2 - 3)(-5) = 3 - 5 = -2$$

$$D_z = \begin{vmatrix} 2 & 5 & 3 \\ 1 & 1 & 0 \\ 3 & -1 & -5 \end{vmatrix} = \begin{vmatrix} 1 & 1 \\ 3 & -1 \end{vmatrix} 3 + \begin{vmatrix} 2 & 5 \\ 1 & 1 \end{vmatrix} (-5)$$

$$= (-1 - 3)3 + (2 - 5)(-5) = -12 + 15 = 3$$

Hence, the values of x, y, and z are

$$x = \frac{D_x}{D} = \frac{-1}{-1} = 1$$

$$y = \frac{D_y}{D} = \frac{-2}{-1} = 2$$

$$z = \frac{D_z}{D} = \frac{3}{-1} = -3$$

Therefore, the solution set is $\{(1, 2, -3)\}$.

EXERCISE 8.3
Cramer's Rule

By use of Cramer's rule, find the solution set of each of the following systems of equations.

1 $2x + y - z = -1$
   $x - 3y + 2z = 8$
   $4x - 2y - 3z = 0$

2 $x + y + z = 2$
   $x - y - z = -4$
   $x + y - z = 0$

3 $x + y + 2z = 1$
   $2x + y - z = 0$
   $x - 2y - 4z = 4$

4 $x - y + z = 0$
   $x + 2y + 4z = 3$
   $2x - 3y - z = 1$

5 $x + y + z = 3$
   $x + 2y + z = 1$
   $2x + 3y - z = 1$

6 $x + y - z = -1$
   $2x + y + 3z = 4$
   $x + 2y + z = 7$

7 $x + 2y - z = -4$
   $2x + 3y + z = 3$
   $x + 4y - 2z = -6$

8 $x - 3y + 2z = 4$
   $2x + y + z = 3$
   $3x + 4y + 3z = -5$

9 $2x - y + 3z = 2$
   $4x + 2y - 5z = 1$
   $6x - 3y + 4z = 1$

10 $2x + 3y + z = 2$
    $4x - 6y + 5z = 5$
    $x + 9y + 2z = -3$

11 $3x + 4y + 2z = -2$
    $2x - y + 8z = 5$
    $x + 5y - 4z = -4$

12 $3x + y - 2z = 2$
    $6x - y + 3z = 3$
    $9x + 4y + z = -3$

13 $2x + 3y + z = 3$
    $6x - y - 2z = -1$
    $2x + 5y + 3z = 6$

14 $5x + 2y + z = 1$
    $x + 4y - 2z = 3$
    $4x - 2y - z = 2$

15
$$x + 2y + z = 4$$
$$x - 6y + 2z = 1$$
$$2x + 4y - 3z = -2$$

16
$$2x - 4y + 3z = -3$$
$$6x + 2y - 3z = 3$$
$$4x - 3y + 6z = -1$$

17
$$x + 3y + 2z = 3$$
$$2x + 3y + 4z = 4$$
$$3x - 6y + 2z = -2$$

18
$$x + 2y + z = 3$$
$$3x - 2y + 5z = 4$$
$$5x - 3y + 3z = 6$$

19
$$2x + 2y - 3z = 2$$
$$x + 4y - 6z = 3$$
$$x - 6y - 3z = -5$$

20
$$x - y + z = 2$$
$$3x + 7y + 5z = 1$$
$$4x + y + 2z = 3$$

21
$$2x + y = 3$$
$$4y + z = -1$$
$$3x - 4z = -6$$

22
$$x + y = 1$$
$$3y - 2z = -2$$
$$5x + 6z = 9$$

23
$$x + 2y = 1$$
$$6y - z = 2$$
$$3x + 4z = -2$$

24
$$x - 2y = 1$$
$$-4y + 3z = 3$$
$$6x + 5z = 4$$

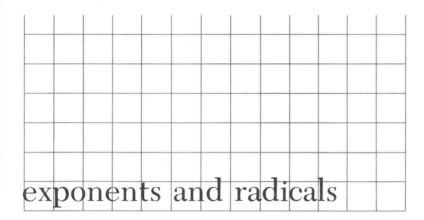

exponents and radicals

In Chap. 2 we defined a positive integral exponent and also the number a^0, $a \neq 0$; furthermore, we derived the laws for the product and the quotient of two positive integral powers of the same number and used these definitions and laws to a limited extent. In fields in which mathematics is used, a broader concept of an exponent is needed. Consequently, in this chapter we shall extend the definition of an exponent so as to include negative and fractional values. These extensions will be made so that the laws developed in Chap. 2 shall hold. Furthermore, we shall develop laws for a power of a product and of a quotient and explain how to use the old and new concepts and laws in more complicated situations than occurred in Chap. 2.

9.1
LAWS OF POSITIVE INTEGRAL EXPONENTS

nth Power, Base, Exponent

We shall begin by repeating the definition of a positive integral exponent and the laws of exponents developed in Chap. 2. If n is a positive integer, then the product $a \cdot a \cdot a \cdots$ to n a's is called the *nth power of a* and is written as a^n. The number a is the *base* and n is the *exponent*. A symbolic form of the definition and the laws developed in Chap. 2 are listed here for the convenience of the reader.

■ $a^n = a \cdot a \cdot a \cdots$ to n a's   (9.1)

■ $a^m a^n = a^{m+n}$ m and n positive integers   (9.2)

■ $\dfrac{a^m}{a^n} = a^{m-n}$ $\begin{cases} m \text{ and } n \text{ positive integers} \\ m > n \\ a \neq 0 \end{cases}$ (9.3)

■ $a^0 = 1$ $a \neq 0$ (9.4)

We shall now develop three more laws for positive integral exponents. If we apply the definition of a positive integral exponent to the number $(a^m)^n$, we have

$$(a^m)^n = a^m \cdot a^m \cdot a^m \cdots \text{ to } n \text{ factors}$$
$$= a^{m+m+m \cdots \text{ to } n \text{ terms}} \qquad \text{by (9.2)}$$
$$= a^{nm}$$

Consequently, we have

■ $(a^m)^n = a^{nm}$ (9.5)

If we apply the definition to $(ab)^n$, we obtain

$$(ab)^n = ab \cdot ab \cdot ab \cdots \text{ to } n \text{ factors}$$
$$= (a \cdot a \cdot a \cdots \text{ to } n \text{ factors})(b \cdot b \cdot b \cdots \text{ to } n \text{ factors}) \qquad \text{by commutative axiom for multiplication}$$
$$= a^n b^n$$

Therefore,

■ $(ab)^n = a^n b^n$ (9.6)

We can show similarly that

■ $\left(\dfrac{a}{b}\right)^n = \dfrac{a^n}{b^n}$ $b \neq 0$ (9.7)

We shall now illustrate the use of Laws (9.1) to (9.7) by several examples.

$$a^5 = a \cdot a \cdot a \cdot a \cdot a \qquad \text{by (9.1)}$$

$$a^7 a^2 = a^{7+2} = a^9 \qquad \text{by (9.2)}$$

$$\frac{a^6 a^5}{a^4} = \frac{a^{6+5}}{a^4} = \frac{a^{11}}{a^4} = a^{11-4} = a^7 \qquad \text{by (9.2) and (9.3)}$$

$$(a^5)^3 = a^{(3)(5)} = a^{15} \qquad \text{by (9.5)}$$

$$(x^2 y^3)^4 = (x^2)^4 (y^3)^4 = x^8 y^{12} \qquad \text{by (9.6) and (9.5)}$$

$$\left(\frac{x^3}{z^4}\right)^2 = \frac{(x^3)^2}{(z^4)^2} = \frac{x^6}{z^8} \qquad \text{by (9.7) and (9.5)}$$

$$(3x^2 y^3)(7x^4 y^5) = 3 \cdot 7x^2 x^4 y^3 y^5 \qquad \text{by commutative axiom}$$

$$= 21x^6 y^8 \qquad \text{by (9.2)}$$

$$\frac{30x^5 y^7}{15x^2 y^3} = \left(\frac{30}{15}\right)\left(\frac{x^5}{x^2}\right)\left(\frac{y^7}{y^3}\right)$$

$$= 2x^3 y^4 \qquad \text{by (9.3)}$$

Simplified

We say that an expression that includes positive integral exponents is *simplified* if all combinations are made that can be made by use of (9.1) to (9.7). The procedures are illustrated in the following example.

EXAMPLE 1 Simplify

$$\left(\frac{3x^2 y^3}{z^4}\right)^3 \left(\frac{2y^2 z^7}{x^3}\right)^2$$

SOLUTION   In order to simplify, we shall raise each product and quotient to the indicated power by use of (9.5), (9.6), and (9.7), then apply (9.2), and finally reduce to lowest terms by use of (9.3).

$$\left(\frac{3x^2 y^3}{z^4}\right)^3 \left(\frac{2y^2 z^7}{x^3}\right)^2 = \frac{3^3 (x^2)^3 (y^3)^3}{(z^4)^3} \frac{2^2 (y^2)^2 (z^7)^2}{(x^3)^2} \qquad \text{by (9.6) and (9.7)}$$

$$= \frac{27x^6 y^9}{z^{12}} \frac{4y^4 z^{14}}{x^6} \qquad \text{by (9.5)}$$

$$= 27(4)\left(\frac{x^6}{x^6}\right)(y^9 y^4)\left(\frac{z^{14}}{z^{12}}\right) \qquad \text{by commutative axiom}$$

$$= 108x^{6-6} y^{9+4} z^{14-12} \qquad \text{by (9.2) and (9.3)}$$

$$= 108x^0 y^{13} z^2$$

$$= 108y^{13} z^2 \qquad \text{since by (9.4), } x^0 = 1$$

If we interchange the members of (9.6) and of (9.7), we get

■ $a^n b^n = (ab)^n$ (9.8)

■ $\dfrac{a^n}{b^n} = \left(\dfrac{a}{b}\right)^n$ (9.9)

This justifies the practice of multiplying or dividing the two bases and then raising the result to the common power.

EXAMPLE 2 Simplify

$$\left[\left(\frac{3a^3 b^5}{13c^4}\right)^2\right]\left[\left(\frac{26c^5}{9a^2 b^3}\right)^2\right]$$

SOLUTION

$$\left[\left(\frac{3a^3 b^5}{13c^4}\right)^2\right]\left[\left(\frac{26c^5}{9a^2 b^3}\right)^2\right] = \left[\left(\frac{3a^3 b^5}{13c^4}\right)\left(\frac{26c^5}{9a^2 b^3}\right)\right]^2 \qquad \text{by (9.8)}$$

$$= \left[\frac{3 \cdot 26}{13 \cdot 9}\frac{a^3}{a^2}\frac{b^5}{b^3}\frac{c^5}{c^4}\right]^2 \qquad \text{by commutative axiom}$$

$$= \left(\frac{2}{3}ab^2 c\right)^2 \qquad \text{by (9.3)}$$

$$= \frac{4}{9}a^2 b^4 c^2 \qquad \text{by (9.6), (9.7), and (9.5)}$$

EXAMPLE 3 Simplify

$$\left(\frac{9x^{3a-4}}{3x^{2a-1}}\right)^3\left(\frac{y^b}{x^{a-1}}\right)^2$$

SOLUTION

$$\left[\frac{9x^{3a-4}}{3x^{2a-1}}\right]^3\left[\frac{y^b}{x^{a-1}}\right]^2 = [3x^{3a-4-(2a-1)}]^3\left[\frac{y^b}{x^{a-1}}\right]^2 \qquad \text{by (9.3)}$$

$$= (3x^{a-3})^3\left(\frac{y^b}{x^{a-1}}\right)^2$$

$$= 27x^{3a-9}\frac{y^{2b}}{x^{2a-2}} \qquad \text{by (9.5) and (9.7)}$$

$$= 27x^{a-7}y^{2b} \qquad \text{by (9.3)}$$

Reminder 1 In applying (9.5), be sure to remember that the exponent inside the parentheses is multiplied by the one outside and not raised to the power indicated by it. Thus,

$$(a^3)^2 = a^{(3)(2)} = a^6 \qquad \text{not} \qquad a^{3^2} = a^9$$

Reminder 2 If x^a is to be multiplied by y^b, we can only indicate the product by writing $x^a y^b$ unless $x = y$ or $a = b$. If the bases are equal, we *use the common base* and add the exponents. If the exponents are equal, we *use the common exponent* and multiply the bases. Thus $a^2 a^3 = a^5$ and $a^2 b^2 = (ab)^2$ as seen from (9.2) and (9.8).

EXERCISE 9.1
Simplification of Exponential Expressions

Perform the indicated operations in each of Probs. 1 to 32.

1 $2^4 2^3$ **2** $3^2 3^5$

3 $4^2 4^4$ **4** $5^0 5^2$

5 $\dfrac{6^5}{6^3}$ **6** $\dfrac{7^4}{7^2}$

7 $\dfrac{5^5}{5^2}$ **8** $\dfrac{8^2}{8^0}$

9 $(3^2)^3$ **10** $(2^3)^3$

11 $(4^2)^2$ **12** $(6^3)^0$

13 $(3^2 2)^2$ **14** $(2^2 3^3)^2$

15 $\left(\dfrac{3^2}{2^3}\right)^2$ **16** $\left(\dfrac{2^2}{3^2}\right)^3$

17 $(3a^3)(2a^4)$ **18** $(2b^2)(3b^0)$

19 $(6b^3)(2b^2)$ **20** $(8c)(3c^5)$

21 $\dfrac{c^8}{c^2}$ **22** $\dfrac{n^7}{n^0}$

23 $\dfrac{s^6}{s^3}$ **24** $\dfrac{d^5}{d^2}$

25 $(x^2)^4$ **26** $(a^2)^3$

27 $(b^0)^2$ **28** $(r^4)^4$

29 $(a^2 b^3)^5$ **30** $(x^4 y^2)^3$

31 $\left(\dfrac{a^4}{b^2}\right)^3$ **32** $\left(\dfrac{x^5}{y^3}\right)^4$

Perform the indicated operations in the following problems and simplify.

33 $\dfrac{32x^2 y^4}{4xy^2}$ **34** $\dfrac{36a^6 b^3}{9a^2 b^2}$

35 $\dfrac{16u^8 v^5}{8u^0 v^3}$ **36** $\dfrac{45r^7 t^9}{5r^5 t^4}$

37 $\left(\dfrac{3a^4 b^7}{7a^2 d^5}\right)\left(\dfrac{14c^5 d^4}{15b^5 c^3}\right)$ **38** $\left(\dfrac{18x^6 y^2}{25x^0 z^3}\right)\left(\dfrac{5y^3 z^2}{6x^4 y^4}\right)$

39 $\left(\dfrac{35r^5 s^6}{24t^0 u^4}\right)\left(\dfrac{9r^0 u^7}{28s^2 t^2}\right)$ **40** $\left(\dfrac{20w^7 y^3}{21x^6 z^2}\right)\left(\dfrac{12x^3 y^2}{35w^5 z^0}\right)$

41 $\left(\dfrac{15a^4 b}{5a^3 b^2}\right)^4$

42 $\left(\dfrac{2c^3 d^2}{3c^4 d^5}\right)^5$

43 $\left(\dfrac{12r^8 s^5}{6r^6 s^2}\right)^6$

44 $\left(\dfrac{24p^5 q^2}{8p^3 q^4}\right)^3$

45 $\left(\dfrac{4a^0 b^3}{3c^5}\right)^2 \left(\dfrac{6a^2 c^2}{8b^3}\right)^2$

46 $\left(\dfrac{6m^2 t^4}{5s^3}\right)^3 \left(\dfrac{10m^4 s^2}{18t^5}\right)^2$

47 $\left(\dfrac{20x^3 y^2}{7z^0}\right)^4 \left(\dfrac{21y^2 z^2}{30x^5}\right)^4$

48 $\left(\dfrac{14r^2 s^4}{21t^3}\right)^3 \left(\dfrac{5r^3 t^2}{10s^4}\right)^4$

49 $\left[\left(\dfrac{3x^2}{2y}\right)^3 \left(\dfrac{8y^3}{9x^2}\right)^2\right]^3$

50 $\left[\left(\dfrac{2a^3}{7b^4}\right)^3 \left(\dfrac{7b^5}{2a^4}\right)^2\right]^2$

51 $\left[\left(\dfrac{4c^3}{3d^4}\right)^4 \left(\dfrac{9d^2}{8c^4}\right)^3\right]^4$

52 $\left[\left(\dfrac{3u^2}{4v^5}\right)^4 \left(\dfrac{16v^7}{9u^3}\right)^2\right]^5$

53 $x^{3a+1} x^{a-2}$

54 $y^{s-3} y^{2s+1}$

55 $w^{x-5} w^{3x+2}$

56 $s^{4x-1} s^{5-3x}$

57 $\dfrac{x^{2a-1} y^{b+5}}{x^{a+2} y^3}$

58 $\dfrac{a^{2n-7} y^{3n+2}}{a^{n-8} y^{n-1}}$

59 $\dfrac{c^{x+3} d^{3y-2}}{c^{2-x} d^{y+1}}$

60 $\dfrac{r^{a+b} s^{3+b}}{r^b s^b}$

61 $\dfrac{(x^{t+2} y^{u-1})^2}{x^3 y^u}$

62 $\dfrac{(u^{a-1} v^{b+3})^3}{u^2 v^{2b}}$

63 $\dfrac{(a^{3+m} b^{m-2})^2}{(a^m b^2)^2}$

64 $\dfrac{(w^{x+y} z^{x+t})^2}{(w^y z^x)^3}$

9.2
NEGATIVE INTEGRAL EXPONENTS

Definition of Negative Exponent

In this section we shall extend the definition of a^n to include an interpretation of a^{-t}, where $t > 0$ and $a \neq 0$. If we disregard the restriction $m > n$ in Law (9.3), we have, since $-t = t - 2t$,

$$a^{-t} = a^{t-2t} = \dfrac{a^t}{a^{2t}} = \dfrac{1}{a^t}$$

Therefore, we define a^{-t} as

■ $\quad a^{-t} = \dfrac{1}{a^t} \qquad a \neq 0$ \hfill (9.10)

We shall next prove that Laws (9.2) and (9.3) hold for this interpretation of negative exponents. To show that Law (9.2) holds, we must prove that $a^{-t} a^{-r} = a^{-t-r}$. By (9.10) we have

$$a^{-t}a^{-r} = \frac{1}{a^t} \cdot \frac{1}{a^r}$$

$$= \frac{1}{a^{t+r}} \qquad \text{by (9.2)}$$

$$= a^{-(t+r)} \qquad \text{by (9.10)}$$

$$= a^{-t-r}$$

To show that (9.3) holds, we must show that $a^{-t}/a^{-r} = a^{-t-(-r)}$. Again using (9.10), we have

$$\frac{a^{-t}}{a^{-r}} = \frac{\dfrac{1}{a^t}}{\dfrac{1}{a^r}}$$

$$= \frac{a^r}{a^t} \qquad \text{multiplying each member of the complex fractions by } a^{t+r}$$

$$= a^{r-t} \qquad \text{by (9.3)}$$

$$= a^{-t+r}$$

$$= a^{-t-(-r)}$$

Since Laws (9.4) to (9.7) were derived from Laws (9.2) and (9.3), these laws hold also for negative exponents. Therefore, we have removed the restrictions on Laws (9.2) to (9.7) that m be greater than n, and the only remaining restriction is that m and n be integers.

It is frequently desirable that a fraction whose numerator or denominator or both include negative exponents should be converted to an equal fraction in which all exponents are positive. We use the fundamental principle of fractions to accomplish this purpose. For example, to convert $a^x b^{-y}/c^z d^{-w}$ to an equal fraction that has no negative exponents, we first notice that $b^{-y} \cdot b^y = b^{-y+y} = b^0 = 1$. Hence, if we multiply the given fraction by b^y/b^y, we obtain

$$\frac{a^x b^{-y}}{c^z d^{-w}} \cdot \frac{b^y}{b^y} = \frac{a^x}{c^z d^{-w} b^y}$$

Similarly, if we multiply the right member by d^w/d^w, we get

$$\frac{a^x d^w}{c^z d^{-w} b^y d^w} = \frac{a^x d^w}{c^z b^y}$$

These two steps can be combined into the single operation of multiplying the given fraction by $b^y d^w/b^y d^w$ and obtaining

$$\frac{a^x b^{-y}}{c^z d^{-w}} = \frac{a^x b^{-y}}{c^z d^{-w}} \cdot \frac{b^y d^w}{b^y d^w} = \frac{a^x d^w}{c^z b^y}$$

Procedure for Eliminating Negative Exponents

The example just given suggests the following procedure by which a fraction whose numerator or denominator or both are monomials that include negative exponents may be expressed as an equal fraction with all exponents positive.

For each negative power of a number that occurs in the numerator or denominator, multiply both numerator and denominator by that number with the numerically equal positive exponent.

EXAMPLE 1 Convert

$$\frac{a^2 b^{-3} c^{-2}}{x^{-1} y^3 z^{-3}}$$

into an equal fraction in which all exponents are positive.

SOLUTION The negative powers of numbers in the numerator and denominator are b^{-3}, c^{-2}, x^{-1}, z^{-3}. Therefore, we multiply each member of the given fraction by $b^3 c^2 x z^3$ and get

$$\frac{a^2 b^{-3} c^{-2}}{x^{-1} y^3 z^{-3}} \cdot \frac{b^3 c^2 x z^3}{b^3 c^2 x z^3} = \frac{a^2 b^{-3+3} c^{-2+2} x z^3}{x^{-1+1} y^3 z^{-3+3} b^3 c^2} \qquad \text{by commutative axiom and (9.2)}$$

$$- \frac{a^2 b^0 c^0 x z^3}{x^0 y^3 z^0 b^3 c^2}$$

$$= \frac{a^2 x z^3}{y^3 b^3 c^2} \qquad \text{since by (9.4), } b^0 = c^0 = z^0 = 1$$

EXAMPLE 2 Express

$$\frac{2 c^{-2} d^{-1}}{3 x^{-1} y^3}$$

as an equal fraction having only positive exponents.

SOLUTION

$$\frac{2 c^{-2} d^{-1}}{3 x^{-1} y^3} = \frac{2 c^{-2} d^{-1}}{3 x^{-1} y^3} \cdot \frac{c^2 d x}{c^2 d x}$$

$$= \frac{2 c^{-2+2} d^{-1+1} x}{3 x^{-1+1} y^3 c^2 d} \qquad \text{by commutative axiom and (9.2)}$$

$$= \frac{2 c^0 d^0 x}{3 x^0 y^3 c^2 d}$$

$$= \frac{2x}{3 y^3 c^2 d} \qquad \text{by (9.4)}$$

If the numerator or denominator or both are polynomials and if either or both the numerator and denominator have negative exponents, we use the principle explained above. For example, to convert $(x^{-1} + y^{-1})/(x^{-2} + y^{-2})$ to an equal fraction in which all exponents are positive, we notice that x appears with exponents -1 and -2 and y likewise appears with exponents -1 and -2. Therefore, if we multiply the fraction by $x^2 y^2/x^2 y^2$, we obtain a fraction in which the exponents of x and of y are positive. The details of the conversion process follow:

$$\frac{x^{-1} + y^{-1}}{x^{-2} - y^{-2}} \cdot \frac{x^2 y^2}{x^2 y^2} = \frac{x^{-1+2} y^2 + x^2 y^{-1+2}}{x^{-2+2} y^2 - x^2 y^{-2+2}} = \frac{xy^2 + x^2 y}{y^2 - x^2}$$

$$= \frac{xy(y + x)}{(y + x)(y - x)} = \frac{xy}{y - x}$$

Since Laws (9.2) to (9.7) now hold for negative as well as positive integral exponents, we can use them for combining powers of the same number regardless of the signs of the exponents. In the following examples we shall make all possible applications of Laws (9.2) to (9.7) and shall express the result without zero or negative exponents. Accordingly,

$$\frac{12a^{-2} b^3 c^{-3}}{4a^3 b^{-1} c^{-2}} = \frac{12a^{-2} b^3 c^{-3}}{4a^3 b^{-1} c^{-2}} \cdot \frac{a^2 bc^3}{a^2 bc^3} \qquad \text{multiplying by } a^2 bc^3/a^2 bc^3$$

$$= \frac{12a^{-2+2} b^{3+1} c^{-3+3}}{4a^{3+2} b^{-1+1} c^{-2+3}} \qquad \text{by (9.2)}$$

$$= \frac{12a^0 b^4 c^0}{4a^5 b^0 c}$$

$$= \frac{3b^4}{a^5 c} \qquad \text{since } a^0 = b^0 = c^0 = 1 \text{ and } \frac{12}{4} = 3$$

$$\left(\frac{2x^{-3} y^2}{x^4 z^3}\right)^{-3} = \frac{2^{-3} x^9 y^{-6}}{x^{-12} z^{-9}} \qquad \text{by (9.7), (9.6), and (9.5)}$$

$$= \frac{2^{-3} x^9 y^{-6}}{x^{-12} z^{-9}} \cdot \frac{2^3 y^6 x^{12} z^9}{2^3 y^6 x^{12} z^9} \qquad \text{multiplying by } 2^3 y^6 x^{12} z^9/2^3 y^6 x^{12} z^9$$

$$= \frac{2^{-3+3} x^{9+12} y^{-6+6} z^9}{2^3 x^{-12+12} y^6 z^{-9+9}} \qquad \text{by commutative axiom and (9.2)}$$

$$= \frac{2^0 x^{21} y^0 z^9}{8x^0 y^6 z^0}$$

$$= \frac{x^{21} z^9}{8y^6} \qquad \text{by (9.4)}$$

$$\frac{2^{-2} + 2^{-3}}{2^{-4}} = \frac{2^{-2} + 2^{-3}}{2^{-4}} \cdot \frac{2^4}{2^4} \qquad \text{multiplying the fraction by } 2^4/2^4$$

$$= \frac{2^{-2+4} + 2^{-3+4}}{2^{-4+4}}$$ by distributive axiom and (9.2)

$$= \frac{2^2 + 2}{2^0}$$

$$= \frac{4 + 2}{1}$$ by (9.4)

$$= 6$$

$$\left(\frac{x^{-1} - y^{-1}}{x^{-1}y^{-1}}\right)^{-2} = \left(\frac{x^{-1} - y^{-1}}{x^{-1}y^{-1}} \frac{xy}{xy}\right)^{-2}$$

multiplying the fraction inside the parentheses by xy/xy, since the negative exponent of x and of y with the greatest absolute value is -1

$$= \left(\frac{x^{-1+1}y - xy^{-1+1}}{x^{-1+1}y^{-1+1}}\right)^{-2}$$ by distributive and commutative axioms and (9.2)

$$= (y - x)^{-2}$$ by (9.4)

$$= \frac{1}{(y - x)^2}$$ by (9.10)

EXERCISE 9.2
Elimination of Negative Exponents

Find the value of each of the following expressions.

1 3^{-2} **2** 5^{-1}

3 9^{-2} **4** 2^{-5}

5 $3^{-3}3^2$ **6** $2^{-3}2^{-2}$

7 $7^{-2}7^3$ **8** $5^{-3}5^0$

9 $\dfrac{4^{-2}}{4}$ **10** $\dfrac{6^{-3}}{6^{-2}}$

11 $\dfrac{7^0}{7^{-2}}$ **12** $\dfrac{3^2}{3^{-3}}$

13 $(2^{-2})^3$ **14** $(3^{-2})^{-3}$

15 $(5^0)^6$ **16** $(4^{-2})^2$

17 $(2^{-1}3^{-2})^{-2}$ **18** $(2^{-5}3^0)^{-1}$

19 $(3^1 3^{-2})^{-2}$ **20** $(1^{-4}1^2)^{-5}$

21 $\left(\dfrac{3^{-1}}{2^{-3}}\right)^{-1}$ **22** $\left(\dfrac{2^{-2}}{3^{-3}}\right)^{-2}$

23 $\left(\dfrac{1^{-3}}{4^{-2}}\right)^2$ **24** $\left(\dfrac{2^{-4}}{3^{-2}}\right)^2$

By use of negative exponents, write the expressions in Probs. 25 to 32 without denominators.

25 $\dfrac{a^2}{2b^2}$

26 $\dfrac{3a^2}{b^3}$

27 $\dfrac{w^3 x^2}{y^{-3} z^4}$

28 $\dfrac{3c^{-2} d^3}{cd^{-2}}$

29 $\dfrac{2^{-1} a^{-3} b^{-4}}{3a^{-2} b^{-5}}$

30 $\dfrac{4a^2 z}{3^0 b^{-2} z^3}$

31 $\dfrac{3r^3 s^{-2} v^2}{2^{-1} r^4 s^{-1} v^{-2}}$

32 $\dfrac{2^{-1} x^2 y^{-4} z^3}{4^0 x^0 y^{-2} z^4}$

In each of the following problems, make all possible combinations, and express each result without zero or negative exponents.

33 $2a^{-3} a^{-4}$

34 $3^{-1} x^{-3} x^0$

35 $2^{-3} y^{-3} y^2$

36 $5^{-1} c^{-3} c^{-2}$

37 $\dfrac{x^{-4}}{x^{-2}}$

38 $\dfrac{c^{-2}}{c^0}$

39 $\dfrac{m^{-4}}{m^{-5}}$

40 $\dfrac{z^{-5}}{z^{-3}}$

41 $\dfrac{a^{-3} b^2 c^0}{a^2 b^{-3} c^{-2}}$

42 $\dfrac{r^{-3} s^3 t^{-2}}{t^3 v^{-2} u^{-1}}$

43 $\dfrac{w^{-2} x^3 y^{-1}}{w^3 x y^2}$

44 $\dfrac{u^{-4} v^{-5} w^{-3}}{u^0 3^{-2} w^{-2}}$

45 $\dfrac{2^{-2} p^{-3} q^2 r^0}{3^{-3} p^{-4} q r^{-3}}$

46 $\dfrac{3^{-1} b^2 c^{-3} d^5}{2^{-4} b^{-4} c d^4}$

47 $\dfrac{2x^0 y^2 z^{-5}}{3^0 x^{-5} y^{-1} z^{-2}}$

48 $\dfrac{4^{-1} r n^{-1} f^{-2}}{2^{-2} r^{-2} n^4 f^{-1}}$

49 $\left(\dfrac{u^2 v^{-3}}{w^{-3} x^2}\right)^{-1}$

50 $\left(\dfrac{p^{-4} q^{-2}}{p^{-1} q^{-5}}\right)^{-2}$

51 $\left(\dfrac{x^{-2} y^{-3}}{x y^{-1}}\right)^{-3}$

52 $\left(\dfrac{2^{-3} m^0 n^{-7}}{4^{-2} m^{-4} n^{-2}}\right)^{-3}$

53 $\left(\dfrac{16^{-1} a^{-3} b^2}{4^{-3} a^2 b^{-1}}\right)^{-2}$

54 $\left(\dfrac{9^{-2} c^{-3} d^{-4}}{3^{-5} c^3 d^0}\right)^{-4}$

55 $\left(\dfrac{16^{-2} x^{-5} y^{-7}}{2^{-7} x^{-6} y^{-4}}\right)^{-5}$

56 $\left(\dfrac{3^{-5} r^{-4} s^{-5}}{9^{-2} r^{-3} s^{-6}}\right)^{-3}$

57 $3a^{-1} + a^{-3}$

58 $c^{-1} d + cd^{-1}$

59 $3b^a - \dfrac{3^{-2}}{b^{-a}}$

60 $x^3 + x^{-3}$

61 $\dfrac{x^{-2} - y^{-2}}{x^{-3} - y^{-3}}$

62 $\dfrac{c}{d^{-2}} - \dfrac{c^{-2}}{d}$

63 $\dfrac{x^{-3} y^{-1} - x^{-2} y^{-2}}{x^{-3} y - y^{-2}}$

64 $\dfrac{r^{-2} s^{-2}}{r^{-2} - s^{-2}}$

65 $-2(x + 2)^2(x - 2)^{-3} + 3(x - 2)^{-2}(x + 2)$

66 $-4(a + 1)^3(a - 2)^{-3} + 2(a - 2)^{-2}(a + 1)^2$

67 $-(c - 4)^{-4}(c - 3)^{-2} + 3(c - 3)^{-3}(c - 4)^{-3}$

68 $-3(x + 4)^2(x + 2)^{-3} + 2(x + 2)^{-2}(x + 4)^2$

69 $3(3a - 1)^{-2}(a + 2)^{-2} - 2(3a - 1)^{-3}(a + 2)^{-1}$

70 $(3c - 4)^{-1}(c + 2) - (c + 2)^2(3c - 4)^{-2}$

71 $-3(2p + 5)^3(3p - 2)^{-2} + 7(3p - 2)^{-1}(2p + 5)^2$

72 $-4(3r + 1)^2(4r - 1)^{-3} + 2(4r - 1)^{-2}(3r + 1)$

73 $3(4a - 3)(3a - 2)^{-3} - 3(3a - 2)^{-2}$

74 $5(2x - 7)^{-3} + 3(4x + 1)(2x - 7)^{-4}$

75 $(4x - 5)^{-2}(3x + 4)^{-2} - 5(3x + 4)^{-1}(4x - 5)^{-3}$

76 $-2(5x - 3)^{-4}(2x + 5)^{-3} - 3(2x + 5)^{-2}(5x - 3)^{-5}$

9.3
FRACTIONAL EXPONENTS

We shall now extend the definition of a^n to include situations in which n is a rational fraction. The extension will be made in such manner that Laws (9.2) to (9.7) hold. If (9.5) holds for $m = 1/k$ and $n = k$, we then have $(a^{1/k})^k = a^{k/k} = a$. Consequently, we define $a^{1/k}$ to be a number whose kth power is a. Unless we add further restrictions, $a^{1/k}$ may have more than one value. For example, $16^{1/2}$ may be 4 or -4 since $4^2 = 16$ and $(-4)^2 = 16$. We shall remove this ambiguity by the following definitions and discussion.

The kth Root

The number b is a *kth root* of a if and only if $b^k = a$.

Principal kth Root

If there is a positive kth root of a, it is called the *principal kth root of a*. If there is no positive kth root of a but there is a negative one, then the negative kth root is called the *principal kth root* of a.

Radical of Order k, Radicand, and Index

It is customary to use $\sqrt[k]{a}$ to indicate the principal kth root of a. This symbol is called a *radical* of order k, a is called the *radicand*, and k the *index* of the radical. If the index is not written, it is understood to be 2.

We are now in a position to define $a^{1/k}$. The definition is

$$■ \quad a^{1/k} = \sqrt[k]{a} \tag{9.11}$$

There is a positive kth root of a if a is positive; hence $a^{1/k}$ is positive for $a > 0$. For example, $64^{1/3} = 4$ and $36^{1/2} = 6$. It can be proved that if a is negative and k is odd, then there is a negative kth root of a. For example, $(-32)^{1/5} = -2$. Finally, there is no real kth root of a if a is negative and k is even since an even power of any nonzero real number is positive. Thus there is no real square root of -4 and no real fourth root of -81. In this chapter *we shall deal only with real numbers; hence we shall not consider even roots of negative numbers.*

If we replace a by b^j in (9.11), we find that $(b^j)^{1/k} = \sqrt[k]{b^j}$. Consequently $(b^j)^{1/k}$ is a kth root of b^j. Now if we raise $(b^{1/k})^j$ to the kth power, we have

$$[(b^{1/k})^j]^k = (b^{1/k})^{jk} \qquad \text{by (9.5)}$$

$$= (b^{1/k})^{kj} \qquad \text{by commutative axiom}$$

$$= [(b^{1/k})^k]^j$$

$$= b^j \qquad \text{by definition of } b^{1/k}$$

Therefore $(b^j)^{1/k}$ and $(b^{1/k})^j$ are both kth roots of b^j, and it can be proved† that, except for the excluded case, the two roots have the same sign and therefore are equal. Consequently, for $\sqrt[k]{b}$ real, we have

$$\blacksquare \quad b^{j/k} = (b^j)^{1/k} = \sqrt[k]{b^j} = (b^{1/k})^j = (\sqrt[k]{b})^j \qquad \sqrt[k]{b} \text{ real} \qquad (9.12)$$

By use of (9.12) and arguments similar to the one just given, we can prove that Laws (9.2) to (9.7) hold for fractional exponents as defined by (9.11). We shall now work several problems that illustrate the procedures to be followed in dealing with fractional exponents.

EXAMPLE 1
Evaluate $9^{1/2}$, $27^{2/3}$, and $(-128)^{5/7}$.

SOLUTION

$$9^{1/2} = 3 \qquad \text{by (9.11)}$$

$$27^{2/3} = (\sqrt[3]{27})^2 \qquad \text{by (9.12)}$$

$$= 3^2 = 9$$

$$(-128)^{5/7} = (\sqrt[7]{-128})^5 \qquad \text{by (9.12)}$$

$$= (-2)^5 = -32$$

EXAMPLE 2
Express $2a^{1/3}b^{2/3}$ and $3a^{2/5}/b^{3/5}$ in radical form.

SOLUTION

$$2a^{1/3}b^{2/3} = 2(ab^2)^{1/3} \qquad \text{by (9.6)}$$

$$= 2\sqrt[3]{ab^2} \qquad \text{by (9.11)}$$

$$\frac{3a^{2/5}}{b^{3/5}} = 3\left(\frac{a^2}{b^3}\right)^{1/5} \qquad \text{by (9.9)}$$

$$= 3\sqrt[5]{\frac{a^2}{b^3}} \qquad \text{by (9.11)}$$

†See P. K. Rees and F. W. Sparks, "Algebra and Trigonometry," 2d ed., p. 107, McGraw-Hill Book Company, New York, 1969.

EXAMPLE 3

Find the product of $3a^{2/5}$ and $2a^{1/3}$ and find the quotient of $8x^{2/3}y^{5/6}$ and $2x^{1/2}y^{1/3}$.

SOLUTION

$$(3a^{2/5})(2a^{1/3}) = 6a^{2/5+1/3}$$

$$= 6a^{(6+5)/15}$$

$$= 6a^{11/15}$$

$$\frac{8x^{2/3}y^{5/6}}{2x^{1/2}y^{1/3}} = \frac{8}{2}x^{2/3-1/2}y^{5/6-1/3}$$

$$= 4x^{(4-3)/6}y^{(5-2)/6}$$

$$= 4x^{1/6}y^{3/6} = 4x^{1/6}y^{1/2}$$

EXAMPLE 4

Use the laws of exponents to make all possible combinations and express the result without zero or negative exponents in

$$\left(\frac{16x^{-1}y^{5/6}z^{3/4}}{9x^3y^{1/3}z^{1/2}}\right)^{1/2}$$

SOLUTION

$$\left(\frac{16x^{-1}y^{5/6}z^{3/4}}{9x^3y^{1/3}z^{1/2}}\right)^{1/2} = \left(\frac{16}{9}\frac{x^{-1}}{x^3}\frac{y^{5/6}}{y^{1/3}}\frac{z^{3/4}}{z^{1/2}}\frac{x}{x}\right)^{1/2} \qquad \text{multiplying by } x/x$$

$$= \left(\frac{16}{9}\frac{x^{-1+1}}{x^{3+1}}y^{5/6-1/3}z^{3/4-1/2}\right)^{1/2} \qquad \text{by (9.2) and (9.3)}$$

$$= \left(\frac{16}{9}\frac{x^0y^{1/2}z^{1/4}}{x^4}\right)^{1/2}$$

$$= \frac{4}{3}\frac{y^{1/4}z^{1/8}}{x^2} \qquad \text{by (9.4), (9.6), and (9.7)}$$

EXERCISE 9.3
Conversion of Radical and Exponential Expressions

Find the value of each of the following without exponents or radicals.

1   $(81)^{1/4}$ 2 $(\frac{1}{4})^{1/2}$

3   $(-8)^{1/3}$ 4 $(-27)^{1/3}$

5   $16^{3/4}$ 6 $(-32)^{3/5}$

7 $(4)^{3/2}$

8 $.064^{2/3}$

9 $\sqrt{49}$

10 $\sqrt[3]{27^2}$

11 $\sqrt[4]{81^3}$

12 $\sqrt[3]{8^4}$

13 $36^{-3/2}$

14 $.25^{-1/2}$

15 $.027^{-2/3}$

16 $32^{-2/5}$

17 $\sqrt{64x^2 y^0}$

18 $\sqrt[3]{27a^6 b^3}$

19 $\sqrt[4]{16r^{12} s^0}$

20 $\sqrt[3]{.008p^6 q^9}$

Make all possible combinations by use of the laws of exponents, and express the results
without zero or negative exponents.

21 $(3x^{1/4} y)(4x^{3/4} y^0)$

22 $(2u^{2/3} v^3)(3u^{-1/3} v^{-2})$

23 $(5a^{1/5} b^{1/2})(3a^{3/5} b^{-1/2})$

24 $(2r^{5/7} s^{-3/4})(4r^{2/7} s^{1/4})$

25 $\dfrac{27^{2/3} a^{2/3} y^{-1/4}}{9^{3/2} a^{1/3} y^{3/4}}$

26 $\dfrac{16^{3/2} x^{1/5} y^{3/7}}{27^{2/3} x^{4/5} y^{1/7}}$

27 $\dfrac{32^{2/5} p^{6/7} q^{4/5}}{16^{1/4} p^{-1/7} q^{-1/5}}$

28 $\dfrac{64^{2/3} x^{-1/6} y^{7/8}}{4^{3/2} x^{-5/3} y^{5/8}}$

29 $\dfrac{(3^{-1} w^{2/3} z^{4/7})}{(16^{3/4} w^{1/3} z^{-3/7})}(4^{3/2} w^{-1/3} z^{-2})$

30 $\dfrac{(27^{-2/3} c^{3/4} d^{-1/3})}{(9^{-1} c^{-3/4} d)}(4^{-1/2} c^{-2} d^{-2/3})$

31 $\dfrac{(8^{4/3} a^{4/9} z^{1/3})}{(32^{4/5} a^{-2/9} z^{1/6})}(27^{-1/3} a^{-2/9} z^{1/2})$

32 $\dfrac{(16^{-3/2} b^{3/5} c^{-1/8})}{(9^{-1/2} b^{1/5} c^{1/4})}(64^{2/3} b^{2/5} c^{1/2})$

33 $(3x^{-1/4} y^{2/3})^2$

34 $(2a^{-1/3} b^{1/6})^3$

35 $(3^{-1} c^{1/3} d^{2/6})^{-3}$

36 $(2u^{1/4} v^{3/4})^{-4}$

37 $(27a^{-6} b^{3/4})^{1/3}$

38 $(64^{-1} b^{6/5} c^{6/7})^{-1/6}$

39 $(9^{3/2} r^{-3/8} s^{3/4})^{1/3}$

40 $(32^{2/3} u^{-5/6} v^5)^{-3/5}$

41 $(2a^{-3} b^{3/5})^{1/3}(4^{-1/2} ab^{2/5})^{-1}$

42 $(9x^{-4/3} y^{5/6})^{3/2}(27x^{3/4} y^{3/4})^{-2/3}$

43 $(64a^{3/5} b^{-2/5})^{5/6}(16a^6 b^{8/3})^{-1/4}$

44 $(27x^{2/5} y^{-3/4})^{5/3}(3^{4/7} x^{-4/7} y^{-3/7})^{-7/2}$

45 $\left(\dfrac{27x^{-2} y^4}{64x^4 y}\right)^{1/3}$

46 $\left(\dfrac{9a^{2/3} b^{-4/6}}{16a^{-2/3} b^0}\right)^{-3/2}$

47 $\left(\dfrac{7^{4/3} c^{5/6} d^{-5/9}}{16^{2/3} c^{3/6} d^{3/9}}\right)^{3/4}$

48 $\left(\dfrac{32x^{1/3} y^{-4/6}}{243x^{-1/2} y}\right)^{-2/5}$

49 $\left(\dfrac{x^{p-q}}{x^{-2q}}\right)^{1/(p+q)}$

50 $\left(\dfrac{x^{r-2s}}{x^{-r-s}}\right)^{s/(2r-s)}$

51 $\left(\dfrac{a^{x+y}}{a^y}\right)^{(x+y)/x}$

52 $\left(\dfrac{d^{x-3y}}{d^{2x-5y}}\right)^{w/(x-2y)}$

53 $(x^{1/2} - y^{1/2})(x^{1/2} + y^{1/2})$

54 $(x^{1/3} - y^{1/3})(x^{2/3} + x^{1/3}y^{1/3} + y^{2/3})$

55 $(x^{1/2} + y^{3/2})(x^{1/2} - y^{3/2})$

56 $(x^{1/2} + x^{-1/2})^2$

57 $\left(\dfrac{x^{-1} - y^{-1}}{x^{-2} - y^{-2}}\right)^{-2}$

58 $\left(\dfrac{a^{-2} - b^{-2}}{a^{-1} + b^{-1}}\right)^{-1}$

59 $\left(\dfrac{a^{-1} + b^{-1}}{a^{-2}b^{-2}}\right)^{-1}$

60 $\left(\dfrac{a^{-2}b^{-2}}{a^{-1} - b^{-1}}\right)^{-2}$

61 $(x - 2)(x + 1)^{-1/2} + 3(x + 1)^{1/2}$

62 $-(a + 1)(a - 2)^{-3/2} + 2(a - 2)^{-1/2}$

63 $(3c + 1)(c - 3)^{-2/3} + 4(c - 3)^{1/3}$

64 $-(z - 2)(2z - 3)^{-5/4} + (2z - 3)^{-1/4}$

65 $-(d + 2)^{1/2}(d - 2)^{-3/2} + (d + 2)^{-1/2}(d - 2)^{-1/2}$

66 $(2b - 3)^{1/2}(b + 5)^{-2/3} + (2b - 3)^{-1/2}(b + 5)^{1/3}$

67 $-(2w + 5)^{1/4}(3w - 4)^{-4/3} + (2w + 5)^{-3/4}(3w - 4)^{-1/3}$

68 $-(4x - 7)^{2/5}(x + 5)^{-3/2} + 3(4x - 7)^{-3/5}(x + 5)^{-1/2}$

9.4
LAWS OF RADICALS

In this section we shall develop and apply three laws of radicals. Since (9.6) and (9.7) are valid for $n = 1/k$, we have

$$(ab)^{1/k} = a^{1/k}b^{1/k} \qquad \text{and} \qquad \left(\frac{a}{b}\right)^{1/k} = \frac{a^{1/k}}{b^{1/k}}$$

Now making use of the relation between fractional exponents and radicals, we have

■ $\sqrt[k]{ab} = \sqrt[k]{a}\,\sqrt[k]{b}$ $\hspace{4cm}$ (9.13)

and

■ $\sqrt[k]{\dfrac{a}{b}} = \dfrac{\sqrt[k]{a}}{\sqrt[k]{b}}$ $\hspace{4cm}$ (9.14)

If we replace m by $1/j$ and n by $1/k$ in (9.5), we have $(a^{1/j})^{1/k} = a^{1/jk}$; *hence, in terms of radicals, we get*

■ $\sqrt[k]{\sqrt[j]{a}} = \sqrt[kj]{a}$ $\hspace{4cm}$ (9.15)

We can use (9.13) to remove rational factors from the radicand, to multiply radicals of the same order, and to insert factors into the radicand as seen in the following examples:

$$\sqrt[3]{108} = \sqrt[3]{(27)(4)}$$
expressing 108 as the product of the perfect cube 27 and 4

$$= \sqrt[3]{27}\ \sqrt[3]{4}$$
by use of (9.13)

$$= 3\sqrt[3]{4}$$

$$\sqrt{5}\ \sqrt{7} = \sqrt{(5)(7)} = \sqrt{35}$$
by use of (9.13) from right to left

$$2\sqrt[4]{3} = \sqrt[4]{2^4}\ \sqrt[4]{3}$$
since $2 = \sqrt[4]{2^4}$

$$= \sqrt[4]{(2^4)(3)} = \sqrt[4]{48}$$
by use of (9.13) from right to left

$$\sqrt[5]{32a^6 b^{13}} = \sqrt[5]{2^5 a^5 a(b^2)^5 b^3}$$
factoring into fifth powers as far as possible

$$= 2ab^2 \sqrt[5]{ab^3}$$
removing the fifth powers from the radicand

$$\sqrt[3]{5ab}\ \sqrt[3]{2ac} = \sqrt[3]{10a^2 bc}$$
by use of (9.13)

$$a\sqrt[4]{2ab^2} = \sqrt[4]{a^4}\ \sqrt[4]{2ab^2}$$
since $a = \sqrt[4]{a^4}$

$$= \sqrt[4]{2a^5 b^2}$$
by (9.13) from right to left

9.6
APPLICATIONS OF $\sqrt[k]{a/b} = \sqrt[k]{a}\ /\ \sqrt[k]{b}$

Law (9.14) is used in obtaining the quotient of two radicals of the same order and in rationalizing monomial denominators as illustrated now.

$$\frac{\sqrt[3]{250a^7 b^5}}{\sqrt[3]{2ab^3}} = \sqrt[3]{\frac{250a^7 b^5}{2ab^3}}$$
by use of (9.14) from right to left

$$= \sqrt[3]{125a^6 b^2}$$

$$= \sqrt[3]{5^3 (a^2)^3 b^2}$$
since $125 = 5^3$ and $a^6 = (a^2)^3$

$$= 5a^2 \sqrt[3]{b^2}$$
since $\sqrt[3]{5^3(a^2)^3} = 5a^2$

Rationalizing a Denominator

We often need to convert a radical with a fractional radicand to an equal fraction with no radicals in the denominator. If this is done, we say the denominator has been *rationalized*. In order to rationalize a monomial denominator, we must multiply the denominator of a radical of order n by the necessary factors so as to make it an nth power and, of course, multiply the numerator by the same factors. Thus, in order to rationalize the denominator of

$$\sqrt[4]{\frac{2a}{3a^3b^2}}$$

we multiply denominator and numerator by 3^3ab^2 since that makes the denominator a fourth power. Hence, we have

$$\sqrt[4]{\frac{2a}{3a^3b^2}} = \sqrt[4]{\frac{2a}{3a^3b^2}\frac{3^3ab^2}{3^3ab^2}}$$

$$= \frac{\sqrt[4]{54a^2b^2}}{3ab} \qquad \text{since } \sqrt[4]{3^4a^4b^4} = 3ab$$

Furthermore, in order to rationalize the denominator of $\sqrt{2x/3y}$, we multiply numerator and denominator by $3y$ and have

$$\sqrt{\frac{2x}{3y}} = \sqrt{\frac{2x}{3y}\frac{3y}{3y}}$$

$$= \frac{\sqrt{6xy}}{3y} \qquad \text{if } y > 0$$

$$= -\frac{\sqrt{6xy}}{3y} \qquad \text{if } y < 0$$

These can be combined into

$$\sqrt{\frac{2x}{3y}} = \frac{\sqrt{6xy}}{3\,|y\,|}$$

9.7
CHANGING THE ORDER OF A RADICAL

If the index of a radical can be factored and if the radicand can be expressed as a power with the exponent of the power equal to one of the factors of the index, then we can decrease the order of the radical. Thus,

$$\sqrt[6]{8x^3} = \sqrt{\sqrt[3]{(2x)^3}} = \sqrt{2x}$$

Another example is

$$\sqrt[15]{a^{10}y^{25}} = \sqrt[3]{\sqrt[5]{(a^2y^5)^5}} = \sqrt[3]{a^2y^5} = \sqrt[3]{a^2y^3y^2} = y\sqrt[3]{a^2y^2}$$

An alternative procedure for obtaining the same result is to change from radical form to fractional-exponent form, reduce the fractional exponents to lowest terms, and then express the result in radical form. If this procedure is used in connection with the last example, we have

$$\sqrt[15]{a^{10}y^{25}} = (a^{10}y^{25})^{1/15} = a^{10/15}y^{25/15}$$

$$= a^{2/3}y^{5/3} \quad \text{since } \tfrac{10}{15} = \tfrac{2}{3} \text{ and } \tfrac{25}{15} = \tfrac{5}{3}$$

$$= \sqrt[3]{a^2 y^5} = y\sqrt[3]{a^2 y^2}$$

Simplify Radicals

In the following exercise, we shall use the word *simplify* to indicate that all possible combinations that can be made by use of the laws of exponents are to be made, the result is to be expressed without zero or negative exponents, all monomial denominators are to be rationalized, and all possible rational factors are to be removed from the radicand.

EXERCISE 9.4
Simplifying Radicals

Simplify the following radicals.

1   $\sqrt{36}$ 2 $\sqrt{49}$

3   $\sqrt[3]{27}$ 4 $\sqrt[6]{64}$

5   $\sqrt{28}$ 6 $\sqrt{45}$

7   $\sqrt[3]{54}$ 8 $\sqrt[5]{96}$

9   $\sqrt{20}\ \sqrt{45}$ 10 $\sqrt{50}\ \sqrt{72}$

11   $\sqrt[3]{32}\ \sqrt[3]{54}$ 12 $\sqrt[5]{108}\ \sqrt[5]{72}$

13   $\dfrac{\sqrt{12}}{\sqrt{27}}$ 14 $\dfrac{\sqrt{48}}{\sqrt{27}}$

15   $\dfrac{\sqrt[3]{54}}{\sqrt[3]{16}}$ 16 $\dfrac{\sqrt[5]{128}}{\sqrt[5]{972}}$

17   $\sqrt[8]{64}$ 18 $\sqrt[4]{64}$

19   $\sqrt[15]{64}$ 20 $\sqrt[12]{64}$

21   $\sqrt{27a^5 b^3}$ 22 $\sqrt{8rs^7}$

23   $\sqrt[3]{16x^4 y^7}$ 24 $\sqrt[3]{54c^6 d^5}$

25   $\sqrt[4]{81x^5 y^9}$ 26 $\sqrt[4]{16w^{12} z^7}$

27   $\sqrt[5]{32a^3 b^{11}}$ 28 $\sqrt[5]{128p^7 q^9}$

29   $\sqrt{8x^3 y^5}\ \sqrt{2xy^3}$ 30 $\sqrt{3a^7 b^2}\ \sqrt{12a^3 b^4}$

31   $\sqrt{3a^2 b}\ \sqrt{27a^4 b^2}$ 32 $\sqrt{18uv^3}\ \sqrt{8u^2 v^4}$

33   $\sqrt[3]{3c^4 d^4}\ \sqrt[3]{9c^2 d^3}$ 34 $\sqrt[3]{49ab^5}\ \sqrt[3]{7a^2 b}$

35   $\sqrt[4]{54r^3 t^2}\ \sqrt[4]{18r^5 t^3}$ 36 $\sqrt[5]{81b^2 c^6}\ \sqrt[5]{18b^3 c^3}$

37   $\dfrac{\sqrt{18a^5 b^3}}{\sqrt{2ab}}$ 38 $\dfrac{\sqrt{72x^7 y^6}}{\sqrt{2x^3 y^2}}$

39   $\dfrac{\sqrt{72u^5 v^9}}{\sqrt{6u^3 v^7}}$ 40 $\dfrac{\sqrt{48s^5 t^7}}{\sqrt{27st^3}}$

41 $\dfrac{\sqrt{7ab^3}}{\sqrt{3a^3b^2}}$

42 $\dfrac{\sqrt{12r^3t}}{\sqrt{15r^4t^2}}$

43 $\dfrac{\sqrt[3]{27a^3b^2}}{\sqrt[3]{2a^2b^6}}$

44 $\dfrac{\sqrt[3]{r^2s^3}}{\sqrt[3]{3r^3s^2}}$

45 $\dfrac{\sqrt{48a^3b^0c^5}}{\sqrt{3a^{-1}bc}}$

46 $\dfrac{\sqrt{98r^{-1}st}}{\sqrt{2rs^{-3}t^0}}$

47 $\dfrac{\sqrt{45w^{-2}y^3z^{-2}}}{\sqrt{15w^2y^2z}}$

48 $\dfrac{\sqrt{12u^{-2}vw^0}}{\sqrt{15uv^{-3}w^{-1}}}$

49 $\dfrac{\sqrt{3a^{-3}b^0c}}{\sqrt{108a^{-2}bc^{-1}}}$

50 $\dfrac{\sqrt{7u^{-2}vw^{-3}}}{\sqrt{21uv^{-2}w^{-4}}}$

51 $\dfrac{\sqrt[3]{4a^2bc^{-2}}}{\sqrt[3]{25a^4bc}}$

52 $\dfrac{\sqrt[3]{9x^0y^{-2}z}}{\sqrt[3]{6x^{-2}yz^{-2}}}$

53 $\dfrac{\sqrt{7x^3y^3z}\,\sqrt{2xyz}}{\sqrt{28x^2y^2z^0}}$

54 $\dfrac{\sqrt{10b^3c^{-1}d^0}\,\sqrt{50b^2c^{-2}d^{-3}}}{\sqrt{20bcd}}$

55 $\dfrac{\sqrt{18r^2s^{-2}t}}{\sqrt{2r^3s^2t}\,\sqrt{75r^{-1}s^0t}}$

56 $\dfrac{\sqrt{108u^2v^3w^0}}{\sqrt{2u^4vw^3}\,\sqrt{30u^3v^2w}}$

57 $\sqrt[6]{27a^3b^9}$

58 $\sqrt[4]{4x^2y^6}$

59 $\sqrt[9]{8x^3y^6}$

60 $\sqrt[15]{64a^{12}b^6}$

61 $\dfrac{\sqrt[4]{36a^2y^6}}{\sqrt[6]{27a^3y^9}}$

62 $\dfrac{\sqrt[9]{8x^{12}z^6}}{\sqrt[6]{4x^8z^4}}$

63 $\dfrac{\sqrt[15]{27x^9y^6}}{\sqrt[20]{16x^{16}y^8}}$

64 $\dfrac{\sqrt[10]{32x^5y^{15}}}{\sqrt[12]{729x^{18}y^6}}$

9.8
RATIONALIZING BINOMIAL DENOMINATORS

The product of the sum and the difference of the same two numbers is the square of the first minus the square of the second; hence, the product of $\sqrt{a}+\sqrt{b}$ and $\sqrt{a}-\sqrt{b}$ is $(\sqrt{a})^2-(\sqrt{b})^2=a-b$. Consequently, if the denominator of a fraction is a binomial that has second-order radicals, we can rationalize it by multiplying each member of the fraction by the binomial obtained by changing the sign between the terms in the denominator.

EXAMPLE Rationalize the denominator in both

$$\dfrac{2+\sqrt{5}}{3-\sqrt{5}} \quad \text{and} \quad \dfrac{3\sqrt{x}-\sqrt{y}}{\sqrt{x}-2\sqrt{y}}$$

SOLUTION

$$\dfrac{2+\sqrt{5}}{3-\sqrt{5}}=\dfrac{2+\sqrt{5}}{3-\sqrt{5}}\dfrac{3+\sqrt{5}}{3+\sqrt{5}}=\dfrac{6+5\sqrt{5}+5}{9-5}=\dfrac{11+5\sqrt{5}}{4}$$

$$\frac{3\sqrt{x}-\sqrt{y}}{\sqrt{x}-2\sqrt{y}} = \frac{3\sqrt{x}-\sqrt{y}}{\sqrt{x}-2\sqrt{y}}\frac{\sqrt{x}+2\sqrt{y}}{\sqrt{x}+2\sqrt{y}} = \frac{3x+5\sqrt{xy}-2y}{x-4y}$$

9.9
ADDITION OF RADICALS

Two or more terms that have a common factor can be combined into a single term by use of the distributive axiom. Thus,

$$2\sqrt{5}-3\sqrt{5}+5\sqrt{5} = (2-3+5)\sqrt{5} \qquad \text{by distributive axiom}$$
$$= 4\sqrt{5}$$

This example illustrates the procedure used in adding radicals. It is advisable to simplify each radical that is to be added since any common factor can then be detected. For example,

$$\sqrt{2}+\sqrt{128}-\sqrt{18} = \sqrt{2}+\sqrt{(64)(2)}-\sqrt{(9)(2)}$$
$$= (1+8-3)\sqrt{2} \qquad \text{by distributive axiom}$$
$$= 6\sqrt{2}$$

It may happen that the terms to be added do not contain a common factor but are such that they can be grouped so that there is a common factor in each group as illustrated below.

$$\sqrt{8}+\sqrt[3]{16}+\sqrt[3]{54}-\sqrt{72}-\sqrt[3]{24}$$
$$= 2\sqrt{2}+2\sqrt[3]{2}+3\sqrt[3]{2}-6\sqrt{2}-2\sqrt[3]{3}$$
$$= (2-6)\sqrt{2}+(2+3)\sqrt[3]{2}-2\sqrt[3]{3} \qquad \text{by distributive axiom}$$
$$= -4\sqrt{2}+5\sqrt[3]{2}-2\sqrt[3]{3}$$

$$\sqrt{2x^3y}+\sqrt[4]{4x^2y^6}+\sqrt[3]{16x^4y} = x\sqrt{2xy}+\sqrt{\sqrt{4x^2y^6}}+2x\sqrt[3]{2xy}$$
$$= x\sqrt{2xy}+\sqrt{2xy^3}+2x\sqrt[3]{2xy}$$
$$= x\sqrt{2xy}+y\sqrt{2xy}+2x\sqrt[3]{2xy}$$
$$= (x+y)\sqrt{2xy}+2x\sqrt[3]{2xy} \qquad \text{by distributive axiom}$$

EXERCISE 9.5
Operations on Radical Expressions

Find the following products.

1 $(\sqrt{5}+\sqrt{3})(\sqrt{5}-\sqrt{3})$

2 $(\sqrt{2}+\sqrt{7})(\sqrt{2}-\sqrt{7})$

3 $(\sqrt{6} + \sqrt{2})(\sqrt{6} - \sqrt{2})$

4 $(\sqrt{7} + \sqrt{6})(\sqrt{7} - \sqrt{6})$

5 $(\sqrt{2} + 3\sqrt{3})(\sqrt{2} - 2\sqrt{3})$

6 $(\sqrt{7} + 2\sqrt{2})(2\sqrt{7} - \sqrt{2})$

7 $(\sqrt{5} - 2\sqrt{3})(2\sqrt{5} - \sqrt{3})$

8 $(\sqrt{6} + 3\sqrt{5})(2\sqrt{6} - \sqrt{5})$

9 $(\sqrt{6} - 3\sqrt{2})(2\sqrt{6} + 2\sqrt{2})$

10 $(2\sqrt{6} - \sqrt{3})(\sqrt{6} + 2\sqrt{3})$

11 $(\sqrt{10} - 3\sqrt{5})(2\sqrt{10} + 2\sqrt{5})$

12 $(2\sqrt{3} - \sqrt{6})(\sqrt{3} + 3\sqrt{6})$

13 $(a + \sqrt{bc})(a - \sqrt{bc})$

14 $(\sqrt{r} + s)(\sqrt{r} - s)$

15 $(x - \sqrt{xy} + y)(\sqrt{x} + \sqrt{y})$

16 $(\sqrt{a} - \sqrt{b})(a + \sqrt{ab} + b)$

17 $(\sqrt{2} + \sqrt{3} - \sqrt{5})(\sqrt{2} - \sqrt{3} + \sqrt{5})$

18 $(\sqrt{6} + \sqrt{2} - \sqrt{3})(\sqrt{6} - \sqrt{2} + \sqrt{3})$

19 $(\sqrt{7} - \sqrt{5} + \sqrt{2})(\sqrt{7} - \sqrt{5} - \sqrt{2})$

20 $(\sqrt{5} - \sqrt{2} + \sqrt{6})(\sqrt{2} - \sqrt{6} + \sqrt{5})$

Rationalize the denominators in the following problems.

21 $\dfrac{\sqrt{2} - 3}{\sqrt{2} + 3}$

22 $\dfrac{\sqrt{5} - 3}{\sqrt{5} + 3}$

23 $\dfrac{\sqrt{5} - \sqrt{6}}{\sqrt{5} + \sqrt{6}}$

24 $\dfrac{\sqrt{7} - \sqrt{3}}{\sqrt{7} + \sqrt{3}}$

25 $\dfrac{2\sqrt{6} - 1}{\sqrt{3} + \sqrt{2}}$

26 $\dfrac{3\sqrt{10} + 2}{\sqrt{5} - \sqrt{2}}$

27 $\dfrac{\sqrt{15} - 5}{\sqrt{5} - \sqrt{3}}$

28 $\dfrac{2\sqrt{14} - 3}{\sqrt{7} - \sqrt{2}}$

29 $\dfrac{\sqrt{a} + \sqrt{b}}{\sqrt{a} - \sqrt{b}}$

30 $\dfrac{x - y}{\sqrt{x} + \sqrt{y}}$

31 $\dfrac{w\sqrt{w} + z\sqrt{z}}{\sqrt{w} - \sqrt{z}}$

32 $\dfrac{c\sqrt{d} - d\sqrt{c}}{\sqrt{c} + \sqrt{d}}$

Simplify the radicals in each of the following problems, and combine coefficients of equal radicands with equal indices.

33 $\sqrt{2} + \sqrt{8} + \sqrt{18}$

34 $\sqrt{3} + \sqrt{12} - \sqrt{75}$

35 $\sqrt{45} + \sqrt{80} - \sqrt{20}$

36 $\sqrt{7} - \sqrt{63} + \sqrt{112}$

37 $\sqrt[3]{24} - \sqrt[3]{81} + \sqrt[3]{375}$

38 $\sqrt[3]{16} + \sqrt[3]{250} - \sqrt[3]{128}$

39 $\sqrt[3]{5} + \sqrt[3]{625} + \sqrt[3]{135}$

40 $\sqrt[3]{32} - \sqrt[3]{108} + \sqrt[3]{500}$

41 $\sqrt{12} + \sqrt{27} - \sqrt[3]{24}$

42 $\sqrt[3]{9} - \sqrt{27} + \sqrt[3]{72}$

43 $\sqrt{5} + \sqrt[3]{40} - \sqrt{20}$

44 $\sqrt[4]{9} + \sqrt{27} - \sqrt[3]{81}$

45 $3\sqrt{3x^4y} - x\sqrt{12x^2y} + x^2\sqrt{75y}$

46 $\sqrt{2c^4d} - c\sqrt{8c^2d} + 2c^2\sqrt{8d}$

47 $3b\sqrt{a^3b} - a\sqrt{4ab^3} + 2\sqrt{9a^3b^3}$

48 $2x^2\sqrt{4xy^5} - xy\sqrt{25x^3y^3} + 3y\sqrt{4x^5y^3}$

49 $ab\sqrt{4a^5b^7} - 5a^2b\sqrt{a^3b^7} + 2b^3\sqrt{4a^7b^3} + b^2\sqrt{9a^7b^5}$

50 $3bc^2\sqrt{4b^5c} + 3b^2c\sqrt{9b^3c^3} - 3bc\sqrt{16b^5c^3} + 7c\sqrt{4b^7c^3}$

51 $2wz^2\sqrt{4w^5z^3} - 2w^2z^2\sqrt{16w^3z^3} + 2w^3\sqrt{9wz^7} + 2z^3\sqrt{25w^7z}$

52 $3p\sqrt{16p^5q^8} - 3q\sqrt{4p^7q^6} - 3pq\sqrt{9p^5q^6} + 3p^2q^2\sqrt{25p^3q^4}$

53 $\dfrac{x}{2}\sqrt{\dfrac{y^3}{9}} + y\sqrt{\dfrac{x^2y}{9}} + \dfrac{1}{2}\sqrt{\dfrac{x^2y^3}{4}} - xy\sqrt{\dfrac{y}{4}}$

54 $3y\sqrt{\dfrac{3x^2}{2}} - 2x\sqrt{\dfrac{8y^2}{3}} + \sqrt{\dfrac{3x^2y^2}{2}}$

55 $\sqrt{\dfrac{4a^2}{3b}} + \dfrac{a}{2b}\sqrt{\dfrac{4b}{3}} - \dfrac{3a}{2}\sqrt{\dfrac{1}{3b}}$

56 $\sqrt{\dfrac{3s}{r}} + \sqrt{\dfrac{4r}{3s}} + \sqrt{\dfrac{3r}{s}} - \sqrt{\dfrac{s}{3r}}$

57 $\sqrt[3]{a^4b} + \sqrt{a^3b} + \sqrt[3]{ab^4} + \sqrt{ab^3}$

58 $\sqrt{3b} + \sqrt[3]{3b^2} - \sqrt{\dfrac{3}{b^3}} - \sqrt[3]{\dfrac{3}{b^4}}$

59 $\sqrt[3]{\dfrac{z^4}{3}} - \sqrt{\dfrac{z^3}{3}} - \sqrt[3]{\dfrac{w^3z}{3}} + \dfrac{1}{3}\sqrt{3w^2z}$

60 $\sqrt[3]{\dfrac{1}{x}} - \sqrt[3]{\dfrac{y^2}{x^2}} + \sqrt[3]{\dfrac{x^5}{y^3}} - \sqrt[3]{\dfrac{x^4}{y}}$

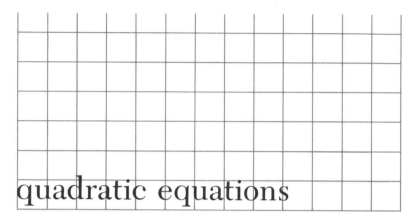

quadratic equations

In Chaps. 5 and 7 we dealt with problems that can be solved by equations that involve only the first power of the variable. Practical problems and theoretical considerations, however, frequently lead to more complicated equations. For example, if the resistance of the air is neglected, the distance s that a compact body falls through the air in t seconds is given by the formula $s = 16.1t^2 + v_0 t$, where s is expressed in feet and v_0 is the vertical velocity in feet per second. Thus, if a body is released from an airplane that is 5000 feet high and is given an initial velocity of 20 feet per second, then the time in seconds required for the body to reach the ground is a root of the equation $5000 = 16.1t^2 + 20t$. This equation, involving the square of the variable t, is called a *quadratic equation* after the Latin word for square. In this chapter we shall explain methods for solving such equations.

10.1
DEFINITION OF A QUADRATIC EQUATION

A quadratic equation is an equation that involves the second, but no higher, power of the variable. The most general form of a quadratic equation is

■ $ax^2 + bx + c = 0$ (10.1)

where a, b, and c are arbitrary constants and $a \neq 0$. We shall frequently refer to c as the *constant term* since it does not involve the variable.

If, in (10.1), $b = 0$, we have

■ $ax^2 + c = 0$ (10.2)

To get the solution set of this equation, we first solve it for x^2 and obtain $x^2 = -c/a$. Then we equate the square roots of the members and have $x = \pm \sqrt{-c/a}$. If $-c/a$ is positive, $\pm \sqrt{-c/a}$ are real numbers. If, however, $-c/a$ is negative, $\sqrt{-c/a}$ is not a real number and Eq. (10.2) has no real roots.

If, in (10.1), $b \neq 0$, the method for solving the equation is more complicated. We shall explain these methods in later sections.

10.2
IMAGINARY NUMBERS

As stated above, if $-c/a$ is negative, Eq. (10.2) has no real roots. It is true, however, that the formal process of solving an equation frequently will yield expressions that involve the square root of a negative number. At present we are not in a position to give such expressions a meaning. We shall, however, explain a notation for dealing with them. We shall define the letter i to be such that $i = \sqrt{-1}$. Next we define $\sqrt{-25}$ as $\sqrt{25(-1)} = \sqrt{25} \sqrt{-1} = 5\sqrt{-1} = 5i$, and in general if $n > 0$, we define $\sqrt{-n^2}$ as ni.

The number ni is called a *pure imaginary number*.

10.3
SOLVING A QUADRATIC EQUATION BY FACTORING

If the left member of (10.1) can be expressed as the product of two linear factors, the roots of the equation can be obtained by use of Theorem (2.26). This theorem states that if a is a real number, then $a \cdot 0 = 0 \cdot a = 0$. We shall next explain how it is used in solving quadratic equations.

We shall first assume that

$ax^2 + bx + c = (dx + e)(fx + g)$

Then (10.1) becomes

$(dx + e)(fx + g) = 0$ (10.3)

By (2.26), $(dx + e)(fx + g) = 0$ if *either* factor is zero. Hence, we can get the roots of Eq. (10.3) by setting each factor of the left member equal to zero, and then by solving the linear equations thus obtained.

Accordingly, we have

$$dx + e = 0 \qquad \text{setting the first factor equal to zero}$$

$$dx = -e \qquad \text{adding } -e \text{ to each member}$$

$$x = \frac{-e}{d} \qquad \text{multiplying each member by } 1/d$$

$$fx + g = 0 \qquad \text{setting the second factor equal to zero}$$

$$fx = -g \qquad \text{adding } -g \text{ to each member}$$

$$x = \frac{-g}{f} \qquad \text{multiplying each member by } 1/f$$

Hence, the solution set of (10.3) is $\{-e/d, -g/f\}$.

Since this method can be applied only when one member of the equation is zero, the first step in applying the method to an equation not in this form is to obtain an equation that is equivalent to the given one and with the right member zero. Obviously, the process of solving cannot be completed if the left member of the equation thus obtained is irreducible.

We shall illustrate the method with the following example.

EXAMPLE   By factoring, solve the equation $2x^2 = x + 6$.

SOLUTION   Since neither member of the above equation is zero, we obtain an equivalent equation with one member zero by adding $-x - 6$ to each member. This operation yields the equation

$$2x^2 - x - 6 = 0$$

Then the method of solving by factoring can be applied. The complete process follows:

$$2x^2 = x + 6 \qquad \text{given equation}$$

$$2x^2 - x - 6 = 0 \qquad \text{adding } -x - 6 \text{ to each member}$$

$$(2x + 3)(x - 2) = 0 \qquad \text{factoring the left member}$$

$$2x + 3 = 0 \qquad \text{setting the first factor equal to zero}$$

$$2x = -3 \qquad \text{adding } -3 \text{ to each member}$$

$$x = -\tfrac{3}{2} \qquad \text{multiplying each member by } \tfrac{1}{2}$$

$$x - 2 = 0 \qquad \text{setting the second factor equal to zero}$$

$$x = 2 \qquad \text{adding 2 to each member}$$

Hence, the solution set is $\{-\tfrac{3}{2}, 2\}$.

Verification   If $x = -\frac{3}{2}$, we have

227

10.3 Solving a
Quadratic
Equation by
Factoring

$$2(-\tfrac{3}{2})^2 = \tfrac{18}{4} = \tfrac{9}{2} \qquad \text{for the left member}$$

and

$$-\tfrac{3}{2} + 6 = \frac{-3 + 12}{2} = \tfrac{9}{2} \qquad \text{for the right member}$$

Hence,

$$2(-\tfrac{3}{2})^2 = -\tfrac{3}{2} + 6$$

Furthermore, if $x = 2$, we have

$$2(2)^2 = 8 \qquad \text{for the left member}$$

$$2 + 6 = 8 \qquad \text{for the right member}$$

Consequently, $2(2)^2 = 2 + 6$

Note   We wish to impress the reader with the fact that this method is applicable *only when the right member of the equation is zero.* If one of the factors of the left member is zero, their product is zero, regardless of the value of the other factor. However, if the right member of the equation is not zero, as in

$$(x - 1)(x + 2) = 6$$

we cannot arbitrarily assign a value to either factor without at the same time fixing the value of the other. For example, if in the preceding example we let $x - 1 = 3$, then surely $x + 2 = 2$ if their product is 6. Obviously, these two conditions cannot be satisfied by the same value of x.

EXERCISE 10.1
Solution by Factoring

Find the solution set of the quadratic equation in each of Probs. 1 to 20.

1	$4x^2 - 9 = 0$	**2**	$16x^2 - 25 = 0$
3	$49x^2 - 64 = 0$	**4**	$16x^2 - 81 = 0$
5	$5x^2 - 80 = 0$	**6**	$3x^2 - 12 = 0$
7	$7x^2 - 63 = 0$	**8**	$13x^2 - 13 = 0$
9	$3x^2 - 4 = 0$	**10**	$9x^2 - 32 = 0$
11	$5x^2 - 8 = 0$	**12**	$6x^2 - 27 = 0$
13	$3x^2 + 12 = 0$	**14**	$5x^2 + 45 = 0$
15	$2x^2 + 32 = 0$	**16**	$3x^2 + 75 = 0$

17 $5x^2 + 15 = 0$

18 $7x^2 + 42 = 0$

19 $2x^2 + 20 = 0$

20 $5x^2 + 35 = 0$

Find the solution set of the equations in Probs. 21 to 44 by the factoring method.

21 $x^2 + 2x - 15 = 0$

22 $x^2 + 2x - 8 = 0$

23 $x^2 + 2x - 3 = 0$

24 $x^2 - 3x - 10 = 0$

25 $x^2 - 3x = 28$

26 $x^2 + 3x = 18$

27 $x^2 = 3x + 40$

28 $x^2 + 5x = 14$

29 $2x^2 = 6 - x$

30 $3x^2 + 7x = 6$

31 $4x^2 + 7x = 2$

32 $5x^2 = 3 + 2x$

33 $3x^2 = 10 + x$

34 $2x^2 + x = 15$

35 $2x^2 = 7 + 5x$

36 $3x^2 = 8x + 16$

37 $6x^2 + 5x = 6$

38 $6x^2 = 4 - 5x$

39 $6x^2 = 7x + 5$

40 $8x^2 = 9 - 6x$

41 $15x^2 - 7x = 2$

42 $15x^2 = 8 + 14x$

43 $8x^2 - 2x = 15$

44 $6x^2 - 5x = 25$

45 Find the set $\{x \mid 15x^2 + 2x - 24 = 0\}$.

46 Find the set $\{x \mid 12x^2 - x - 35 = 0\}$.

47 Find the set $\{x \mid 20x^2 + 3x - 56 = 0\}$.

48 Find the set $\{x \mid 12x^2 - 8x - 15 = 0\}$.

49 Find the set $\{x \mid 6x^2 + 11x = 35\}$.

50 Find the set $\{x \mid 8x^2 + 2x = 45\}$.

51 Find the set $\{x \mid 42x^2 = 2 - 5x\}$.

52 Find the set $\{x \mid 6x^2 + x = 35\}$.

53 Find the set $\{x \mid 20x^2 + 3x = 56\}$.

54 Find the set $\{x \mid 4x^2 + 4x = 63\}$.

55 Find the set $\{x \mid 12x^2 - 5x = 72\}$.

56 Find the set $\{x \mid 12x^2 + 7x = 49\}$.

57 Find the set $\{x \mid 30x^2 + 19x = 63\}$.

58 Find the set $\{x \mid 20x^2 + 13x = 72\}$.

59 Find the set $\{x \mid 42x^2 = 45 - 19x\}$.

60 Find the set $\{x \mid 42x^2 + x = 56\}$.

61 Find the set $\{x \mid 3x^2 + 2b^2 = 5bx\}$.

62 Find the set $\{x \mid 2x^2 - ax = 10a^2\}$.

63 Find the set $\{x \mid 2x^2 + 21t^2 = 13tx\}$.

64 Find the set $\{x \mid 3x^2 + 7dx = 40d^2\}$.

65 Find the set $\{x \mid x^2 + 2ax - bx - 2ab = 0\}$.

66 Find the set $\{x \mid px^2 + 3pqx - qx = 3q^2\}$.

67 Find the set $\{x \mid 3bx^2 - 3cx + bcx = c^2\}$.

68 Find the set $\{x \mid acx^2 + bcx - adx - bd = 0\}$.

229

10.4 Solving a
Quadratic
Equation by
Completing the
Square

10.4
SOLVING A QUADRATIC EQUATION BY
COMPLETING THE SQUARE

Although the method of solving an equation by use of factoring is efficient, it can be applied only when one member of the equation is zero and the other member is reducible. In this section we shall explain a method that can be used for solving any quadratic equation. We shall first consider the equation

$$(x + d)^2 = k \tag{1}$$

Evidently, if $x + d = \pm\sqrt{k}$, then $(x + d)^2 = k$. Hence the roots of the two linear equations $x + d = \sqrt{k}$ and $x + d = -\sqrt{k}$ are also roots of Eq. (1). These roots are $-d + \sqrt{k}$ and $-d - \sqrt{k}$. For example, to get the roots of $(x - 2)^2 = 9$, we solve the linear equations $x - 2 = \pm 3$ and obtain $x = 2 \pm 3 = 5$ and -1.

Hence, if a quadratic equation can be converted to an equivalent equation of the type of (1), the process of solving can be completed by this method. It is therefore essential at this point to review the conditions under which a trinomial is a perfect square. These conditions as given in Sec. 3.5 require that two of the terms be perfect squares and that the third term be twice the product of the square roots of these two. Consequently, the trinomial

$$x^2 \pm px + q^2 \tag{2}$$

is a perfect square if $2\sqrt{x^2 q^2} = \pm px$. From this statement it follows that $4x^2 q^2 = (\pm px)^2 = (\pm p)^2 x^2$; hence q^2 must be equal to $(\pm\frac{1}{2}p)^2$. Therefore, the trinomial (2) is a perfect square if *the third term is equal to the square of one-half the coefficient of* x. Furthermore, $x^2 \pm px + (\frac{1}{2}p)^2 = (x \pm \frac{1}{2}p)^2$. For example, $x^2 + 6x + 9$ is a perfect square, since $9 = [\frac{1}{2}(6)]^2$ and it is the square of $x + 3$. Similarly, $x^2 - 5x/3 + \frac{25}{36}$ is a perfect square, since $\frac{25}{36} = [\frac{1}{2}(\frac{5}{3})]^2$. Also $x^2 - 5x/3 + \frac{25}{36} = (x - \frac{5}{6})^2$.

We shall now solve the equation

$$x^2 + 6x = 27 \tag{3}$$

by use of the above information. If we add the square of one-half the coefficient of x to each member of Eq. (3), we obtain an equivalent equation in which the left member is a perfect square. The square of one-half the coefficient of x is $[\frac{1}{2}(6)]^2 = 3^2 = 9$. Hence, if we add 9 to each member of Eq. (3), we get

$$x^2 + 6x + 9 = 27 + 9$$

or

$(x + 3)^2 = 36$ since $x^2 + 6x + 9 = (x + 3)^2$ and $27 + 9 = 36$

Now we can complete the solution as follows:

$x + 3 = \pm 6$ equating the square roots of the members and giving the right member both the plus and minus signs

$x = \pm 6 - 3$ adding -3 to each member

$x = 3$ and -9 since $6 - 3 = 3$ and $-6 - 3 = -9$

Consequently, the solution set is $\{3, -9\}$.

Verification   If $x = 3$, then

$3^2 + 6(3) = 9 + 18 = 27$

If $x = -9$, then

$(-9)^2 + 6(-9) = 81 - 54 = 27$

The above procedure is called the *method of solving a quadratic equation by completing the square*. We shall illustrate it with three examples.

EXAMPLE 1   By the method of completing the square, solve the equation

$$3x - 9 = -x^2 \tag{1}$$

SOLUTION   Since our first objective is to convert Eq. (1) into an equivalent equation in which the left member is the square of a binomial of the type $x + d$ and the right member is a constant, we must first convert (1) to an equivalent equation in which the first term is x^2, the second term involves x, and the right member is a constant. Therefore, we add $x^2 + 9$ to each member and get

$3x - 9 + x^2 + 9 = -x^2 + x^2 + 9$

or

$x^2 + 3x = 9$

The next step is to complete the square by adding $[\frac{1}{2}(3)]^2 = \frac{9}{4}$ to each member. Thus, we get

$x^2 + 3x + \frac{9}{4} = 9 + \frac{9}{4}$

or

$(x + \frac{3}{2})^2 = \frac{45}{4}$ since $x^2 + 3x + \frac{9}{4} = (x + \frac{3}{2})^2$ and $9 + \frac{9}{4} = (36 + 9)/4 = \frac{45}{4}$

Now we equate the principal† square root of the left member to both square roots of the right and get

231

10.4 Solving a
Quadratic
Equation by
Completing the
Square

$$x + \tfrac{3}{2} = \pm \sqrt{45/4}$$

$$= \pm \frac{3\sqrt{5}}{2} \qquad \text{since } \sqrt{45/4} = \sqrt{(9/4)(5)} = \frac{3\sqrt{5}}{2}$$

Finally, we solve the two linear equations and get

$$x = \frac{-3}{2} \pm \frac{3\sqrt{5}}{2}$$

$$= \tfrac{3}{2}(-1 \pm \sqrt{5})$$

Hence, the solution set is

$$\{\tfrac{3}{2}(-1 + \sqrt{5}), \quad \tfrac{3}{2}(-1 - \sqrt{5})\}$$

We next check these roots by substituting them for x in each member of Eq. (1). For the left member, we get

$$3[\tfrac{3}{2}(-1 \pm \sqrt{5})] - 9 = -\frac{9}{2} \pm \frac{9\sqrt{5}}{2} - \frac{18}{2} = -\frac{27}{2} \pm \frac{9\sqrt{5}}{2}$$

and for the right,

$$-[\tfrac{3}{2}(-1 \pm \sqrt{5})]^2 = -\tfrac{9}{4}(1 \mp 2\sqrt{5} + 5) = -\tfrac{9}{4}(6 \mp 2\sqrt{5}) = -\frac{27}{2} \pm \frac{9\sqrt{5}}{2}$$

EXAMPLE 2 By the method of completing the square, solve the equation

$$8x - 2 = 3x^2 \qquad\qquad (1)$$

SOLUTION Again the first objective is to convert Eq. (1) to an equivalent equation in which the left member is the square of a binomial of the type $x + d$. Consequently, we must first convert (1) to an equivalent equation in which the first term of the left member is x^2, the second term involves x, and the right member is a constant. Therefore, we add $-3x^2 + 2$ to each member and divide the resulting equation by -3. Thus, we obtain

$$-3x^2 + 8x = 2 \qquad \text{adding } -3x^2 + 2 \text{ to each member of Eq. (1)}$$

$$x^2 - \frac{8x}{3} = -\frac{2}{3} \qquad \text{dividing each member by } -3$$

†The question arises here: Why not use both square roots of the left member? The answer is that such procedure will yield two linear equations that are equivalent to those obtained by use of the principal square root, since $-(x + \tfrac{3}{2}) = \pm 3\sqrt{5}/2$ is equivalent to $x + \tfrac{3}{2} = \mp 3\sqrt{5}/2$.

We now add the square of one-half the coefficient of x to each member and complete the process as follows:

$$x^2 - \frac{8x}{3} + \frac{16}{9} = -\frac{2}{3} + \frac{16}{9} \qquad \text{adding } [\tfrac{1}{2}(-\tfrac{8}{3})]^2 = \tfrac{16}{9} \text{ to each member}$$

$$= \frac{-6 + 16}{9}$$

$$= \frac{10}{9}$$

Now we express

$$x^2 - \frac{8x}{3} + \frac{16}{9} \qquad \text{as} \qquad (x - \tfrac{4}{3})^2$$

and have

$$(x - \tfrac{4}{3})^2 = \tfrac{10}{9}$$

$$x - \tfrac{4}{3} = \pm \sqrt{\tfrac{10}{9}}$$

$$x = \frac{4}{3} \pm \frac{\sqrt{10}}{3} \qquad \text{since } \sqrt{\tfrac{10}{9}} = \sqrt{10}/3$$

$$x = \tfrac{1}{3}(4 \pm \sqrt{10})$$

Consequently, the solution set is

$$\{\tfrac{1}{3}(4 + \sqrt{10}), \tfrac{1}{3}(4 - \sqrt{10})\}$$

We check the solution set by substituting $\tfrac{1}{3}(4 \pm \sqrt{10})$ in each member of Eq. (1). The substitution is as follows:

Left Member

$$8[\tfrac{1}{3}(4 \pm \sqrt{10})] - 2$$

$$= \tfrac{32}{3} \pm \tfrac{8}{3}\sqrt{10} - 2$$

$$= \frac{32 \pm 8\sqrt{10} - 6}{3}$$

$$= \frac{26 \pm 8\sqrt{10}}{3}$$

Right Member

$$3[\tfrac{1}{3}(4 \pm \sqrt{10})]^2$$

$$= 3[\tfrac{1}{9}(16 \pm 8\sqrt{10} + 10)]$$

$$= \tfrac{1}{3}(26 \pm 8\sqrt{10})$$

$$= \frac{26 \pm 8\sqrt{10}}{3}$$

Consequently, the two members of the equation are equal if $x = \tfrac{1}{3}(4 \pm \sqrt{10})$.

EXAMPLE 3

Solve the equation $2x^2 + 9 = 3x$.

SOLUTION

233

10.4 Solving a
Quadratic
Equation by
Completing the
Square

$$2x^2 + 9 = 3x \qquad \text{given equation copied}$$

$$2x^2 - 3x = -9 \qquad \text{adding } -3x - 9 \text{ to each member}$$

$$x^2 - \frac{3x}{2} = -\frac{9}{2} \qquad \text{dividing each member by the coefficient of } x^2$$

$$x^2 - \frac{3x}{2} + \frac{9}{16} = -\frac{9}{2} + \frac{9}{16} \qquad \text{adding } \left[\frac{1}{2}\left(\frac{3}{2}\right)\right]^2 \text{ to each member}$$

$$= \frac{-72 + 9}{16}$$

$$= -\frac{63}{16} \qquad \text{simplifying right member}$$

$$\left(x - \frac{3}{4}\right)^2 = -\frac{63}{16} \qquad \text{expressing left member as } \left(x - \frac{3}{4}\right)^2$$

$$x - \frac{3}{4} = \pm\sqrt{-\frac{63}{16}} \qquad \text{equating the square roots of the members}$$

$$= \pm\frac{3\sqrt{-7}}{4} \qquad \text{since } \sqrt{-63} = \sqrt{9(-7)} = 3\sqrt{-7} \text{ and } \sqrt{16} = 4$$

$$= \pm\frac{3i\sqrt{7}}{4} \qquad \text{since } \sqrt{-7} = i\sqrt{7}$$

$$x = \frac{3}{4} \pm \frac{3i\sqrt{7}}{4} \qquad \text{solving for } x$$

$$= \frac{3}{4}(1 \pm i\sqrt{7})$$

Therefore, the solution set is

$$\left\{\frac{3}{4}(1 + i\sqrt{7}), \quad \frac{3}{4}(1 - i\sqrt{7})\right\}$$

To check the solution, we substitute $\frac{3}{4}(1 \pm i\sqrt{7})$ for x in each member of the original equation and get the following results:

Left Member

$$2\left[\frac{3}{4}(1 \pm i\sqrt{7})\right]^2 + 9$$

$$= 2\left[\frac{9}{16}(1 \pm 2i\sqrt{7} + 7i^2)\right] + 9$$

$$= \frac{9}{8}(1 \pm 2i\sqrt{7} - 7) + 9$$

$$= \frac{9}{8}(-6 \pm 2i\sqrt{7}) + 9$$

$$= -\frac{54}{8} \pm \frac{18i\sqrt{7}}{8} + 9$$

$$= \frac{18}{8} \pm \frac{18i\sqrt{7}}{8}$$

$$= \frac{9}{4} \pm \frac{9i\sqrt{7}}{4}$$

Right Member

$$3\left[\frac{3}{4}(1 \pm i\sqrt{7})\right]$$

$$= \frac{9}{4}(1 \pm i\sqrt{7})$$

$$= \frac{9}{4} \pm \frac{9i\sqrt{7}}{4}$$

Consequently, the two members are equal when $x = \frac{3}{4}(1 \pm i\sqrt{7})$.

Steps in Solving a Quadratic Equation by Completing the Square

Examples 1 to 3 illustrate the following formal steps in solving a quadratic equation by completing the square:

1. Convert the given equation to an equivalent equation of the form $ax^2 + bx = k$.
2. Divide each member by the coefficient of x^2.
3. Add the square of one-half the coefficient of x to each member.
4. Simplify the right member of the equation obtained in step 3.
5. Equate the square roots of the two members obtained in step 4, and give the square root of the right member both the plus and the minus signs.
6. Solve the two linear equations obtained in step 5 for x.
7. Substitute the values of x obtained in step 6 in the original equation in order to verify that each is a root.

10.5
COMPLEX NUMBERS

The solution of Example 3 of the previous section introduces a type of number which we have not previously met. We shall describe these numbers in the definition:

■ A number of the type $a + bi$, where a and b are real and $i = \sqrt{-1}$, is called a *complex number.*

If $b = 0$, the complex number is equal to the real number a. Consequently, the set of complex numbers includes the set of real numbers as a subset. The complex number $a + bi$ is called *imaginary* if and only if $b \neq 0$. The imaginary numbers $2 + 3i$, $\frac{3}{4} + \frac{1}{4}i$, and $2i$ and the real number 3 are all examples of complex numbers.

The name *imaginary* for such numbers is unfortunate; it was adopted when such numbers were not understood and were regarded as fictitious. Later the mathematician Gauss invented a way to give such numbers a geometrical meaning, and since this invention, they have become very useful and important in many theoretical and practical fields.

EXERCISE 10.2
Solution by Completing the Square

Find the solution sets of the quadratic equations in Probs. 1 to 28 by the method of completing the square.

1. $x^2 + 6x - 27 = 0$
2. $x^2 - 2x - 8 = 0$
3. $x^2 + 10x + 21 = 0$
4. $x^2 - 4x - 12 = 0$

5 $x^2 + 5x = 24$

6 $x^2 - x = 20$

7 $x^2 = 54 + 3x$

8 $70 - 3x = x^2$ $+ 3x - 70 = 0$

9 $4x^2 + 4x = 35$

10 $9x^2 = 10 - 9x$

11 $4x^2 + 8x = 21$

12 $27x - 8 = 9x^2$

13 $2x^2 + 3x = 2$

14 $3x^2 = 5x - 2$

15 $2x^2 + 5x = 3$

16 $3x^2 + 7x + 2 = 0$

17 $5x - 1 = 6x^2$

18 $8x^2 = 15 - 2x$

19 $10x^2 = 7x - 1$

20 $6x^2 + 7x = 5$

21 $6x^2 + 5x = 4$

22 $6x^2 + 3 = 11x$

23 $8x^2 - 45 = 2x$

24 $15x^2 + 4x - 3 = 0$

25 $x^2 - 4x - 2 = 0$

26 $x^2 = 2x + 4$

27 $x^2 + 7 = 6x$

28 $x^2 + 10x + 23 = 0$

29 Find the set $\{x \mid 4x^2 + 8x - 1 = 0\}$.

30 Find the set $\{x \mid 4x^2 + 7 = 12x\}$.

31 Find the set $\{x \mid 9x^2 = 5 + 15x\}$.

32 Find the set $\{x \mid 12x - 1 = 9x^2\}$.

33 Find the set $\{x \mid 4x^2 + 5 = 10x\}$.

34 Find the set $\{x \mid 9x^2 + 8x + 1 = 0\}$.

35 Find the set $\{x \mid 4x^2 + 11 = -14x\}$.

36 Find the set $\{x \mid 18x - 17 = 4x^2\}$.

37 Find the set $\{x \mid x^2 + 25 = 8x\}$.

38 Find the set $\{x \mid x^2 = 10x - 26\}$.

39 Find the set $\{x \mid 25 + x^2 = 6x\}$.

40 Find the set $\{x \mid x^2 = 12x - 40\}$.

41 Find the set $\{x \mid 2x^2 + 13 = 2x\}$.

42 Find the set $\{x \mid 4x^2 + 8x = -13\}$.

43 Find the set $\{x \mid 9x^2 = 6x - 10\}$.

44 Find the set $\{x \mid 9x^2 + 5 = 12x\}$.

45 Find the set $\{x \mid 4x^2 = 8x - 9\}$.

46 Find the set $\{x \mid 9x^2 - 18x + 15 = 0\}$.

47 Find the set $\{x \mid 4x^2 + 16x + 21 = 0\}$.

48 Find the set $\{x \mid 9x^2 = 6x - 7\}$.

49 Find the set $\{x \mid x^2 + 4ax = 5a^2\}$.

50 Find the set $\{x \mid x^2 = 6m^2 - mx\}$.

51 Find the set $\{x \mid x^2 + (3p - 5q)x = 15pq\}$.

52 Find the set $\{x \mid x^2 + 2wx = -7xz - 14wz\}$.

53 Find the set $\{x \mid bx^2 - 3bx + 2a + 2b = ax\}$.

54 Find the set $\{x \mid rtx^2 - (rt^2 + 1)x + t = 0\}$.

55 Find the set $\{x \mid 4x^2 + b^2 + bc = (4b + 2c)x\}$.

56 Find the set $\{x \mid uvx^2 - (u^2 + v^2)x + uv = 0\}$.

57 Find the set $\{x \mid (a - b)x^2 + (2a - 3b)x = 3a\}$.

58 Find the set $\{x \mid (c - d)x^2 = c + d + 2dx\}$.

59 Find the set $\{x \mid px^2 + (3 - 5p)x = 15\}$.

60 Find the set $\{x \mid wzx^2 - (w^2 + 2wz - z^2)x + w^2 - z^2 = 0\}$.

10.6
THE QUADRATIC FORMULA $\frac{-b \sqrt{b^2\ 4ac}}{2a}$

If we solve the equation

■ $ax^2 + bx + c = 0$ (10.4)

for x, we get a formula that can be used for obtaining the roots of any quadratic equation. We shall derive the formula by solving (10.4) by completing the square, and then we shall explain its use.

$ax^2 + bx + c = 0$ Eq. (10.4) copied

$ax^2 + bx = -c$ adding $-c$ to both members

$x^2 + \dfrac{bx}{a} = -\dfrac{c}{a}$ dividing each member by a

$x^2 + \dfrac{bx}{a} + \left(\dfrac{b}{2a}\right)^2 = -\dfrac{c}{a} + \dfrac{b^2}{4a^2}$ adding $\left[\frac{1}{2}(b/a)\right]^2$ to each member

$\left(x + \dfrac{b}{2a}\right)^2 = \dfrac{b^2 - 4ac}{4a^2}$

$x + \dfrac{b}{2a} = \pm\dfrac{\sqrt{b^2 - 4ac}\dagger}{2a}$

$x = -\dfrac{b}{2a} \pm \dfrac{\sqrt{b^2 - 4ac}}{2a}$

    $\dagger$Here we have tacitly assumed that a is positive. If, however, a is negative, then $\sqrt{4a^2} = -2a$, since by definition the principal square root of a number is positive. Then we have

$x + \dfrac{b}{2a} = \pm\sqrt{\dfrac{b^2 - 4ac}{4a^2}} = \pm\dfrac{\sqrt{b^2 - 4ac}}{\sqrt{4a^2}} = \mp\dfrac{\sqrt{b^2 - 4ac}}{2a}$

Hence,

$x = \dfrac{-b \mp \sqrt{b^2 - 4ac}}{2a}$

This equation yields the same values of x as (10.5), but the values are in reverse order.

Since the two denominators in the right member are the same, we can write

$$\blacksquare \quad x = \frac{-b \pm \sqrt{b^2 - 4ac}}{2a} \tag{10.5}$$

Equation (10.5) is known as the *quadratic formula*. It expresses the roots of Eq. (10.4) in terms of the constant term and the coefficients of x^2 and x. By properly matching the coefficients and the constant term in any quadratic equation with those in (10.4), we can determine the values of a, b, and c, and then we can use the formula to get the roots of the equation. We shall illustrate the use of (10.5) with two examples.

EXAMPLE 1

Solve the equation $3x^2 - 5x + 2 = 0$ by the quadratic formula.

SOLUTION   To solve $3x^2 - 5x + 2 = 0$ with the quadratic formula, we compare the equation with (10.4) and see that $a = 3$, $b = -5$, and $c = 2$. Hence, if we substitute these values in (10.5), we get

$$x = \frac{-(-5) \pm \sqrt{(-5)^2 - 4(3)(2)}}{2(3)}$$

$$= \frac{5 \pm \sqrt{25 - 24}}{6}$$

$$= \frac{5 \pm 1}{6}$$

$$= \tfrac{6}{6} \quad \text{and} \quad \tfrac{4}{6}$$

$$= 1 \quad \text{and} \quad \tfrac{2}{3}$$

Hence, the solution set is $\{1, \tfrac{2}{3}\}$.

The reader should verify that this is true.

If the terms in a given equation do not occur in the same order as those in (10.4), we convert the equation to an equivalent equation in which the terms have the desired order before applying the quadratic formula.

EXAMPLE 2

By the quadratic formula, solve the equation $4x^2 = 8x - 7$.

SOLUTION   The first step in solving the given equation is to convert it to an equivalent equation in which the terms are in the same order as in (10.4). For this purpose we add $-8x + 7$ to each member and obtain $4x^2 - 8x + 7 = 0$. By comparing this equation with (10.4), we see that $a = 4$, $b = -8$, and

$c = 7$. If we substitute these values in (10.5), we get

$$x = \frac{-(-8) \pm \sqrt{(-8)^2 - 4(4)(7)}}{2(4)}$$

$$= \frac{8 \pm \sqrt{64 - 112}}{8}$$

$$= \frac{8 \pm \sqrt{-48}}{8}$$

$$= \frac{8 \pm 4i\sqrt{3}}{8}$$

$$= \tfrac{1}{2}(2 \pm i\sqrt{3})$$

Therefore, the solution set is

$$\{\tfrac{1}{2}(2 + i\sqrt{3}), \quad \tfrac{1}{2}(2 - i\sqrt{3})\}$$

The reader should verify that these two numbers are roots of the given equation.

EXERCISE 10.3

Solution by Formula

Find the solution set of each of the equations in Probs. 1 to 28 by using the quadratic formula.

1	$x^2 + 9x + 14 = 0$	**2**	$x^2 + x - 30 = 0$
3	$x^2 - 4x - 21 = 0$	**4**	$x^2 - 4x - 45 = 0$
5	$2x^2 - x = 10$	**6**	$3x^2 = 5x + 12$
7	$5x^2 + 3x = 14$	**8**	$4x^2 = 11x + 20$
9	$6x^2 + x = 35$	**10**	$20 = 9x^2 - 3x$
11	$12x^2 - 7x = 45$	**12**	$10x^2 = 9 - 9x$
13	$42x^2 + x = 30$	**14**	$25x^2 + 10x = 63$
15	$56x^2 = 3x + 20$	**16**	$42 - 5x = 63x^2$
17	$x^2 - 4x = -1$	**18**	$x^2 - 10x + 18 = 0$
19	$x^2 = 8x - 13$	**20**	$6x = 3 + x^2$
21	$x^2 - 3x + 1 = 0$	**22**	$3x^2 + 4x - 1 = 0$
23	$25x^2 - 50x + 22 = 0$	**24**	$9x^2 = 24x - 14$
25	$9x^2 = 12x + 14$	**26**	$4x^2 - 11 = 12x$
27	$9x^2 - 24x = 47$	**28**	$24x + 41 = 16x^2$

29 Find the set $\{x \mid 25x^2 + 37 = 70x\}$.

30 Find the set $\{x \mid 9x^2 - 21x + 1 = 0\}$.

31 Find the set $\{x \mid 7x^2 + 10x + 1 = 0\}$.

32 Find the set $\{x \mid 9x^2 + 8 = -36x\}$.

33 Find the set $\{x \mid x^2 + 6x + 10 = 0\}$.

34 Find the set $\{x \mid x^2 + 20 = 8x\}$.

35 Find the set $\{x \mid x^2 - 4x + 29 = 0\}$.

36 Find the set $\{x \mid x^2 - 10x + 34 = 0\}$.

37 Find the set $\{x \mid 2x^2 - 6x + 5 = 0\}$.

38 Find the set $\{x \mid 2x^2 + 17 = 10x\}$.

39 Find the set $\{x \mid 9x^2 + 34 = 18x\}$.

40 Find the set $\{x \mid 9x^2 = 30x - 29\}$.

41 Find the set $\{x \mid 20x = 25x^2 + 9\}$.

42 Find the set $\{x \mid 4x^2 + 12x + 15 = 0\}$.

43 Find the set $\{x \mid 4x^2 + 27 = 20x\}$.

44 Find the set $\{x \mid 3x^2 = 14x - 17\}$.

45 Find the set $\{x \mid 4x^2 + 12x + 15 = 0\}$.

46 Find the set $\{x \mid 3x^2 + 10x + 9 = 0\}$.

47 Find the set $\{x \mid 2x^2 + 12x + 19 = 0\}$.

48 Find the set $\{x \mid 24x = 9x^2 + 23\}$.

49 Find the set $\{x \mid x^2 + 2ab = (2a + b)x\}$.

50 Find the set $\{x \mid x^2 + 5uv = (5 + uv)x\}$.

51 Find the set $\{x \mid 2c^2x^2 = 3bcx + 2b^2\}$.

52 Find the set $\{x \mid pqx^2 + 4pq = 2p^2x + 2q^2x\}$.

53 Find the set $\{x \mid x^2 - (2r - 2t)x + r^2 = 2rt + 3t^2\}$.

54 Find the set $\{x \mid 3d^2x^2 + 2c^2 = 5cdx\}$.

55 Find the set $\{x \mid 5abx^2 + 3ab = 15x + a^2b^2x\}$.

56 Find the set $\{x \mid x^2 + w^2 + 2wz = 2wx + 2zx\}$.

57 Find the set $\{x \mid 3x^2 + 3cd + 6d^2 = cx + 11dx\}$.

58 Find the set $\{x \mid (2m^2 + 3mn + n^2)x^2 + (m - n)x - 6 = 0\}$.

59 Find the set $\{x \mid bx^2 + 7x + b^2x + 7b = 0\}$.

60 Find the set $\{x \mid 6x^2 + a^2 = 5ax + 2bx + 4b^2\}$.

10.7
NATURE OF THE ROOTS OF A QUADRATIC EQUATION

The roots of the equation in Example 1 of Sec. 10.3 were two unequal rational numbers. In the first two examples of Sec. 10.4 the roots were unequal irrational numbers, and in Example 2 of Sec. 10.6 the roots were imaginary. In this section we shall discuss a method that enables us to determine, without solving, the nature of the roots of any quadratic equation. We shall use the quadratic formula (10.5) for this purpose. Remember that the quadratic formula expresses the roots of Eq. (10.4) in terms of the coefficients in the equation. For convenient reference, we shall repeat this equation. It is

$$ax^2 + bx + c = 0 \tag{10.4}$$

Discriminant of a Quadratic Equation

We shall first assume that a, b, and c are rational numbers. Furthermore, we shall let r represent the root when the positive sign is used before the radical in the quadratic formula, and we shall let s represent the root when the negative sign is used. Accordingly, we have

$$\blacksquare \quad r = \frac{-b + \sqrt{b^2 - 4ac}}{2a} \tag{10.6}$$

and

$$\blacksquare \quad s = \frac{-b - \sqrt{b^2 - 4ac}}{2a} \tag{10.7}$$

We shall let $D = b^2 - 4ac$, and we shall next demonstrate that the nature of the roots of (10.4) depends upon the value of D. For this reason D is called the *discriminant* of the equation.

Tests for Determining the Nature of the Roots

The following conclusions can be derived from Formulas (10.6) and (10.7) for a, b, and c rational:

1 If $D = 0$, then $r = s = -b/2a$ and the roots of (10.4) are rational and equal.
2 If D is a perfect square, then $\sqrt{D}$ is rational and the roots $r = (1/2a)(-b + \sqrt{D})$ and $s = (1/2a)(-b - \sqrt{D})$ are therefore rational and unequal.
3 If $D > 0$ but is not a perfect square, then $\sqrt{D}$ is irrational and consequently the roots are irrational and unequal.
4 If $D < 0$, $\sqrt{D}$ is imaginary and therefore the roots are imaginary.

EXAMPLE 1

Compute the value of the discriminant, and then determine the nature of the roots of each of the following four equations:

$$4x^2 - 12x + 9 = 0, \qquad 3x^2 - 7x - 6 = 0, \qquad 5x^2 + 2x - 9 = 0,$$

and $x^2 + 3x + 5 = 0$

SOLUTION

$4x^2 - 12x + 9 = 0$ $\qquad D = (-12)^2 - 4(4)(9)$ $\qquad$ roots are rational

$\qquad\qquad\qquad\qquad\qquad = 144 - 144 = 0$ $\qquad$ and equal

$3x^2 - 7x - 6 = 0$ $D = (-7)^2 - 4(3)(-6)$ roots are rational and unequal

$$= 49 + 72 = 121 = 11^2$$

$5x^2 + 2x - 9 = 0$ $D = (2)^2 - 4(5)(-9)$ roots are irrational and unequal

$$= 4 + 180 = 184$$

$x^2 + 3x + 5 = 0$ $D = (3)^2 - 4(1)(5)$ roots are imaginary

$$= 9 - 20 = -11$$

If, in (10.4), a, b, and c are real but not necessarily rational, the information we derive from D is less specific:

1. If $D = 0$, then the roots are real and equal. (In this case $r = s = -b/2a$, but we know only that b and a are real.)
2. If $D > 0$, then the roots are real and unequal.
3. If $D < 0$, then the roots are imaginary.

EXAMPLE 2

Compute the value of the discriminant, and determine the nature of the roots in each of the following three equations:

$$4x^2 - 4\sqrt{5}x + 5 = 0, \qquad \sqrt{3}x^2 - 6x + \sqrt{12} = 0,$$

and $\qquad \sqrt{2}x^2 + 3x + \sqrt{5} = 0$

SOLUTION

$4x^2 - 4\sqrt{5}x + 5 = 0$ $D = 80 - 80 = 0$ roots are real and equal

$\sqrt{3}x^2 - 6x + \sqrt{12} = 0$ $D = 36 - 24 = 12$ roots are real and unequal

$\sqrt{2}x^2 + 3x + \sqrt{5} = 0$ $D = 9 - 4\sqrt{10} < 0$ roots are imaginary

10.8
SUM AND PRODUCT OF THE ROOTS

Sum of the Roots of a Quadratic Equation

If we add the roots r and s of $ax^2 + bx + c = 0$ given by Formulas (10.6) and (10.7), we get

$$r + s = \frac{-b + \sqrt{b^2 - 4ac}}{2a} + \frac{-b - \sqrt{b^2 - 4ac}}{2a}$$

$$= \frac{-2b}{2a} = -\frac{b}{a}$$

Product of the Roots of a Quadratic Equation

Similarly, the product of the roots is

$$rs = \left(\frac{-b + \sqrt{b^2 - 4ac}}{2a}\right)\left(\frac{-b - \sqrt{b^2 - 4ac}}{2a}\right)$$

$$= \frac{b^2 - (b^2 - 4ac)}{4a^2} = \frac{4ac}{4a^2} = \frac{c}{a}$$

Consequently, we have the following theorem:

■ In the quadratic equation $ax^2 + bx + c = 0$, the sum of the roots is $-b/a$ and the product of the roots is c/a.

The application of this theorem is shown in the following example.

EXAMPLE 1

Find the sum of the roots and the product of the roots in each of the following three equations:

$$x^2 - 3x + 2 = 0, \quad 2x^2 + 8x - 5 = 0, \quad \text{and} \quad \sqrt{2}x^2 + 5x - \sqrt{8} = 0$$

SOLUTION In the following tabulation $-b/a$ is the sum of the roots and c/a is the product of the roots.

$$x^2 - 3x + 2 = 0 \qquad -\frac{b}{a} = \frac{-(-3)}{1} = 3 \qquad \frac{c}{a} = \frac{2}{1} = 2$$

$$2x^2 + 8x - 5 = 0 \qquad -\frac{b}{a} = \frac{-8}{2} = -4 \qquad \frac{c}{a} = \frac{-5}{2}$$

$$\sqrt{2}x^2 + 5x - \sqrt{8} = 0 \qquad -\frac{b}{a} = \frac{-5}{\sqrt{2}} \qquad \frac{c}{a} = \frac{-\sqrt{8}}{\sqrt{2}}$$

$$= -\sqrt{4}$$

$$= -2$$

The theorem provides a rapid method for checking the roots of an equation. For example, suppose the equation is $6x^2 - x - 12 = 0$ and the roots $\frac{3}{2}$ and $-\frac{4}{3}$ are obtained. By the theorem the sum of the roots is $\frac{1}{6}$ and the product is $-\frac{12}{6} = -2$. Therefore, since

$$\frac{3}{2} - \frac{4}{3} = \frac{9 - 8}{6} = \frac{1}{6} \quad \text{and} \quad \frac{3}{2} \times \frac{-4}{3} = -2$$

the roots are correct.

As we shall next demonstrate, we can use this theorem to form a quadratic equation that has two specified numbers as roots. We first divide each member of the equation $ax^2 + bx + c = 0$ by a and obtain $x^2 + bx/a + c/a = 0$.

If we now replace b/a by $-(r+s)$ and c/a by rs, the equation becomes

$$x^2 - (r+s)x + rs = 0$$

EXAMPLE 2

Obtain quadratic equations that have the following pairs of numbers as roots:
$5, -7; -\frac{1}{2}, -\frac{2}{3}$; and $2 + \sqrt{3}, 2 - \sqrt{3}$.

SOLUTION   In each of the following examples the equation at the right is formed so that it has the roots indicated at the left:

Roots are $5, -7$ $\qquad$ $x^2 - (5 - 7)x + (5)(-7) = 0$

$\qquad\qquad\qquad\qquad$ $x^2 + 2x - 35 = 0$

Roots are $-\frac{1}{2}, -\frac{2}{3}$ $\qquad$ $x^2 - (-\frac{1}{2} - \frac{2}{3})x + (-\frac{1}{2})(-\frac{2}{3}) = 0$

$\qquad\qquad\qquad\qquad$ $x^2 + \dfrac{7x}{6} + \dfrac{2}{6} = 0$ $\qquad$ simplifying

$\qquad\qquad\qquad\qquad$ $6x^2 + 7x + 2 = 0$ $\qquad$ multiplying each member by 6

Roots are $2 + \sqrt{3}$, $\qquad$ $x^2 - (2 + \sqrt{3} + 2 - \sqrt{3})x + (2 + \sqrt{3})(2 - \sqrt{3}) = 0$

$2 - \sqrt{3}$ $\qquad\qquad\qquad$ $x^2 - 4x + (4 - 3) = 0$

$\qquad\qquad\qquad\qquad$ $x^2 - 4x + 1 = 0$

EXERCISE 10.4
Sum, Product, and Nature of Roots

Without solving, determine the nature of the roots of each of the following equations, and also find the sum and the product of the roots.

1	$x^2 - 6x - 15 = 0$	**2**	$x^2 - 4x - 16 = 0$
3	$x^2 + 10x + 15 = 0$	**4**	$x^2 + 8x + 9 = 0$
5	$x^2 - 10x + 25 = 0$	**6**	$x^2 - 3x - 28 = 0$
7	$x^2 + 6x - 16 = 0$	**8**	$x^2 - 16x + 64 = 0$
9	$2x^2 + 7x - 15 = 0$	**10**	$4x^2 - 4x + 1 = 0$
11	$9x^2 - 12x + 4 = 0$	**12**	$5x^2 + 7x - 6 = 0$
13	$5x^2 - 6x + 2 = 0$	**14**	$2x^2 + 10x + 13 = 0$
15	$3x^2 - 4x + 2 = 0$	**16**	$2x^2 - 14x + 25 = 0$
17	$9x^2 - 12x + 4 = 0$	**18**	$5x^2 + 6x - 8 = 0$
19	$9x^2 - 24x + 16 = 0$	**20**	$7x^2 + 5x - 2 = 0$
21	$9x^2 - 18x + 1 = 0$	**22**	$2x^2 - 6x - 7 = 0$
23	$4x^2 - 10x + 5 = 0$	**24**	$3x^2 - 8x - 4 = 0$
25	$5x^2 + 10x + 6 = 0$	**26**	$x^2 + 5x + 7 = 0$

27 $5x^2 - 14x + 10 = 0$ **28** $3x^2 + 18x + 28 = 0$

29 $5x^2 - 14x + 5 = 0$ **30** $3x^2 + 16x + 12 = 0$

31 $4x^2 - 18x + 9 = 0$ **32** $5x^2 + 18x + 5 = 0$

33 $4x^2 - 4\sqrt{3}\,x - 1 = 0$ **34** $9x^2 + 6\sqrt{6}\,x + 5 = 0$

35 $3x^2 + 4\sqrt{3}\,x + 4 = 0$ **36** $2x^2 - 14\sqrt{2}\,x + 49 = 0$

Form an equation that has each of the following sets as its solution set.

37 $\{\frac{3}{4}, -\frac{1}{3}\}$ **38** $\{\frac{2}{7}, -7\}$

39 $\{\frac{2}{3}, -\frac{3}{2}\}$ **40** $\{\frac{1}{4}, \frac{1}{5}\}$

41 $\{2 + \sqrt{5}, 2 - \sqrt{5}\}$ **42** $\{-3 + \sqrt{7}, -3 - \sqrt{7}\}$

43 $\left\{\dfrac{1 + \sqrt{2}}{3}, \dfrac{1 - \sqrt{2}}{3}\right\}$ **44** $\left\{\dfrac{-5 + \sqrt{6}}{2}, \dfrac{-5 - \sqrt{6}}{2}\right\}$

10.9
RADICAL EQUATIONS

Principle for Solving Radical Equations

A radical equation is an equation in which either or both members contain a radical that has an unknown in the radicand. If the radicals are of the second order, we can solve the equation by the method explained in this section. The method depends upon the following principle.

■ Any root of a given equation is also a root of the equation obtained by equating the squares of the members of the given equation.

To justify this statement, we shall assume that r is a root of the equation $f(x) = a$ and shall prove by the following argument that r is also a root of $[f(x)]^2 = a^2$.

$$f(r) = a \qquad \text{since } r \text{ is a root of } f(x) = a$$

$$a[f(r)] = a^2 \qquad \text{by (2.5)}$$

$$[f(r)][f(r)] = a^2 \qquad \text{replacing } a \text{ by } f(r)$$

$$[f(r)]^2 = a^2 \qquad \text{performing the indicated multiplication}$$

Consequently, r is a root of $[f(x)]^2 = a^2$.

The converse of the above principle is not true. For example, if we equate the squares of the members of

$$\sqrt{x^2 - 4x} = 3x$$

we get

$$x^2 - 4x = 9x^2$$

The roots of $x^2 - 4x = 9x^2$ are 0 and $-\frac{1}{2}$. If $x = 0$ is substituted in $\sqrt{x^2 - 4x} = 3x$, each member becomes zero, and $x = 0$ is therefore a root. If, however, $x = -\frac{1}{2}$ is substituted in $\sqrt{x^2 - 4x} = 3x$, the right member is equal to $-\frac{3}{2}$, and we know that this is not a root because $\sqrt{x^2 - 4x}$ is the principal square root of $x^2 - 4x$ and is therefore positive and hence cannot be equal to $-\frac{3}{2}$. If we keep in mind that a radical of the second order denotes the principal root and is zero or positive, we may frequently tell in advance that a radical equation has no roots. For example, in the equation $\sqrt{x^2 - 5x} = -3$ the left member is positive or zero and the right member is negative. Therefore the equation has no root.

Steps in Solving a Radical Equation

The process of solving a radical equation in which all radicals are of the second order consists of the following steps:

1 Obtain an equation that is equivalent to the given equation and that has one radical and no other term in one member. This process is called *isolating a radical*.
2 Equate the squares of the members of the equation obtained in step 1.
3 If the equation obtained in step 2 contains one or more radicals, repeat the process until an equation free of radicals is obtained. The equation obtained is called the *rationalized equation*.
4 Solve the rationalized equation.
5 Substitute the roots obtained in step 4 in the original equation in order to determine which of these numbers satisfy the original equation.

Note Step 5 is an essential step in solving a radical equation since the rationalized equation may have roots that are not roots of the given equation.

We shall illustrate the process with four examples.

EXAMPLE 1 Solve the equation

$$x = \sqrt{5x - 1} - 1 \tag{1}$$

SOLUTION

$$x = \sqrt{5x - 1} - 1 \qquad \text{given Eq. (1)}$$

$$x + 1 = \sqrt{5x - 1} \qquad \text{isolating radical by adding 1 to each member of Eq. (1)}$$

$$x^2 + 2x + 1 = 5x - 1 \qquad \text{equating squares of the members}$$

$$x^2 + 2x + 1 - 5x + 1 = 5x - 1 - 5x + 1 \qquad \text{adding } -5x + 1 \text{ to each member}$$

$$x^2 - 3x + 2 = 0 \qquad \text{combining similar terms}$$

$$(x - 2)(x - 1) = 0 \qquad \text{factoring left member}$$

$$x = 2 \qquad \text{setting } x - 2 = 0 \text{ and solving}$$

$$x = 1 \qquad \text{setting } x - 1 = 0 \text{ and solving}$$

We check these roots by substituting each of them for x in each member of Eq. (1) and obtain

Value of x	Left Member	Right Member
2	2	$\sqrt{10-1}-1=3-1=2$
1	1	$\sqrt{5-1}-1=2-1=1$

Therefore, since the left and right members of Eq. (1) are equal for $x=2$ and for $x=1$, the solution set of Eq. (1) is $\{2, 1\}$.

EXAMPLE 2 Solve the equation

$$\sqrt{5x-11} - \sqrt{x-3} = 4 \tag{1}$$

SOLUTION

$\sqrt{5x-11} - \sqrt{x-3} = 4$	given Eq. (1)
$\sqrt{5x-11} = \sqrt{x-3} + 4$	isolating $\sqrt{5x-11}$
$5x-11 = x-3+8\sqrt{x-3}+16$	equating squares of the members
$5x-11-x+3-16 = 8\sqrt{x-3}$	isolating $8\sqrt{x-3}$
$4x-24 = 8\sqrt{x-3}$	combining similar terms
$x-6 = 2\sqrt{x-3}$	dividing each member by 4
$x^2-12x+36 = 4(x-3)$	equating squares of the members
$x^2-12x+36 = 4x-12$	by distributive axiom
$x^2-16x+48 = 0$	adding $-4x+12$ to each member
$(x-12)(x-4) = 0$	factoring left member
$x = 12$	setting $x-12=0$ and solving
$x = 4$	setting $x-4=0$ and solving

$$\sqrt{60-11}-\sqrt{12-3}=\sqrt{49}-\sqrt{9} \qquad \text{substituting 12 for } x \text{ in left}$$
member of Eq. (1)

$$=7-3$$

$$=4$$

Therefore, since the right member of Eq. (1) is also 4, then 12 is a root.

$$\sqrt{20-11}-\sqrt{4-3}=\sqrt{9}-\sqrt{1} \qquad \text{substituting 4 for } x \text{ in left}$$
member of Eq. (1)

$$=3-1=2$$

Consequently, since the right member of Eq. (1) is not 2, then 4 is not a root. Therefore, the solution set of Eq. (1) is $\{12\}$.

EXAMPLE 3 Solve the equation

$$\sqrt{x+1}+\sqrt{2x+3}-\sqrt{8x+1}=0 \qquad\qquad\qquad (1)$$

SOLUTION

$$\sqrt{x+1}+\sqrt{2x+3}-\sqrt{8x+1}=0 \qquad \text{given Eq. (1)}$$

$$\sqrt{x+1}+\sqrt{2x+3}=\sqrt{8x+1} \qquad \text{isolating } \sqrt{8x+1}$$

$$x+1+2\sqrt{(x+1)(2x+3)}+2x+3=8x+1 \qquad \text{equating the squares of the members}$$

$$2\sqrt{(x+1)(2x+3)}=8x+1-x-1-2x-3$$
$$\text{isolating } 2\sqrt{(x+1)(2x+3)}$$

$$2\sqrt{2x^2+5x+3}=5x-3 \qquad \text{simplifying radicand and combining similar terms}$$

$$4(2x^2+5x+3)=25x^2-30x+9 \qquad \text{equating squares of the members}$$

$$8x^2+20x+12=25x^2-30x+9 \qquad \text{by distributive axiom}$$

$$17x^2 - 50x - 3 = 0$$

adding $-25x^2$ $+ 30x - 9$ to each member and dividing by -1

$$x = \frac{50 \pm \sqrt{2500 + 204}}{34}$$

by quadratic formula

$$= \frac{50 \pm \sqrt{2704}}{34}$$

$$= \frac{50 \pm 52}{34}$$

$$= \tfrac{102}{34} \quad \text{and} \quad \tfrac{-2}{34}$$

$$= 3 \quad \text{and} \quad \tfrac{-1}{17}$$

Check

$$\sqrt{3+1} + \sqrt{6+3} - \sqrt{24+1} = \sqrt{4} + \sqrt{9} - \sqrt{25}$$

substituting 3 for x in left member of Eq. (1)

$$= 2 + 3 - 5 = 0$$

Therefore since the right member of Eq. (1) is also zero, 3 is a root.

$$\sqrt{\tfrac{-1}{17}+1} + \sqrt{\tfrac{-2}{17}+3} - \sqrt{\tfrac{-8}{17}+1} = \sqrt{\tfrac{16}{17}} + \sqrt{\tfrac{49}{17}} - \sqrt{\tfrac{9}{17}}$$

substituting $\tfrac{-1}{17}$ for x in Eq. (1)

$$= \frac{4}{\sqrt{17}} + \frac{7}{\sqrt{17}} - \frac{3}{\sqrt{17}}$$

$$= \frac{8}{\sqrt{17}}$$

Consequently, since the right member of Eq. (1) is not $8/\sqrt{17}$, then $\tfrac{-1}{17}$ is not a root. Therefore, the solution set of Eq. (1) is {3}.

EXAMPLE 4 Solve the equation

$$\sqrt{x+3} - \sqrt{x-2} = 5 \tag{1}$$

SOLUTION

$$\sqrt{x+3} - \sqrt{x-2} = 5$$

given Eq. (1)

$$\sqrt{x+3} = \sqrt{x-2} + 5$$

isolating $\sqrt{x+3}$

$$x + 3 = x - 2 + 10\sqrt{x - 2} + 25 \quad \text{equating squares of the members}$$

$$-10\sqrt{x - 2} = 20 \quad \text{isolating } -10\sqrt{x - 2} \text{ and}$$
combining terms

$$\sqrt{x - 2} = -2 \quad \text{dividing by } -10$$

$$x - 2 = 4 \quad \text{equating squares of the members}$$

$$x = 6 \quad \text{adding 2 to each member}$$

Check

$$\sqrt{6 + 3} - \sqrt{6 - 2} = \sqrt{9} - \sqrt{4} \quad \text{replacing } x \text{ by 6}$$
in left member of Eq. (1)

$$= 3 - 2 = 1$$

Consequently, 6 is not a root of Eq. (1) since the right member of Eq. (1) is 5. Furthermore, we conclude that Eq. (1) has no roots since the set of roots of the rationalized equation includes the set of roots of the given equation.

EXERCISE 10.5
Radical Equations

Find the solution set of each of the following equations.

1 $\sqrt{x + 3} = \sqrt{5x - 1}$ 2 $\sqrt{x + 4} = \sqrt{2x - 1}$

3 $\sqrt{x + 6} = \sqrt{4x - 3}$ 4 $\sqrt{2x - 3} = \sqrt{x + 1}$

5 $\sqrt{8x - 1} - \sqrt{4x + 1} = 0$ 6 $\sqrt{3x + 6} - \sqrt{4x - 4} = 0$

7 $\sqrt{2x - 9} - \sqrt{x + 8} = 0$ 8 $\sqrt{2x + 2} = \sqrt{3x - 1}$

9 $\sqrt{x^2 - 5x + 1} - \sqrt{1 - 8x} = 0$ 10 $\sqrt{x^2 - 3x - 3} - \sqrt{3x + 4} = 0$

11 $\sqrt{x^2 + 3x - 3} - \sqrt{6x + 7} = 0$ 12 $\sqrt{2x^2 + 5x + 1} - \sqrt{2x + 6} = 0$

13 $x = \sqrt{x + 3} - 3$ 14 $2x = \sqrt{2x + 5} + 1$

15 $\sqrt{9 - 2x} + 3 = x$ 16 $2x = \sqrt{-2x + 5} - 1$

17 $\sqrt{5x^2 - 4x + 3} - x = 1$ 18 $3 + \sqrt{x^2 - 2x + 6} = 2x$

19 $\sqrt{2x^2 + 5x + 4} - 2x = 1$ 20 $\sqrt{2x^2 - 3x + 1} - 2x = -4$

21 $\sqrt{2x - 1} - \sqrt{x - 1} = 1$ 22 $\sqrt{2x + 3} - \sqrt{x + 1} = 1$

23 $\sqrt{2x - 7} - 1 = \sqrt{x - 4}$ 24 $\sqrt{5x + 1} - \sqrt{2x + 2} = 2$

25 $\sqrt{x + 3} + \sqrt{2x - 1} = \sqrt{7x + 2}$ 26 $\sqrt{3x + 1} - \sqrt{2x - 1} = \sqrt{x - 4}$

27 $\sqrt{3x + 4} + \sqrt{2x + 6} = \sqrt{x + 10}$ 28 $\sqrt{2x + 2} + \sqrt{2x - 5} = \sqrt{6x + 7}$

29 $\sqrt{3x + 4} - \sqrt{2x - 4} = \sqrt{3x - 8}$ 30 $\sqrt{7x + 8} = \sqrt{3x + 1} + \sqrt{x + 1}$

31 $\sqrt{2x - 1} + \sqrt{3x - 3} = \sqrt{10x - 9}$ 32 $\sqrt{x + 1} + \sqrt{3x - 5} = \sqrt{5x + 1}$

33 $\sqrt{x^2 + 3x - 1} - \sqrt{x^2 + x - 2} = 1$ 34 $\sqrt{x^2 - x - 4} - \sqrt{x^2 - 4x - 1} = 2$

35 $\sqrt{x^2 - x + 3} - \sqrt{x^2 - 3x + 4} = 1$ 36 $\sqrt{x^2 - 6x - 2} - \sqrt{x^2 - 7x - 2} = 1$

37 $\dfrac{\sqrt{2x+1}-1}{\sqrt{2x-4}-1} = 2$ **38** $\dfrac{\sqrt{2x-3}-2}{\sqrt{3x-2}-3} = 1$

39 $\dfrac{\sqrt{3x+4}+2}{\sqrt{x+5}-1} = 3$ **40** $\dfrac{\sqrt{3x-2}+1}{\sqrt{x+2}-1} = 3$

41 $\sqrt{bx+b^2} = \sqrt{bx-2b^2}+b,\ b>0$

42 $\sqrt{ax+3a^2} = \sqrt{8ax+a^2}-a,\ a>0$

43 $\sqrt{bx+2b^2} = \sqrt{3bx+3b^2}-b,\ b>0$

44 $\sqrt{3ax+a^2} - \sqrt{3ax-2a^2} = a,\ a>0$

10.10
EQUATIONS IN QUADRATIC FORM

An equation of the type

■ $a[f(x)]^2 + b[f(x)] + c = 0 \qquad a \neq 0$ (10.8)

is said to be in *quadratic form*. The symbol $f(x)$ stands for an expression in x, and it should be noted that this expression appears in both brackets. For example, the equation

$$4(x^2-x)^2 - 11(x^2-x) + 6 = 0$$

is in quadratic form since $x^2 - x$ appears in both parentheses.
  If we let

$$z = f(x)$$ (10.9)

and replace $f(x)$ by z in Eq. (10.8), we get

$$az^2 + bz + c = 0$$ (10.10)

which is a quadratic equation in z. We can solve Eq. (10.10) for z and then replace z in (10.9) by the roots of Eq. (10.10). If the equations thus obtained are linear or quadratic in x, we can solve each of them for x. We shall illustrate the method with two examples.

EXAMPLE 1 Solve the equation

$$4(x^2-x)^2 - 11(x^2-x) + 6 = 0$$ (1)

SOLUTION In order to solve Eq. (1), we shall let

$$x^2 - x = z$$ (2)

and then replace $x^2 - x$ by z in Eq. (1). Thus, we obtain

$$4z^2 - 11z + 6 = 0 \qquad\qquad\qquad\qquad (3)$$

which is a quadratic equation in z. We next solve Eq. (3) for z, replace z in Eq. (2) by each of the roots thus obtained, and solve the two resulting equations for x. The details of the solution follow.

$4(x^2 - x)^2 - 11(x^2 - x) + 6 = 0 \qquad$ given Eq. (1)

$\qquad\qquad 4z^2 - 11z + 6 = 0 \qquad$ replacing $x^2 - x$ by z in Eq. (1)

$z = \dfrac{-(-11) \pm \sqrt{(-11)^2 - 4 \cdot 4 \cdot 6}}{2 \cdot 4} \qquad$ by the quadratic formula

$\quad = \dfrac{11 \pm \sqrt{121 - 96}}{8}$

$\quad = \dfrac{11 \pm \sqrt{25}}{8}$

$\quad = \dfrac{11 \pm 5}{8}$

$\quad = 2 \quad$ and $\quad \frac{3}{4}$

$\qquad\qquad x^2 - x = 2 \qquad$ replacing z by 2 in Eq. (2)

$\qquad\quad x^2 - x - 2 = 0 \qquad$ adding -2 to each member

$\quad (x - 2)(x + 1) = 0 \qquad$ factoring left member

$\qquad\qquad\qquad x = 2 \qquad$ setting $x - 2 = 0$ and solving

$\qquad\qquad\quad x = -1 \qquad$ setting $x + 1 = 0$ and solving

Similarly, replacing z by $\frac{3}{4}$ in Eq. (2), we get

$$x^2 - x = \tfrac{3}{4}$$

which has the roots $\frac{3}{2}$ and $-\frac{1}{2}$.

Hence, the solution set of Eq. (1) is $\{2, -1, \frac{3}{2}, -\frac{1}{2}\}$.

Check

Value of x	Left Member of Eq. (1)
2	$4(4 - 2)^2 - 11(4 - 2) + 6 = 16 - 22 + 6 = 0$
-1	$4(1 + 1)^2 - 11(1 + 1) + 6 = 16 - 22 + 6 = 0$
$\frac{3}{2}$	$4(\frac{9}{4} - \frac{3}{2})^2 - 11(\frac{9}{4} - \frac{3}{2}) + 6 = 4(\frac{3}{4})^2 - 11(\frac{3}{4}) + 6$
	$\qquad\qquad = \frac{9}{4} - \frac{33}{4} + 6 = 0$
$-\frac{1}{2}$	$4(\frac{1}{4} + \frac{1}{2})^2 - 11(\frac{1}{4} + \frac{1}{2}) + 6 = 4(\frac{3}{4})^2 - 11(\frac{3}{4}) + 6$
	$\qquad\qquad = \frac{9}{4} - \frac{33}{4} + 6 = 0$

EXAMPLE 2

Solve the equation

$$3x^4 = 2x^2 + 1 \tag{1}$$

SOLUTION

$3x^4 = 2x^2 + 1$ given Eq. (1)

Let

$x^2 = z$

Then,

$3z^2 = 2z + 1$ replacing x^2 by z in Eq. (1)

$3z^2 - 2z - 1 = 0$ adding $-2z - 1$ to each member

$z = \dfrac{2 \pm \sqrt{4 + 12}}{6}$ by quadratic formula

$ = \dfrac{2 \pm 4}{6}$

$ = 1 \text{ and } -\tfrac{1}{3}$

$x^2 = 1$ replacing z by 1 in Eq. (2)

$x = \pm 1$ solving for x

$x^2 = -\tfrac{1}{3}$ replacing z by $-\tfrac{1}{3}$ in Eq. (2)

$x = \pm \sqrt{-\tfrac{1}{3}}$

$ = \pm \dfrac{1}{\sqrt{3}} i$ since $\sqrt{-\tfrac{1}{3}} = \sqrt{\tfrac{1}{3}(-1)} = \dfrac{1}{\sqrt{3}} i$

Therefore, the solution set of Eq. (1) is

$$\left\{ 1, -1, \frac{i}{\sqrt{3}}, -\frac{i}{\sqrt{3}} \right\}$$

Check

Value of x	Left Member of Eq. (1)	Right Member of Eq. (1)
1	3	$2 + 1 = 3$
-1	3	$2 + 1 = 3$
$\dfrac{1}{\sqrt{3}} i$	$3\left(\dfrac{1}{\sqrt{3}} i\right)^4 = \tfrac{1}{3}(i^2)^2$	$2\left(\dfrac{1}{\sqrt{3}} i\right)^2 + 1 = \tfrac{2}{3}(-1) + 1 = \tfrac{1}{3}$

$$= \tfrac{1}{3}(-1)^2 = \tfrac{1}{3}$$

$-\dfrac{1}{\sqrt{3}}i \qquad 3\left(-\dfrac{1}{\sqrt{3}}i\right)^4 = \tfrac{3}{9} = \tfrac{1}{3} \qquad 2(-\tfrac{1}{3}i)^2 + 1 = \tfrac{2}{3}(-1) + 1 = \tfrac{1}{3}$

Equations in Quadratic Form

Find the solution set of each of the following equations.

1 $x^4 - 5x^2 + 4 = 0$

2 $x^4 - 10x^2 + 9 = 0$

3 $x^4 - 20x^2 + 64 = 0$

4 $x^4 - 26x^2 + 25 = 0$

5 $x^4 = 5x^2 + 36$

6 $x^4 = 144 - 7x^2$

7 $x^4 - 21x^2 = 100$

8 $x^4 + 9x^2 = 400$

9 $4x^4 = 5x^2 - 1$

10 $4x^4 + 4 = 17x^2$

11 $36x^4 + 1 = 13x^2$

12 $36x^4 - 25x^2 + 4 = 0$

13 $x^4 = x^2 + 2$

14 $x^4 + x^2 = 12$

15 $2x^4 = 4 - 7x^2$

16 $27x^4 = 6x^2 + 1$

17 $x^{-2} - 5x^{-1} + 4 = 0$

18 $x^{-2} - 4x^{-1} + 3 = 0$

19 $x^{-4} - 10x^{-2} + 9 = 0$

20 $4x^{-4} - 17x^{-2} + 4 = 0$

21 $8x^6 + 7x^3 - 1 = 0$

22 $27x^6 - 19x^3 - 8 = 0$

23 $x^6 - 7x^3 - 8 = 0$

24 $x^6 - 35x^3 + 216 = 0$

25 $(x^2 + 6)^2 - 17(x^2 + 6) + 70 = 0$

26 $(x^2 - 2)^2 - 9(x^2 - 2) + 14 = 0$

27 $(x^2 + 3)^2 - 16(x^2 + 3) + 48 = 0$

28 $(x^2 - 5)^2 - 15(x^2 - 5) + 44 = 0$

29 $(2x^2 - 5)^2 - 39 = 10(2x^2 - 5)$

30 $(3x^2 - 8)^2 = 23(3x^2 - 8) - 76$

31 $(2x^2 + 1)^2 = 12(2x^2 + 1) - 27$

32 $(2x^2 - 3)^2 - 10(2x^2 - 3) = 75$

33 $(x^2 + x)^2 - 5(x^2 + x) + 6 = 0$

34 $(x^2 + x)^2 + 72 = 18(x^2 + x)$

35 $(x^2 + 2x)^2 = 8 + 7(x^2 + 2x)$

36 $(3x^2 + 5x)^2 + 24 = 14(3x^2 + 5x)$

37 $\left(\dfrac{x}{x+1}\right)^2 - 3\left(\dfrac{x}{x+1}\right) - 18 = 0$

38 $\left(\dfrac{x-3}{x+1}\right)^2 - 6\left(\dfrac{x-3}{x+1}\right) + 8 = 0$

39 $\left(\dfrac{x+7}{2x-1}\right)^2 + 3 = 4\left(\dfrac{x+7}{2x-1}\right)$

40 $3\left(\dfrac{2x-1}{1-3x}\right)^2 = 2\left(\dfrac{2x-1}{1-3x}\right) + 1$

41 $4\left(\dfrac{x-3}{2x+1}\right) - \left(\dfrac{2x+1}{x-3}\right) = 3$ HINT: Let $z = \dfrac{x-3}{2x+1}$; then $\dfrac{1}{z} = \dfrac{2x+1}{x-3}$

42 $\left(\dfrac{x-1}{x+3}\right) + 2 = 15\left(\dfrac{x+3}{x-1}\right)$

43 $\dfrac{2x-3}{x-1} - 4\left(\dfrac{x-1}{2x-3}\right) = 3$

44 $3\left(\dfrac{2x-1}{x+3}\right) + 2\left(\dfrac{x+3}{2x-1}\right) = 5$

45 $2x + 1 - 5\sqrt{2x+1} + 6 = 0$

46 $3x - 1 - 2\sqrt{3x-1} = 3$

47 $4(x^2 - 3) - 5\sqrt{x^2 - 3} + 1 = 0$

48 $2x^2 + 3x - 1 - 3\sqrt{2x^2 + 3x - 1} + 2 = 0$

49 $2x - 2 = -6\sqrt{2x+5}$ HINT: Add 7 to each member, and
then let $z = \sqrt{2x+5}$

50 $2x + 6 = 4\sqrt{2x+3}$

51 $6x - 1 = 5\sqrt{3x-2}$

52 $3x + 6 = 8\sqrt{x+3}$

10.11
QUADRATIC INEQUALITIES

An open sentence of the type

$$ax^2 + bx + c > 0 \qquad\qquad (1)$$

or

$$ax^2 + bx + c < 0 \qquad\qquad (2)$$

is a *quadratic inequality*.

The method we shall use to get the solution sets of such inequalities is based on the nature and the location of the graph of the equation

$$y = \pm ax^2 + bx + c \qquad a > 0 \qquad\qquad (3)$$

It is proved in analytic geometry that the graph of Eq. (3) is a parabola that opens upward if a is preceded by the plus sign and downward if a is preceded by the minus sign. These curves are illustrated in Figs. 10.1 and 10.2.

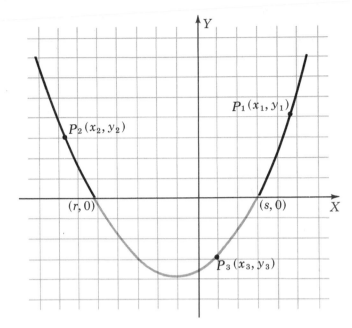

FIGURE 10.1

If $\{r, s\}$ is the solution set of the equation

$$ax^2 + bx + c = 0 \qquad\qquad (4)$$

the graph of $y = ax^2 + bx + c$ intersects the X axis at points $(r, 0)$ and $(s, 0)$. We shall assume that $r < s$. Then since the curve opens upward, it is above the X axis if $x < r$ or if $x > s$, and it is below it if $r < x < s$, as illustrated in Fig. 10.1. Hence, if $x_1 > s$ and $x_2 < r$, then $P_1(x_1, y_1)$ and $P_2(x_2, y_2)$ are on the portions of the graph above the X axis, and it follows that y_1 and y_2 are each greater than zero. Therefore,

$$ax_1{}^2 + bx_1 + c > 0 \qquad \text{if } x_1 > s$$

and

$$ax_2{}^2 + bx_2 + c > 0 \qquad \text{if } x_2 < r$$

Hence, the solution set of $ax^2 + bx + c > 0$ is

$$\{x \mid x < r\} \cup \{x \mid x > s\}$$

Likewise, if $r < x_3 < s$, then $P_3(x_3, y_3)$ is on the portion of the graph below the X axis, and therefore, $y_3 < 0$. Consequently, $ax_3{}^2 + bx_3 + c < 0$ if

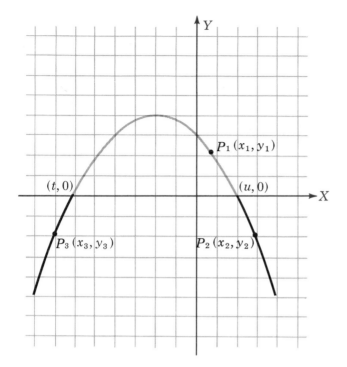

FIGURE 10.2

$r < x_3 < s$, and thus the solution set of

$$ax^2 + bx + c < 0 \quad \text{is} \quad \{x \mid r < x < s\}$$

Figure 10.2 shows the nature of the graph of

$$y = -ax^2 + bx + c \qquad a > 0$$

where $\{t, u\}$ is the solution set of $-ax^2 + bx + c = 0$. From the figure we see that if $t < x_1 < u$, then $P_1(x_1, y_1)$ is on the portion of the graph above the X axis and that $y_1 > 0$. Hence the solution set of

$$-ax^2 + bx + c > 0 \quad \text{is} \quad \{x \mid t < x < u\}$$

Similarly, if $x_2 > u$ or $x_3 < t$, then $P_2(x_2, y_2)$ and $P_3(x_3, y_3)$ are on the portion of the graph that is below the X axis and thus y_2 and y_3 are each less than zero. Therefore, the solution set of

$$-ax^2 + bx + c < 0 \quad \text{is} \quad \{x \mid x < t\} \cup \{x \mid x > u\}$$

EXAMPLE 1 Find the solution set of

$$x^2 > x + 6 \tag{5}$$

$$x^2 - x - 6 > 0 \tag{6}$$

that is equivalent to Eq. (5) by adding $-x - 6$ to each member. Since inequality (6) is of the same type as (1), we can apply the information derived in the preceding discussion. We first solve the equation

$$x^2 - x - 6 = 0$$

by the quadratic formula and get

$$x = \frac{1 \pm \sqrt{1 + 24}}{2}$$

$$= 3 \quad \text{and} \quad -2$$

The graph of

$$y = x^2 - x - 6$$

shown in Fig. 10.3 is a parabola opening upward, since the coefficient of x^2 is

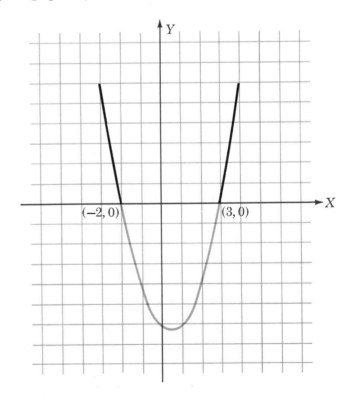

$(-2, 0)$ $(3, 0)$

FIGURE 10.3

positive, and the curve crosses the X axis at $(-2, 0)$ and $(3, 0)$. Furthermore, the curve is above the X axis at the left of $(-2, 0)$ and at the right of $(3, 0)$. Hence, $x^2 - x - 6 > 0$ if $x < -2$ or $x > 3$. Therefore, the required solution set is $\{x \mid x < -2\} \cup \{x \mid x > 3\}$.

EXAMPLE 2 Find the solution set of

$$-2x^2 - 5x > -3 \tag{7}$$

and of

$$-2x^2 - 5x < -3 \tag{8}$$

SOLUTION We add 3 to each member of Eqs. (7) and (8) and get the inequalities

$$-2x^2 - 5x + 3 > 0 \tag{9}$$

and

$$-2x^2 - 5x + 3 < 0 \tag{10}$$

that are equivalent to (7) and (8), respectively.

We now get the solution set of $-2x^2 - 5x + 3 = 0$ by the quadratic formula as follows:

$$x = \frac{5 \pm \sqrt{25 + 24}}{-4}$$

$$= \frac{5 \pm 7}{-4}$$

$$= -3 \quad \text{and} \quad \tfrac{1}{2}$$

Hence, since the coefficient of x^2 is negative, the graph of

$$y = -2x^2 - 5x + 3$$

is a parabola opening downward, and it crosses the X axis at $(-3, 0)$ and $(\tfrac{1}{2}, 0)$. Furthermore, as shown in Fig. 10.4, the graph is above the X axis if $-3 < x < \tfrac{1}{2}$, and below if $x < -3$ or $x > \tfrac{1}{2}$. Consequently, since $y > 0$ if the graph is above the X axis and $y < 0$ if the graph is below it, the solution set of inequality (9) is

$$S_1 = \{x \mid -3 < x < \tfrac{1}{2}\}$$

and the solution set of inequality (10) is

$$S_2 = \{x \mid x < -3\} \cup \{x \mid x > \tfrac{1}{2}\}$$

Therefore, since (9) and (10) are equivalent to (7) and (8), respectively, the solution set of (7) is S_1 and the solution set of (8) is S_2.

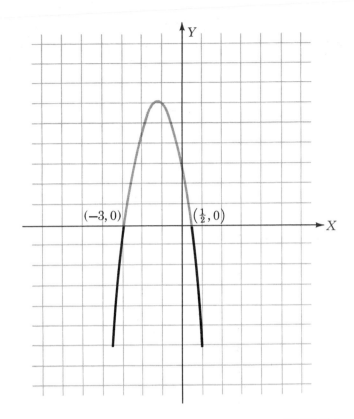

FIGURE 10.4

EXAMPLE 3 Find the solution set of

$$x^2 - 4x + 5 > 0 \qquad\qquad (11)$$

and of

$$x^2 - 4x + 5 < 0 \qquad\qquad (12)$$

SOLUTION Since the coefficient of x^2 is positive, the graph of

$$y = x^2 - 4x + 5 \qquad\qquad (13)$$

is a parabola that opens upward. However, if we solve the equation

$$x^2 - 4x + 5 = 0$$

by the quadratic formula, we have

$$x = \frac{4 \pm \sqrt{16 - 20}}{2}$$

$$= 2 \pm i$$

Since these values of x are imaginary, the graph does not cross the X axis and is therefore entirely above it. Consequently, the solution set of (11) is the set of all real numbers, and the solution set of (12) is the empty set $\varnothing$.

EXERCISE 10.7
Quadratic Inequalities

Find the solution set of each of the following inequalities.

1	$x^2 - x - 6 > 0$	**2**	$x^2 - 7x + 10 > 0$
3	$2x^2 - 3x - 2 > 0$	**4**	$3x^2 - 10x + 3 > 0$
5	$2x^2 + 3 < 5x$	**6**	$x^2 + 10x < -21$
7	$2x^2 + x < 3$	**8**	$2x^2 < x + 1$
9	$-2x^2 - 5x + 3 > 0$	**10**	$-3x^2 - 4x - 1 > 0$
11	$-3x^2 - 4x + 4 > 0$	**12**	$-2x^2 + x + 10 > 0$
13	$-6x^2 + x + 2 < 0$	**14**	$-2x^2 - 5x + 3 < 0$
15	$-4x^2 - 5x + 6 < 0$	**16**	$-5x^2 - 13x + 6 < 0$
17	$4x^2 + 9x - 20 > 2x - 2x^2$	**18**	$16x^2 + 3x < 4x^2 - 2x + 3$

19 $4x^2 - 9x - 5 < 10x^2 + 10x + 5$

20 $2x^2 - 2x + 3 > 5x^2 + 6x$

21 $2x^2 - x + 2 > x^2 + x + 1$

22 $3x^2 - 5x < 4x^2 - x + 4$

23 $2x^2 - 3x < x^2 + 3x - 9$

24 $3x^2 + 5x > 4x^2 - 3x + 16$

25 $x^2 - 4x + 5 > 0$

26 $-x^2 - 6x - 13 > 0$

27 $-2x^2 + 2x - 1 < 0$

28 $8x^2 - 4x + 1 < 0$

10.12
PROBLEMS THAT LEAD TO QUADRATIC EQUATIONS

Many stated problems, especially those which deal with products or quotients involving the unknown, lead to quadratic equations. The method of obtaining the equation for solving such problems is the same as that in Sec. 5.9, and the reader should review that article at this point. It should be noted here that often a problem which can be solved by the use of a quadratic equation has only one solution, while the equation has two solutions. In such cases the root which does not satisfy the conditions of the problem is discarded.

EXAMPLE 1

261

10.12 Problems
that Lead to
Quadratic
Equations

A rectangular building whose depth is twice its frontage is divided into two parts by a partition that is 30 feet from, and parallel to, the front wall. If the rear portion of the building contains 3500 square feet, find the dimensions of the building.

SOLUTION Let

x = the frontage of the building in feet

Then,

$2x$ = the depth in feet

Also,

$2x - 30$ = the length of the rear portion in feet   since partition
is 30 feet from
front wall

and

x = the width of the rear portion

Consequently,

$x(2x - 30)$ = the area of the rear portion in square feet   since area of a
rectangle is equal to
the product of
length and width

Therefore,

$x(2x - 30) = 3500$ since area of the rear portion is
3500 square feet

We solve this equation as follows:

$2x^2 - 30x - 3500 = 0$ performing indicated operations and
adding -3500 to each member

$(x - 50)(2x + 70) = 0$ factoring left member

$x = 50$ setting $x - 50 = 0$ and solving

$x = -35$ setting $2x - 70 = 0$ and solving

Since, however, no dimension of the building can be negative, we discard -35 and have

$x = 50$ frontage

$2x = 100$ depth

EXAMPLE 2

The periods of time required by two painters to paint a square yard of floor differ by 1 minute. Together they can paint 27 square yards in 1 hour. How long does it take each to paint 1 square yard?

SOLUTION   Let

x = the number of minutes required by the faster painter to paint 1 square yard

Then,

$x + 1$ = the number of minutes required by the other

Consequently,

$\dfrac{1}{x}$ = the fraction of a square yard the first man paints in 1 minute

and

$\dfrac{1}{x + 1}$ = the fraction of a square yard the other paints in 1 minute

Hence,

$\dfrac{1}{x} + \dfrac{1}{x + 1}$ = the fraction of a square yard painted by both men in 1 minute

However, since together they painted 27 square yards in 60 minutes, they covered $\frac{27}{60} = \frac{9}{20}$ square yard in 1 minute. Therefore,

$$\frac{1}{x} + \frac{1}{x + 1} = \frac{9}{20}$$

Solving this equation, we have

$20(x + 1) + 20x = 9x(x + 1)$ multiplying each member by $20x(x + 1)$

$20x + 20 + 20x = 9x^2 + 9x$ by distributive axiom

$-9x^2 + 31x + 20 = 0$ adding $-9x^2 - 9x$ to each member

$$x = \frac{-31 \pm \sqrt{(31)^2 - 4(-9)(20)}}{2(-9)}$$ by quadratic formula

$$= \frac{-31 \pm \sqrt{961 + 720}}{-18}$$

$$= \frac{-31 \pm \sqrt{1681}}{-18}$$

$$= \frac{-31 \pm 41}{-18}$$

$$= -\frac{5}{9} \quad \text{and} \quad 4$$

We discard $-\frac{5}{9}$, since a negative time has no meaning in this problem. Hence,

$x = 4$

and

$x + 1 = 5$

Thus, the painters require 4 and 5 minutes, respectively, to paint 1 square yard.

EXERCISE 10.8
Problems Solvable by Quadratic Equations

1 **Find two consecutive positive integers whose product exceeds their sum by 19.**

2 Find a number that is 56 less than its square.

3 Find two consecutive even integers whose product is 30 more than half the square of the larger.

4 Two numbers differ by 15, and their product is four times their sum. Find the numbers.

5 Find two integers that differ by 5, and the sum of whose squares is 25 times the smaller.

6 The sum of a number and its reciprocal is $\frac{13}{6}$. Find the number.

7 Two numbers differ by 5, and their reciprocals differ by $\frac{1}{10}$. Find the numbers.

8 The sum of two numbers is 17, and the sum of their squares is 169. Find the numbers.

9 The number of square yards in the area of a square exceeds the number of linear yards in the perimeter by 2300. Find the dimensions of the square.

10 The length and width of a rectangle differ by 5 ft, and the length of the diagonal is 25 ft. Find the dimensions.

11 A circular flower garden is surrounded by a walk 2 ft wide. If the combined areas of the walk and garden are 1.44 times the area of the garden, find the radius of the latter.

12 The length of a rectangular fish pond exceeds the width by 4 ft, and the pond is surrounded by a walk 4 ft wide. If the combined areas of the walk and the pond are 672 sq ft, find the dimensions of the pond.

13 Airfield A is 660 miles due north of B. A plane flew from A to B and returned in $6\frac{2}{3}$ hr of flying time. If the wind was blowing from the south at 20 miles per hour during both flights, find the airspeed of the plane.

14 Airfield B is 480 miles north of field A, and field C is 400 miles east of B. A pilot flew from A to B, delayed an hour, and then flew to C. If the wind blew from the north at 20 miles per hour during the flight from A to B, but changed to the west at 20 miles per hour during the delay, and the entire trip required 6 hr, find the airspeed of the plane.

15 A highway patrolman left his headquarters and cruised at a constant speed for 28 miles, and then was notified of an accident. He drove to the scene of the accident 8 miles away at a speed that was 45 miles per hour faster than his cruising speed. If he had been on duty 54 min when he reached the accident, find his cruising speed.

16 A rancher drove 100 miles to a city to accept delivery of a new car, and returned in the new car. His average speed to the city was 10 miles per hour more than his returning speed, and the entire trip required $3\frac{2}{3}$ hr of driving time. Find the speed on each part of the trip.

17 A carpenter and his helper can build a garage in $2\frac{2}{5}$ days. Each of the men working alone built similar garages, and the helper required 2 more days than the carpenter. How many days were required for each construction?

18 Two brothers washed the walls of their room in 3 hr. How long will it take each boy working alone to wash the walls of a similar room if the older boy can do the job in $2\frac{1}{2}$ hr less time than the younger?

19 A farmer and his helper, each driving a tractor, ploughed a tract of land in 6 days. Two years before, the helper ploughed the tract alone with the smaller tractor. The next year, the farmer ploughed the tract in 5 days less time with the larger tractor. How many days were required on each of the 2 years?

20 The outlet pipe drains a full irrigation reservoir in 2 hr less time than it takes the inlet pipe to fill it. One day, at the start of an irrigation job when the reservoir was full, the farmer opened both pipes, and the reservoir was empty at the end of 24 hr. In how many hours can the intake pipe refill the reservoir if the outlet is closed?

21 A square piece of linoleum was 4 ft too short to cover the floor of a rectangular room with an area of 192 sq ft. Find the dimensions of a supplementary piece necessary to complete the coverage.

22 A clothesline 35 ft long is stretched diagonally between the corners of a rectangular service yard. If the yard is enclosed by 98 ft of fencing, find its dimensions.

23 The outside dimensions of a framed picture are 20 by 18 in. Find the width of the frame if its area is one-fourth of the area enclosed by it.

24 A man built a garage door whose width exceeded its height by 11 ft, and in the construction he used 252 linear ft of lumber 6 in. wide. Find the dimensions of the door.

25 A builder used 54 linear ft of lumber to make the form for a concrete sidewalk 4 in. thick. If the walk contained 24 cu ft of concrete, find the dimensions.

26 A grocer sold two lots of eggs for $18 and $26, respectively, and there were 10 dozen more eggs in the latter than in the former. If the price per dozen of the second lot was 5 cents more than that of the first, and the price of each was more than 50 cents, how many dozen were in each lot?

27 The brothers in a family bought a used car for $1200 and shared equally in the cost. After 6 months one of the brothers left home and sold his share to the others for $240. If each of the remaining brothers' shares in this purchase was $220 less than his share in the original price, find the number of brothers in the family.

28 The driver for a car pool estimated that the cost of driving his car to a plant was $6, which was divided equally among the passengers, including himself. When two additional men joined the pool, the cost per person per trip was reduced by 50 cents, how many people, including the driver were in the original pool?

29 A man bought two apartment buildings for $48,000 apiece. There were 14 apartments in all, and the cost per apartment in one building was $2000 more than the cost in the other. How many apartments were in each building?

30 The expense of the annual club party is shared equally by the members attending. In 2 consecutive years the expense was $100 and $105, respectively, and the share of each member was 25 cents less the second year. How many members attended each year if the attendance the second year was 10 more than the first?

31 A swimming pool that holds 1800 cu ft of water can be drained at the rate of 15 cu ft per min faster than it can be filled. If it takes 20 min longer to fill it than to drain it, find the drainage rate.

32 A contractor who owned two power shovels with different capacities agreed to do three excavation jobs, each of which required moving the same number of cubit feet of soil. He started two of the jobs at the same time, using the larger shovel on the first and the smaller on the second. When the first job was finished, he started the third job with the larger machine. When the second job was finished 7 days later, he moved the smaller machine to the the third job, and with the two machines operating, the excavation was finished in 8 days. How many days were required for the first excavation?

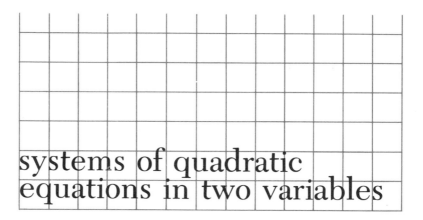

systems of quadratic equations in two variables

Chapter

11

General Form of a Quadratic Equation in Two Variables

The general quadratic equation in two variables is an equation of the type

$$Ax^2 + Bxy + Cy^2 + Dx + Ey + F = 0$$

where at least one of A, B, and C is not zero.

Conic Sections

It is proved in analytic geometry that the graph of a quadratic equation in two variables is a member of the important class of curves called *conic sections*, which includes the circle, the ellipse, the hyperbola, and the parabola (see Fig. 11.1). In the next two sections we shall discuss the graphs of special cases of a quadratic equation in two variables. In the remainder of the chapter we discuss methods for solving systems of two equations, at least one of which is a quadratic equation in two variables.

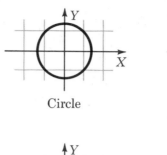

Circle   Ellipse

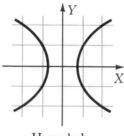

Hyperbola   Parabola

11.1
GRAPHS OF QUADRATIC EQUATIONS IN
TWO VARIABLES

We shall discuss the graphs of the following special cases of the general quadratic:

1   $x^2 + y^2 = r^2$
2   $ax^2 + by^2 = c$ a, b, and c positive
3   $ax^2 - by^2 = c$ a and b positive
4   $y = ax^2 + bx + c$ $a \neq 0$
5   $x = ay^2 + by + c$ $a \neq 0$

As in Chap. 6, the steps followed in constructing the graph are as follows:

1   Solve† the equation for y in terms of x.
2   Assign several values to x, compute each corresponding value of y, and arrange the associated pairs of values in tabular form.
3   Plot the points determined by the pairs of values, and then draw a smooth curve through them.

We shall illustrate the method with several examples.

†If the equation is easier to solve for x than for y, we solve it for x. Then in reading the succeeding steps, we interchange x and y.

Graphs of Equations of the Type $x^2 + y^2 = r^2$

EXAMPLE 1   Construct the graph of

$$x^2 + y^2 = 25 \qquad (1)$$

SOLUTION

1   If we perform the operations suggested by the steps listed above, we have

$$y = \pm \sqrt{25 - x^2}$$

2   Assign to x the integers from -5 to 5, inclusive, since that is the domain, and compute each corresponding value of y. For example, if $x = -5$, then

$$y = \pm \sqrt{25 - (-5)^2} = \pm \sqrt{25 - 25} = 0$$

Similarly, if $x = 2$, then

$$y = \pm \sqrt{25 - (2)^2} = \pm \sqrt{25 - 4} = \pm \sqrt{21} = \pm 4.6$$

When a similar computation is performed for each of the other values assigned to x and the results are arranged in tabular form, we have the following:

x	-5	-4	-3	-2	-1	0	1	2	3	4	5
y	0	± 3	± 4	± 4.6	± 4.9	± 5	± 4.9	± 4.6	± 4	± 3	0

3   Note that in this table we have two values of y for each x except $x = -5$ and $x = 5$. The pair of values $x = 3$, $y = \pm 4$ determines the two points $(3, 4)$ and $(3, -4)$.

With this understanding, if we plot the points determined by the table and join them by a smooth curve, we have the graph in Fig. 11.2.

We can readily see that the curve is a circle, since the coordinates (x, y) of any point P on it satisfy Eq. (1); that is, the sum of their squares is 25. Furthermore, by looking at the figure, we see that the square of the distance OP of P from the center is $x^2 + y^2$. Hence, any point whose coordinates satisfy Eq. (1) is at a distance of 5 from the origin.

In general, by similar reasoning, we conclude that the graph of $x^2 + y^2 = r^2$ is a circle of radius r, and the graph of $ax^2 + ay^2 = c$ is a circle of radius $\sqrt{c/a}$ if a and c have the same sign.

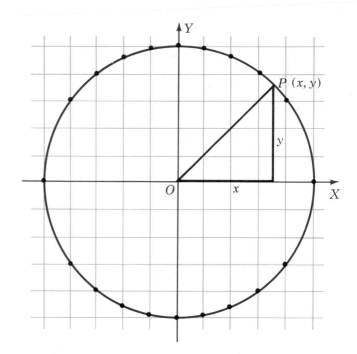

FIGURE 11.2

Equations of the Type $ax^2 + by^2 = c$

EXAMPLE 2

Construct the graph of $4x^2 + 9y^2 = 36$.

SOLUTION

1   Solving for y, we have

$$y = \pm\sqrt{\frac{36 - 4x^2}{9}}$$

$$= \pm\tfrac{2}{3}\sqrt{9 - x^2}$$

2   We note here that if $x^2 > 9$, the radicand is negative and y is imaginary. Hence, the graph exists only for values of x from -3 to 3, inclusive. Therefore, we assign to x the integers, $0, \pm 1, \pm 2, \pm 3$, compute each corresponding value of y, arrange the results in a table, and get

x	-3	-2	-1	0	1	2	3
y	0	± 1.5	± 1.9	± 2	± 1.9	± 1.5	0

269

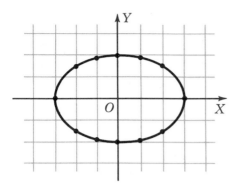

FIGURE 11.3

3   When we construct the graph determined by this table, we get the curve in Fig. 11.3.

By referring to Fig. 11.1, we surmise that this curve is an ellipse. The proof that the equation

$$ax^2 + by^2 = c$$

with a, b, and c positive always defines an ellipse is beyond the scope of this book. However, the statement is true, and it is helpful to remember this fact when dealing with such an equation.

Equations of the Type $ax^2 - by^2 = c$

EXAMPLE 3
Construct the graph of the equation $3x^2 - 4y^2 = 12$.

SOLUTION

1   Proceeding as before, we have

$$y = \pm\sqrt{\frac{3x^2 - 12}{4}}$$
$$= \pm\tfrac{1}{2}\sqrt{3(x^2 - 4)}$$

2   In this case, we notice that if $x^2 < 4$, the radicand is negative and y is imaginary. Hence, the graph does not exist between $x = -2$ and $x = 2$. If, however, x is either 2 or -2, y is zero. Thus, the curve must extend to the right from $(2, 0)$ and to the left from $(-2, 0)$. Hence, we assign the values ±2, ±3, ±4, ±5, ±7, ±9, to x, proceed as in the previous example, and get the following table:

x	-9	-7	-5	-4	-3	-2	2	3	4	5	7	9
y	±7.6	±5.8	±4	±3	±1.9	0	0	±1.9	±3	±4	±5.8	±7.6

3 When the above points are plotted and the graph is drawn, we obtain the curve in Fig. 11.4.

Again, by referring to Fig. 11.1, we conclude that this curve is a hyperbola.

This example illustrates the fact that if $a > 0$ and $b > 0$, an equation of the type

$$ax^2 - by^2 = c$$

defines a hyperbola. If c is positive, the curve is in the same general position as that in Fig. 11.4. However, if c is negative, the two branches of the curve cross the Y axis instead of the X axis and open upward and downward.

Equations of the Type $y = ax^2 + bx + c$ or $x = ay^2 + by + c$

EXAMPLE 4

Construct the graph of $x^2 - 4x - 4y - 4 = 0$.

SOLUTION By solving the equation

$$x^2 - 4x - 4y - 4 = 0$$

for y, we have

$$y = \tfrac{1}{4}x^2 - x - 1$$

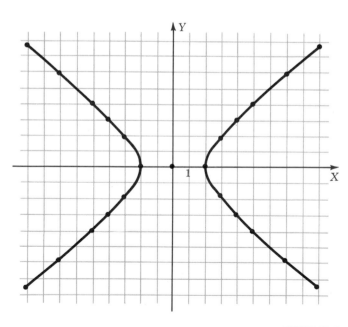

FIGURE 11.4

and this is the first type mentioned above. We avoid fractions here if we substitute only even values for x. If we use the values $-4, -2, 0, 2, 4, 6, 8$ for x and proceed as before, we get the following table of corresponding values of x and y:

x	-4	-2	0	2	4	6	8
y	7	2	-1	-2	-1	2	7

Plotting the above points and drawing the graph, we get the curve in Fig. 11.5.

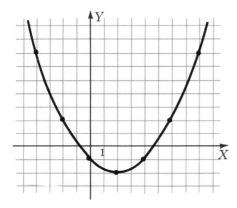

FIGURE 11.5

EXAMPLE 5

Construct the graph of $2y^2 + 1 = x + 4y$.

SOLUTION Since this equation contains only one term in x, the algebra is easier if we solve for x in terms of y and get

$$x = 2y^2 - 4y + 1$$

Now we assign values to y and compute each corresponding value of x. The following table was obtained by using the values $-2, -1, 0, 1, 2, 3, 4$ for y:

x	17	7	1	-1	1	7	17
y	-2	-1	0	1	2	3	4

Now we plot the graph and obtain the curve in Fig. 11.6.

The curves in Figs. 11.5 and 11.6 are parabolas. It is proved in analytic geometry that an equation of the type

$$y = ax^2 + bx + c$$

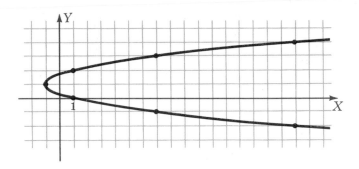

FIGURE 11.6

defines a parabola opening upward if a is positive and a parabola opening downward if a is negative. Furthermore, an equation of the type

$$x = ay^2 + by + c$$

defines a parabola opening to the right if a is positive and a parabola opening to the left if a is negative.

11.2
GRAPHICAL SOLUTION OF TWO QUADRATIC EQUATIONS IN TWO VARIABLES

A simultaneous solution set of a system of two quadratic equations in two variables is the set of all pairs of corresponding values of x and y that satisfies both equations. Now the coordinates of each point on the graph of an equation satisfy the equation. Therefore, if the graphs of two quadratic equations in two unknowns are constructed with respect to the same axes, the coordinates of each point of intersection is a simultaneous solution pair of the equations, since these points are on both graphs. Consequently, to obtain the simultaneous solution set of a system of two quadratic equations in two variables graphically, we construct the graphs of the equations with respect to the same axes and then estimate the coordinates of their points of intersection.

EXAMPLE
Obtain the simultaneous solution set of the system of equations

$$y = x^2 - 4 \tag{1}$$

$$9x^2 + 25y^2 = 225 \tag{2}$$

by using the graphical method.

SOLUTION   We construct the graph of Eq. (1) by use of the following table:

x	0	±1	±2	±3	±4
y	−4	−3	0	5	12

We thus obtain the parabola in Fig. 11.7.

The graph of Eq. (2) is an ellipse, and we obtain this graph by use of the following table:

x	−5	−4	−2	−1	0	1	2	4	5
y	0	±1.8	±2.7	±2.9	±3	±2.9	±2.7	±1.8	0

The ellipse is also shown in Fig. 11.7.

The two graphs intersect at the points whose coordinates to one decimal place appear to be $(2.6, 2.5)$, $(−2.6, 2.5)$, $(1, −2.9)$, and $(−1, −2.9)$. Therefore, the simultaneous solution set of (1) and (2) is $\{(2.6, 2.5), (−2.6, 2.5), (1, −2.9), (−1, −2.9)\}$ where the elements are accurate to one decimal place.

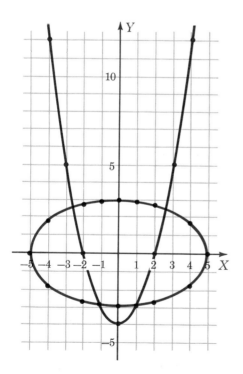

FIGURE 11.7

Graphical Methods

Construct the graph of the equation in each of Probs. 1 to 20.

1 $y = x^2 - 3x - 4$ **2** $y = \frac{1}{2}x^2 + 2x - 1$

3 $x = y^2 - 2y - 4$ **4** $x = \frac{1}{3}y^2 - 5y + 3$

5 $x^2 + y^2 = 36$ **6** $x^2 + y^2 = 49$

7 $x^2 - y^2 = 25$ **8** $x^2 - y^2 = 16$

9 $9x^2 + y^2 = 36$ **10** $4x^2 + 25y^2 = 100$

11 $9x^2 + 16y^2 = 144$ **12** $9x^2 + 4y^2 = 324$

13 $2x^2 - 9y^2 = 18$ **14** $9x^2 - 4y^2 = -36$

15 $4x^2 - 25y^2 = -400$ **16** $x^2 - 4y^2 = 64$

17 $3x + xy = 12$ **18** $4x^2 + 3xy = 9$

19 $2x^2 - 3xy = 4$ **20** $xy = 15$

By the graphical method, find the solution set of the pair of equations in each of Probs. 21 to 36.

21 $3x - y = 1$
    $y = 2x^2 - 5$

22 $x^2 + 3y = 7$
    $y = 2x + 4$

23 $x^2 + 4y^2 = 25$
    $x - 2y = -1$

24 $4x^2 + y^2 = 64$
    $3x - y = 3$

25 $2y = x^2 - 12$
    $x = y^2$

26 $y = x^2 - 2$
    $y^2 = 2x$

27 $y = x^2 - x - 1$
    $(y - 3)^2 = 2x$

28 $y = x^2 - 2x$
    $x = -y^2 + 4y$

29 $x^2 + y^2 = 169$
    $y^2 = 5x - 35$

30 $4x^2 - 5y^2 = -16$
    $y^2 = 4x$

31 $3x^2 + 2y^2 = 35$
    $y^2 = 2(x + 5)$

32 $x^2 - 3y^2 = -11$
    $x^2 = 2(y + 5)$

33 $2x^2 + y^2 = 11$
    $9x^2 - 4y^2 = -27$

34 $4x^2 + y^2 = 100$
    $12x^2 - 5y^2 = 12$

35 $4x^2 + y^2 = 36$
    $x^2 + y^2 = 16$

36 $25x^2 + 4y^2 = 100$
    $4x^2 + 9y^2 = 36$

11.3
ALGEBRAIC SOLUTIONS

Solution Set

As we stated in Sec. 11.2, a simultaneous solution of a system of two quadratic equations in two variables is an ordered pair of numbers (x, y) that satisfies each equation. We shall call the set of pairs of numbers that satisfy both equations in the system the *solution set* of the system. As the example in Sec. 11.2 illustrates, there are usually four pairs in the solution set of a system if both equations are quadratic. If, however, one equation is linear and the other is quadratic, the graphs of the equations are a straight line and one of the conic sections. Since these two graphs can intersect in, at most, two points, there are at most two pairs of numbers in the solution set.

Eliminating a Variable

The general method for solving a system of equations in two variables at least one of which is quadratic is the same as that used in Chap. 7. The first step in the method is to combine the equations in such a way as to obtain one equation in one variable each of whose roots is one member of a pair in the solution set. This process is called *eliminating a variable*. After one number in each pair of the solution set is obtained, the other is determined by substitution. Frequently, the elimination of a variable yields an equation of the fourth degree. In such cases the completion of the process of solving requires more advanced methods than those presented in this book. We shall confine our discussion and exercises, however, to problems that can be solved by the available methods. In the remainder of this chapter we shall discuss the algebraic methods generally used.

11.4
ELIMINATION BY SUBSTITUTION

If in a system of two equations in two variables, one equation can be solved for one variable in terms of the other, this variable can be eliminated by substitution. We shall assume that the variables are x and y, that one equation can be solved for y in terms of x, and that the solution is in the form $y = f(x)$. We then replace y in the other equation by $f(x)$ and obtain an equation including x only. We shall designate this equation by $F(x) = 0$. Next, we solve $F(x) = 0$ for x, and the roots obtained will be the first number in each pair of numbers in the solution set. We complete the process by substituting each root of $F(x) = 0$ in $y = f(x)$ and thus obtain the corresponding values of y. Finally, we arrange the corresponding values of x and y in pairs in this way, $(x_1, y_1), (x_2, y_2), (x_3, y_3), (x_4, y_4)$, and thus obtain the solution set.

We shall illustrate the method with three examples, the first of which shows the process of solving a system containing *a linear equation in two variables and a quadratic equation in two variables*. The second and third illustrate the method for solving a system of *two quadratic equations in two variables where one equation is easily solvable for one variable in terms of the other.*

EXAMPLE 1 Solve the following system of equations:

$$x^2 + 2y^2 = 54 \tag{1}$$

$$2x - y = -9 \tag{2}$$

SOLUTION We first solve Eq. (2) for y and get

$$y = 2x + 9 \tag{3}$$

Next, we replace y in Eq. (1) by $2x + 9$ and get

$$x^2 + 2(2x + 9)^2 = 54 \tag{4}$$

which we solve as follows:

$$x^2 + 2(4x^2 + 36x + 81) = 54 \quad \text{squaring } 2x + 9$$

$$x^2 + 8x^2 + 72x + 162 = 54 \quad \text{by distributive axiom}$$

$$x^2 + 8x^2 + 72x + 162 - 54 = 0 \quad \text{adding } -54 \text{ to each member}$$

$$9x^2 + 72x + 108 = 0 \quad \text{combining similar terms}$$

$$x^2 + 8x + 12 = 0 \quad \text{dividing by 9}$$

$$(x + 6)(x + 2) = 0 \quad \text{factoring}$$

$$x = -6 \quad \text{setting } x + 6 = 0 \text{ and solving}$$

$$x = -2 \quad \text{setting } x + 2 = 0 \text{ and solving}$$

We now replace x in Eq. (3) by -6 and get

$$y = 2(-6) + 9$$

$$= -12 + 9$$

$$= -3$$

Similarly, replacing x in Eq. (3) by -2, we have

$$y = 2(-2) + 9$$

$$= -4 + 9$$

$$= 5$$

Therefore, the solution set of the given system is $\{(-6, -3), (-2, 5)\}$.

The final step is to verify that each of the above pairs of numbers satisfies the given equations. This verification is as follows.

Verification

Replacement for (x, y)	Equation Number	Left Member	Right Member
$(-6, -3)$	(1)	$(-6)^2 + 2(-3)^2$ $= 36 + 18 = 54$	54
	(2)	$2(-6) - (-3)$ $= -12 + 3 = -9$	-9
$(-2, 5)$	(1)	$(-2)^2 + 2(5^2)$ $= 4 + 50 = 54$	54
	(2)	$2(-2) - 5$ $= -4 - 5 = -9$	-9

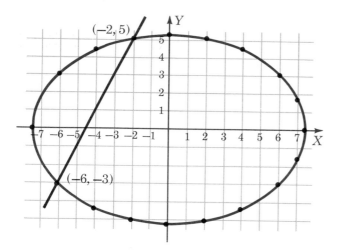

FIGURE 11.8

The graphs of Eqs. (1) and (2) together with the geometrical significance of the solution set are shown in Fig. 11.8.

EXAMPLE 2   Solve the following system of equations:

$$4x^2 - 2xy - y^2 = -5 \qquad (1)$$

$$y + 1 = -x^2 - x \qquad (2)$$

SOLUTION   We first solve Eq. (2) for y and get

$$y = -x^2 - x - 1 \qquad (3)$$

Next, we substitute the right member of Eq. (3) for y in Eq. (1) and have

$$4x^2 - 2x(-x^2 - x - 1) - (-x^2 - x - 1)^2 = -5$$

By performing the indicated operations in the last equation and then adding 5 to each member of the resulting equation, we get

$$-x^4 + 3x^2 + 4 = 0 \qquad (4)$$

This is an equation in quadratic form, and we solve it as follows:

$x^4 - 3x^2 - 4 = 0$	dividing each member of Eq. (4) by -1	(5)
$(x^2 - 4)(x^2 + 1) = 0$	factoring left member of Eq. (5)	
$x^2 - 4 = 0$	setting first factor equal to zero	
$x^2 = 4$	adding 4 to each member	
$x = \pm 2$	solving for x	

$$x^2 + 1 = 0 \qquad \text{setting second factor equal to zero}$$

$$x^2 = -1 \qquad \text{adding} -1 \text{ to each member}$$

$$x = \pm\sqrt{-1} \qquad \text{solving for } x$$

$$x = \pm i \qquad \text{since, by the definition in Sec. 10.2, } i = \sqrt{-1}$$

Finally, we substitute each of the above values of x in Eq. (3) and obtain the corresponding value of y. This procedure yields the following results.

$$y = -2^2 - 2 - 1 \qquad \text{replacing } x \text{ by 2 in Eq. (3)}$$

$$= -4 - 2 - 1$$

$$= -7$$

$$y = -(-2)^2 - (-2) - 1 \qquad \text{replacing } x \text{ by } -2 \text{ in Eq. (3)}$$

$$= -4 + 2 - 1$$

$$= -3$$

$$y = -i^2 - i - 1 \qquad \text{replacing } x \text{ by } i \text{ in Eq. (3)}$$

$$= -(-1) - i - 1 \qquad \text{since } i^2 = -1$$

$$= 1 - i - 1$$

$$= -i$$

$$y = -(-i)^2 - (-i) - 1 \qquad \text{replacing } x \text{ by } -i \text{ in Eq. (3)}$$

$$= -(-1) + i - 1 \qquad \text{since } (-i)^2 = i^2 = -1$$

$$= 1 + i - 1$$

$$= i$$

Consequently, the possible solution set is $\{(2, -7), (-2, -3), (i, -i), (-i, i)\}$.

Check

Replacement for (x, y)	Equation Number	Left Member	Right Member
$(2, -7)$	(1)	$4(2^2) - 2(2)(-7) - (-7)^2$ $= 16 + 28 - 49 = -5$	-5
	(2)	$-7 + 1 = -6$	$-2^2 - 2 = -4 - 2 = -6$
$(-2, -3)$	(1)	$4(-2)^2 - 2(-2)(-3) - (-3)^2$ $= 16 - 12 - 9 = -5$	-5
	(2)	$-3 + 1 = -2$	$-(-2)^2 - (-2)$ $= -4 + 2 = -2$

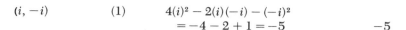

$$(i, -i) \qquad (1) \qquad \begin{aligned} &4(i)^2 - 2(i)(-i) - (-i)^2 \\ &= -4 - 2 + 1 = -5 \end{aligned} \qquad\qquad -5$$

$$\qquad\qquad\qquad (2) \qquad -i + 1 \qquad\qquad\qquad -i^2 - i = 1 - i$$

$$(-i, i) \qquad (1) \qquad \begin{aligned} &4(-i)^2 - 2(-i)(i) - (-i)^2 \\ &= -4 - 2 + 1 = -5 \end{aligned} \qquad\qquad -5$$

$$\qquad\qquad\qquad (2) \qquad i + 1 \qquad\qquad\qquad \begin{aligned} &-(-i)^2 - (-i) \\ &= 1 + i \end{aligned}$$

Since the coordinates of any point in a Cartesian plane are real numbers, $(i, -i)$ and $(-i, i)$ do not represent points on the graph of either equation, although they satisfy each equation. The graphs of Eqs. (1) and (2) are shown in Fig. 11.9, and in the figure we see that $(2, -7)$ and $(-2, -3)$ are the points of intersection of the graphs, but $(i, -i)$ and $(-i, i)$ are not points on either graph.

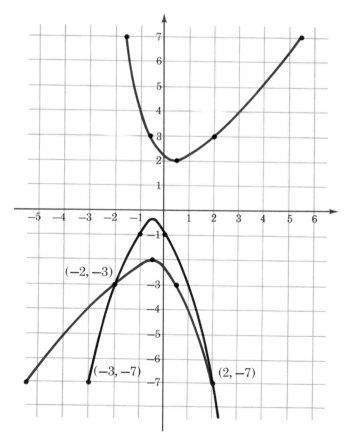

FIGURE 11.9

EXAMPLE 3 Solve the following system of equations by substitution:

$$xy = 2 \tag{1}$$

$$15x^2 + 4y^2 = 64 \tag{2}$$

SOLUTION Equation (1) can be solved easily for y in terms of x, and the solution is

$$y = \frac{2}{x} \tag{3}$$

We may now eliminate y by replacing it with $2/x$ in Eq. (2) and thus obtain

$$15x^2 + 4\left(\frac{2}{x}\right)^2 = 64 \tag{4}$$

which we solve for x as follows:

$$15x^2 + 4\left(\frac{4}{x^2}\right) = 64 \qquad \text{squaring } 2/x$$

$$15x^4 + 16 = 64x^2 \qquad \text{multiplying each member by } x^2$$

$$15x^4 - 64x^2 + 16 = 0 \qquad \text{adding } -64x^2 \text{ to each member and arranging terms}$$

$$(x^2 - 4)(15x^2 - 4) = 0 \qquad \text{factoring}$$

$$x^2 - 4 = 0 \qquad \text{setting first factor equal to zero}$$

$$x^2 = 4$$

$$x = \pm 2$$

$$15x^2 - 4 = 0 \qquad \text{setting second factor equal to zero}$$

$$15x^2 = 4$$

$$x^2 = \tfrac{4}{15}$$

$$x = \pm \frac{2}{\sqrt{15}}$$

$$= \pm \frac{2\sqrt{15}}{15}$$

We next substitute each of the above values for x in Eq. (3) and thereby obtain the corresponding value of y. This procedure yields:

with $x = 2$, $\qquad y = \tfrac{2}{2} = 1$

with $x = -2$, $\qquad y = \tfrac{2}{-2} = -1$

with $x = \dfrac{2\sqrt{15}}{15}$, $y = \sqrt{15}$

with $x = -\dfrac{2\sqrt{15}}{15}$, $y = -\sqrt{15}$

Consequently, the solution set is

$$\left\{ (2,\,1),\, (-2,\,-1),\, \left(\frac{2\sqrt{15}}{15},\, \sqrt{15}\right),\, \left(-\frac{2\sqrt{15}}{15},\, -\sqrt{15}\right) \right\}$$

The solution set may be checked by replacing x and y in Eqs. (1) and (2) by the appropriate member of each pair of numbers in the solution set, as done in Examples 1 and 2.

The graphs of Eqs. (1) and (2) are shown in Fig. 11.10.

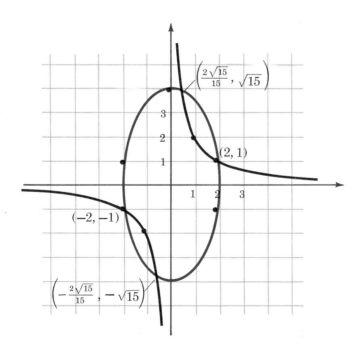

FIGURE 11.10

Elimination by Substitution

Find the simultaneous solution set of each of the following problems by the method of substitution.

1 $y^2 = 4x$
$x + y = 3$

2 $x^2 = 9y$
$x - y = 2$

3 $y^2 = 2x$
$y = x$

4 $y^2 = x$
$y + 2x = 1$

5 $x^2 + y^2 = 5$
$y = x - 1$

6 $x^2 + y^2 = 13$
$2x + y = 1$

7 $4x^2 + 4y^2 = 5$
$2x = 2y + 1$

8 $x^2 + 4y^2 = 4$
$y = x - 2$

9 $2x - y = -1$
$16x^2 - 3y^2 = -11$

10 $8x^2 + 5y^2 = 77$
$4x - y = 11$

11 $3x^2 + 16y^2 = 91$
$x + 8y = 13$

12 $3x^2 - 2y^2 = 25$
$x + 2y = 1$

13 $3x^2 - 16y^2 = 11$
$3x - 8y = 1$

14 $3x^2 - 20y^2 = 28$
$3x - 10y = 2$

15 $5x^2 + 9y^2 = 6$
$2x - 3y = 1$

16 $9x^2 + 4y^2 = 5$
$3x + 6y = -1$

17 $2x^2 - 3y^2 = -a^2$
$2x + y = 3a$

18 $ax^2 - by^2 = ab^2 - a^2 b$
$ax - by = 0$

19 $m^2 x^2 + y^2 = b^2$
$y = mx + b$

20 $x^2 + a^2 y^2 = 10b^2$
$x + ay = 2b$

21 $x^2 - y = 3$
$x^2 + 2x - 3y = 5$

22 $2x^2 - y = 3$
$x^2 + 3x - 2y = 6$

23 $x^2 + y = 11$
$x^2 - x - 2y = 2$

24 $3x^2 + y = 0$
$x^2 + 2x + 2y = -3$

25 $x^2 + 20xy + y^2 = 100$
$x^2 - 10x = y$

26 $x^2 - 2xy + y^2 = 144$
$y = 3x^2 + x$

27 $x^2 - 2xy + y^2 = 1$
$x + 2 = y^2 + y$

28 $y^2 + 2y + 4xy + x + x^2 = -6$
$x = y^2 - 2y - 2$

29 $x^2 - 8xy - 4y + x = -14$
$x = y^2 + 4y$

30 $x^2 + 2xy + y + x = 3$
$x = 2y^2 - y$

31 $x^2 + 2xy + y + x = 0$
$x = 4y^2 - y$

32 $4x^2 - 8xy + 3y - 3x = 5$
$x = -y^2 + y$

33 $x^2 + 3y^2 = 7$
$xy = 2$

34 $2x^2 + 3y^2 = 21$
$xy = 3$

35 $3x^2 - 4y^2 = 32$
$xy = -8$

36 $3x^2 - y^2 = 8$
$xy = -4$

37 $x^2 + 2xy - y^2 = 1$
$xy = 2$

38 $4x^2 - 6xy + y^2 = 1$
$2xy = 3$

39 $2x^2 + 2xy + y^2 = 17$
$xy = -5$

40 $3x^2 - 3xy + 3y^2 = 1$
$9xy = 2$

11.5

ELIMINATION BY ADDITION OR SUBTRACTION

In this section we shall discuss two classes of pairs of equations in which the first step is to eliminate one of the variables by addition or subtraction. We justify the method by the following argument:

If $x = p$ and $y = q$ satisfy each of the equations

$$Ax^2 + Bxy + Cy^2 = F \tag{11.1}$$

$$ax^2 + bxy + cy^2 = f \tag{11.2}$$

then (p, q) also satisfies the equation

$$m(Ax^2 + Bxy + Cy^2) \pm n(ax^2 + bxy + cy^2) = mF \pm nf \tag{11.3}$$

This statement follows from the fact that since (p, q) satisfies (11.1) and (11.2), we have

$$m(Ap^2 + Bpq + Cq^2) = mF \qquad \text{and} \qquad n(ap^2 + bpq + nq^2) = nf$$

Consequently, if (x, y) is replaced by (p, q), the left member of (11.3) is equal to $mF \pm nf$, and therefore (p, q) satisfies the equation. The object, then, is to choose first the variable to be eliminated. Then, if possible, we shall so choose m and n that (11.3) will not contain that variable.

Two Equations of the Type $Ax^2 + Cy^2 = F$

We solve two equations of the type $Ax^2 + Cy^2 = F$ simultaneously by first eliminating one of the variables by addition or subtraction and then solving the resulting equation for the remaining variable. The value of the other variable is then found by substitution.

EXAMPLE 1 Solve the following system of equations:

$$2x^2 + 3y^2 = 21 \tag{1}$$

$$3x^2 - 4y^2 = 23 \tag{2}$$

SOLUTION

$$
\begin{array}{lll}
8x^2 + 12y^2 = 84 & \text{multiplying Eq. (1) by 4} & (3) \\
9x^2 - 12y^2 = 69 & \text{multiplying Eq. (2) by 3} & (4) \\
\hline
17x^2 = 153 & \text{adding corresponding members of Eqs. (3) and (4)} & \\
x^2 = 9 & \text{dividing by 17} & \\
x = \pm 3 & &
\end{array}
$$

If either 3 or -3 is substituted for x in Eq. (1), we have

$$18 + 3y^2 = 21$$
$$3y^2 = 3$$
$$y^2 = 1$$
$$y = \pm 1$$

Therefore, if x is equal to either 3 or -3, then y is equal to ± 1. Hence, the solution set is

$$\{(3, 1), (3, -1), (-3, 1), (-3, -1)\}$$

Figure 11.11 shows the graph of the two given equations and the coordinates of the points of intersection.

Two Equations of the Type $Ax^2 + Cy^2 + Dx = F$ or $Ax^2 + Cy^2 + Ey = F$

The first step in solving simultaneously two equations of the type $Ax^2 + Cy^2 + Dx = F$ is to eliminate y^2 by addition or subtraction. Then the process of solving can be completed by the method illustrated in Example 2.

EXAMPLE 2

Solve the following system of equations:

$$3x^2 - 2y^2 - 6x = -23 \tag{1}$$
$$x^2 + y^2 - 4x = 13 \tag{2}$$

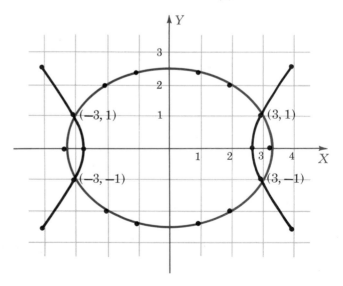

FIGURE 11.11

SOLUTION Since each of the given equations contains y^2 and no other term involving y, we shall eliminate y^2 and complete the solution as follows:

$$3x^2 - 2y^2 - 6x = -23 \qquad \text{Eq. (1) copied} \qquad (1)$$
$$\underline{2x^2 + 2y^2 - 8x = 26} \qquad \text{multiplying Eq. (2) by 2} \qquad (3)$$
$$5x^2 - 14x = 3 \qquad \begin{array}{l}\text{adding corresponding members of Eqs.(1)}\\ \text{and (3)}\end{array}$$

$$5x^2 - 14x - 3 = 0 \qquad \text{adding } -3 \text{ to each member}$$

$$x = \frac{14 \pm \sqrt{196 + 60}}{10} \qquad \text{by quadratic formula}$$

$$= \frac{14 \pm \sqrt{256}}{10}$$

$$= \frac{14 \pm 16}{10}$$

$$= 3 \quad \text{and} \quad -\tfrac{1}{5}$$

$$9 + y^2 - 4(3) = 13 \qquad \text{replacing } x \text{ by 3 in Eq. (2)}$$

$$9 + y^2 - 12 = 13$$

$$y^2 = 13 + 3 \quad \text{adding 3 to each member}$$

$$y^2 = 16$$

$$y = \pm 4$$

Hence, if $x = 3$, then $y = \pm 4$.

$$(-\tfrac{1}{5})^2 + y^2 - 4(-\tfrac{1}{5}) = 13 \qquad \text{replacing } x \text{ by } -\tfrac{1}{5} \text{ in Eq. (2)}$$

$$\tfrac{1}{25} + y^2 + \tfrac{4}{5} = 13 \qquad \text{since } (-\tfrac{1}{5})^2 = \tfrac{1}{25} \text{ and } -4(-\tfrac{1}{5}) = \tfrac{4}{5}$$

$$1 + 25y^2 + 20 = 325 \qquad \text{multiplying each member by 25}$$

$$25y^2 = 325 - 21 \quad \text{adding } -21 \text{ to each member}$$

$$25y^2 = 304$$

$$y^2 = \tfrac{304}{25}$$

$$y = \pm \sqrt{\frac{304}{25}} = \pm \frac{4\sqrt{19}}{5}$$

Consequently, if $x = -\tfrac{1}{5}$, then

$$y = \pm \frac{4\sqrt{19}}{5}$$

Therefore, the solution set is

$$\left\{ (3, 4), (3, -4), \left(-\frac{1}{5}, \frac{4\sqrt{19}}{5}\right), \left(-\frac{1}{5}, -\frac{4\sqrt{19}}{5}\right) \right\}$$

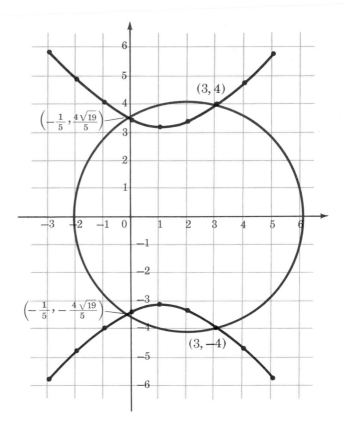

FIGURE 11.12

The graphs of Eqs. (1) and (2) are shown in Fig 11.12.

EXERCISE 11.3
Elimination by Addition or Subtraction

Solve the following pairs of equations simultaneously.

1 $3x^2 + y^2 = 12$
$x^2 + 2y^2 = 19$

2 $x^2 - 3y^2 = 13$
$3x^2 - y^2 = 71$

3 $x^2 + y^2 = 5$
$2x^2 - 3y^2 = 5$

4 $x^2 + 4y^2 = 45$
$3x^2 - y^2 = 18$

5 $2x^2 + 4y^2 = 19$
$3x^2 - 8y^2 = 25$

6 $3x^2 - 5y^2 = 70$
$2x^2 + y^2 = 51$

7 $3x^2 + 2y^2 = 14$
$2x^2 + 3y^2 = 11$

8 $2x^2 + 5y^2 = 53$
$4x^2 + 3y^2 = 43$

9 $3x^2 - 5y^2 = -8$
$2x^2 - y^2 = 4$

10 $4x^2 - 3y^2 = -39$
$5x^2 - 4y^2 = -55$

11 $2x^2 + 5y^2 = 70$
$3x^2 - 7y^2 = 47$

12 $6x^2 + 5y^2 = 59$
$4x^2 - 3y^2 = 33$

13 $x^2 + 4y^2 - 2x = 1$
$3x^2 + 8y^2 - 6x = 2$

14 $9x^2 - 8y^2 - 21x = -14$
$x^2 + 3y^2 - 14x = -21$

15 $28x^2 + 12y^2 - 24x = 7$
$6x^2 + 4y^2 - 3x = 4$

16 $28x^2 - 15y^2 - 8x = -12$
$44x^2 - 23y^2 - 16x = -20$

17 $3x^2 + y^2 - 8x = 0$
$5x^2 + 3y^2 - 4x = 24$

18 $2x^2 - 7x + 2y^2 = 5$
$5x^2 - 22x + 4y^2 = -5$

19 $2x^2 + 24x - y^2 = 18$
$8x^2 + 151x - 6y^2 = 111$

20 $3x^2 + 98x - 3y^2 = 65$
$9x^2 + 191x - 6y^2 = 128$

21 $3x^2 - 2y^2 - 2y = -12$
$4x^2 - 5y^2 + 16y = 19$

22 $3y^2 - 9x^2 + 19y = 13$
$-3y^2 + 6x^2 - 11y = -8$

23 $125x^2 + 100y^2 - 75y = -11$
$125x^2 + 75y^2 - 50y = -17$

24 $9y^2 - 9x^2 - 3y = -1$
$27y^2 - 18x^2 - 15y = -4$

25 $x^2 + 3xy + 3x = -8$
$3x^2 + xy + x = 8$

26 $3x^2 - 5xy + x = 4$
$2x^2 - 4xy + x = 2$

27 $y^2 + 2xy - y = 4$
$2y^2 - xy + 3y = 3$

28 $5y^2 + 5xy + 4y = -18$
$48y^2 + 25xy - 3y = 48$

29 $4x^2 + 5xy + 2x = 0$
$2x^2 + xy + 3x = 3$

30 $x^2 - 2xy - x = 12$
$x^2 - xy - 3x = 3$

31 $3y^2 - 5xy + y = 4$
$2y^2 - 4xy + y = 2$

32 $y^2 + 3xy - 3y = -5$
$y^2 + 2xy - 2y = -2$

11.6
ELIMINATION BY A COMBINATION OF METHODS

Frequently, the computation involved in solving the system of equations

$$F(x, y) = K \tag{11.4}$$

$$f(x, y) = k \tag{11.5}$$

by substitution is very tedious. In such cases it may be more efficient to use the method explained in this section.

The method depends upon the fact that the solution set of the system composed of (11.4) and (11.5) is the same as the solution set of the system composed of either (11.4) or (11.5) and the equation

$$mF(x, y) \pm nf(x, y) = mK \pm nk \tag{11.6}$$

In order to prove this statement, we shall first show that any pair of numbers in the solution set of (11.4) and (11.5) is also in the solution set of (11.4) and (11.6) and that, conversely, any pair of numbers in the solution set of (11.4) and (11.6) is in the solution set of (11.4) and (11.5). For this purpose we shall assume that (p, q) satisfies (11.4) and (11.5). Consequently, $F(p, q) = K$ and $f(p, q) = k$, and therefore we have $mF(p, q) \pm nf(p, q) = mK \pm nk$. Therefore, (p, q) satisfies (11.6).

To prove the converse, we shall assume that (p, q) satisfies (11.4) and (11.6). Then we have

$$mF(p, q) \pm nf(p, q) = mK \pm nk \qquad \text{since } (p, q) \text{ satisfies (11.6)}$$

and

$$F(p, q) = K \qquad \text{since } (p, q) \text{ satisfies (11.4)}$$

Now, replacing $F(p, q)$ by K in the above equation, we get

$$mK \pm nf(p, q) = mK \pm nk$$

Finally, we add $-mK$ to each member of the above equation, divide each member of the result by $\pm n$, and have

$$f(p, q) = k$$

Hence, (p, q) satisfies Eq. (11.5).

By use of a similar argument we can show that if (p, q) satisfies (11.5) and (11.6), it also satisfies (11.4).

Now if we can so determine m and n in (11.6) that the resulting equation is easily solvable for either variable in terms of the other, we can obtain the solution set of (11.4) and (11.5) by solving (11.6) with either (11.4) or (11.5) by substitution. We shall discuss two classes of systems of quadratic equations in two variables that can be solved by this method.

Two Equations of the Type $Ax^2 + Bxy + Cy^2 = D$

The first step in solving a system of two equations of the type $Ax^2 + Bxy + Cy^2 = D$ is to eliminate the constant terms by addition or subtraction. That is, we combine the two given equations in such a way as to obtain an equivalent equation in which the constant term is zero. We then solve this equation for one variable in terms of the other and finally complete the process by substitution. For example, if we solve the equation obtained by eliminating the constants for y in terms of x, we usually obtain the two equations $y = rx$ and $y = tx$ where r and t are constants. We then replace y in one of the given equations successively by rx and tx and find each corresponding value of x. We shall illustrate the process with the following example.

EXAMPLE 1 Solve the following system of equations:

$$3x^2 - 4xy + 2y^2 = 3 \tag{1}$$

$$2x^2 - 6xy + y^2 = -6 \tag{2}$$

SOLUTION If we multiply Eq. (1) by 2 and add each member of the resulting equation to the corresponding member of (2), we obtain an equation in which the constant term is zero. Then the process of solving can be completed as now indicated:

$$6x^2 - 8xy + 4y^2 = 6 \qquad \text{multiplying Eq. (1) by 2} \qquad (3)$$
$$\underline{2x^2 - 6xy + y^2 = -6 \qquad \text{Eq. (2) copied}} \qquad (2)$$
$$8x^2 - 14xy + 5y^2 = 0 \qquad (4)$$

We now solve Eq. (4) for y in terms of x, using the quadratic formula with $a = 5$, $b = -14x$, and $c = 8x^2$, and obtain

$$y = \frac{14x \pm \sqrt{196x^2 - 160x^2}}{10}$$

$$= \frac{14x \pm \sqrt{36x^2}}{10}$$

$$= \frac{14x \pm 6x}{10} = \frac{20x}{10} \quad \text{and} \quad \frac{8x}{10}$$

Consequently,

$$y = 2x \qquad (5)$$

and

$$y = \frac{4x}{5} \qquad (6)$$

We continue the solving process by replacing y in either Eq. (1) or Eq. (2) separately by $2x$ and $4x/5$ and solving the resulting equations for x. Using Eq. (1) and $y = 2x$, we get

$$3x^2 - 4x(2x) + 2(2x)^2 = 3$$
$$3x^2 - 8x^2 + 8x^2 = 3 \qquad \text{performing indicated operations}$$
$$3x^2 = 3 \qquad \text{combining similar terms}$$
$$x^2 = 1$$
$$x = \pm 1$$

Now we replace x in Eq. (5) by ± 1 and get $y = 2(\pm 1) = \pm 2$. Hence, two number pairs of the solution set are $(1, 2)$ and $(-1, -2)$. We continue the process by replacing y by $4x/5$ in Eq. (1) and solving the resulting equation for x. Thus, we obtain

$$3x^2 - 4x\left(\frac{4x}{5}\right) + 2\left(\frac{4x}{5}\right)^2 = 3$$

$$3x^2 - \frac{16x^2}{5} + \frac{32x^2}{25} = 3 \qquad \text{performing indicated operations}$$

$$75x^2 - 80x^2 + 32x^2 = 75 \qquad \text{multiplying each member by 25}$$

$$27x^2 = 75$$

$$x^2 = \frac{75}{27} = \frac{25}{9}$$

$$x = \pm\frac{5}{3}$$

Finally, we replace x in Eq. (6) by $\pm\frac{5}{3}$ and obtain $y = (\frac{4}{5})(\pm\frac{5}{3}) = \pm\frac{4}{3}$. Therefore, two additional members of the solution set are $(\frac{5}{3}, \frac{4}{3})$ and $(-\frac{5}{3}, -\frac{4}{3})$. Consequently, the complete solution set is

$$\{(1, 2), (-1, -2), (\tfrac{5}{3}, \tfrac{4}{3}), (-\tfrac{5}{3}, -\tfrac{4}{3})\}$$

The graphs of Eqs. (1) and (2) together with the coordinates of their points of intersection are shown in Fig. 11.13.

If the constant term in either of the given equations is zero, as in the system

$$2x^2 - 3xy - 2y^2 = 0$$

$$x^2 + 2xy + 5y^2 = 17$$

it is unnecessary to perform the first step in Example 1. We proceed directly to solve the first equation for y in terms of x, or for x in terms of y, and then complete the process by the method of substitution.

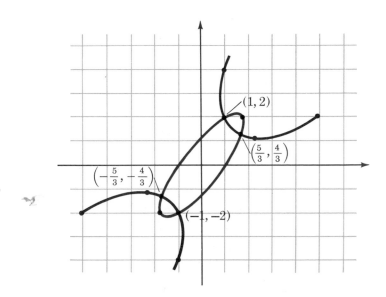

FIGURE 11.13

Two Equations of the Type $Ax^2 + Ay^2 + Dx + Ey = F$

If each of the given equations is of the type $Ax^2 + Ay^2 + Dx + Ey = F$, we can eliminate the second-degree terms by addition or subtraction and obtain a linear equation in x and y. We can then solve this equation with one of the given equations by the method illustrated by Example 1 of Sec. 11.4. We shall illustrate the method with the following example.

EXAMPLE 2 Solve the following system of equations:

$$3x^2 + 3y^2 + \ x - 2y = 20 \tag{1}$$

$$2x^2 + 2y^2 + 5x + 3y = 9 \tag{2}$$

SOLUTION

$6x^2 + 6y^2 + \ 2x - \ 4y = 40$	multiplying Eq. (1) by 2	(3)
$3x^2 + 6y^2 + 15x + \ 9y = 27$	multiplying Eq. (2) by 3	(4)
$\qquad\quad - 13x - 13y = 13$	subtracting corresponding members	(5)
	of Eqs. (3) and (4)	

It is proved in analytic geometry that the graphs of Eqs. (1) and (2) are circles and that the graph of Eq. (5) is a straight line that passes through the intersections, if any, of the two circles. Therefore, we can obtain the coordinates of the points of intersection of Eqs. (1) and (2) by solving either of them simultaneously with (5).

$$y = -x - 1 \qquad \text{solving Eq. (5) for } y \tag{6}$$

$$2x^2 + 2(-x - 1)^2 + 5x + 3(-x - 1) = 9 \qquad \text{replacing } y \text{ by } -x - 1 \text{ in Eq. (2)}$$

$$2x^2 + 2x^2 + 4x + 2 + 5x - 3x - 3 - 9 = 0 \qquad \text{performing indicated operations and adding } -9 \text{ to each member}$$

$$x^2(2 + 2) + x(4 + 5 - 3) + 2 - 3 - 9 = 0 \qquad \text{by commutative and distributive axioms}$$

$$4x^2 + 6x - 10 = 0 \qquad \text{combining terms}$$

$$2x^2 + 3x - 5 = 0 \qquad \text{dividing by 2}$$

$$(x - 1)(2x + 5) = 0 \qquad \text{factoring}$$

$$x - 1 = 0$$

$$x = 1$$

$$2x + 5 = 0$$

$$x = -\frac{5}{2}$$

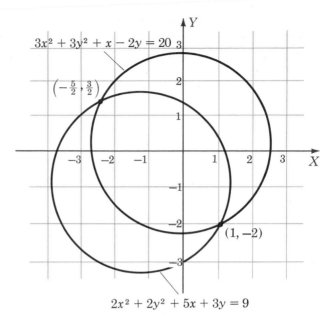

$3x^2 + 3y^2 + x - 2y = 20$

$\left(-\frac{5}{2}, \frac{3}{2}\right)$

$(1, -2)$

$2x^2 + 2y^2 + 5x + 3y = 9$

FIGURE 11.14

We complete the solving process by replacing x in Eq. (6) successively by 1 and by $-\frac{5}{2}$ and then computing each corresponding value of y. Thus, we obtain

$y = -1 - 1 = -2$ replacing x by 1 in Eq. (6)

$y = -(-\frac{5}{2}) - 1 = \frac{3}{2}$ replacing x by $-\frac{5}{2}$ in Eq. (6)

Consequently, the solution set is

$\{(1, -2), (-\frac{5}{2}, \frac{3}{2})\}$

The solution set can be checked by replacing (x, y) successively by $(1, -2)$ and $(-\frac{5}{2}, \frac{3}{2})$ in Eqs. (1) and (2).

The graphs of Eqs. (1) and (2) together with their points of intersection are shown in Fig. 11.14.

11.7
SYMMETRIC EQUATIONS

An equation in two variables is symmetric if the equation is not altered when the variables are interchanged. For example, the equation

$$Ax^2 + Bxy + Ay^2 + Dx + Dy = F \tag{11.7}$$

is symmetric since interchanging x and y does not alter the equation.

If we transform (11.7) by means of the **equations**

$$x = u + v$$
$$y = u - v \tag{11.8}$$

we get

$$(2A + B)u^2 + (2A - B)v^2 + 2Du = F \tag{11.9}$$

Therefore, the transformation (11.8) will convert two symmetric equations into two equations of the type of (11.9), and then the method of Example 2, Sec. 11.5, can be used. We shall illustrate the procedure with the following example:

EXAMPLE Solve the following system of equations:

$$7x^2 + 2xy + 7y^2 - 8x - 8y = 108 \tag{1}$$

$$3x^2 - 2xy + 3y^2 + 4x + 4y = 68 \tag{2}$$

SOLUTION We first replace x by $u + v$ and y by $u - v$, divide each resulting equation by the common factor of its terms, and get

$$4u^2 + 3v^2 - 4u = 27 \qquad \text{from Eq. (1)} \tag{3}$$

$$u^2 + 2v^2 + 2u = 17 \qquad \text{from Eq. (2)} \tag{4}$$

Since each of these equations contains one term including v^2 and no other term with v, we can eliminate v^2 by addition or subtraction and complete the procedure of solving as follows:

$$\begin{array}{ll} 8u^2 + 6v^2 - \ 8u = 54 & \text{multiplying Eq. (3) by 2} \tag{5} \\ 3u^2 + 6v^2 + \ 6u = 51 & \text{multiplying Eq. (4) by 3} \tag{6} \\ \hline 5u^2 \qquad\quad - 14u = \ 3 & \text{Eq. (5) minus Eq. (6)} \tag{7} \end{array}$$

$$u = \frac{14 \pm \sqrt{196 + 60}}{10} \qquad \text{solving (7) by quadratic formula}$$

$$u = \frac{14 \pm 16}{10} = 3 \quad \text{and} \quad -\tfrac{1}{5}$$

We now obtain the corresponding values of v by replacing u in Eq. (4) by each of the above values as now shown:

$$9 + 2v^2 + 6 = 17 \qquad \text{replacing } u \text{ by 3 in Eq. (4)}$$

$$2v^2 = 2$$

$$v^2 = 1$$

$$v = \pm 1$$

Therefore, if $u = 3$, then $v = \pm 1$.

Now we replace u by 3 and v by ± 1 in Eqs. (11.8) and get

$$\cdot = 3 \pm 1 = 4 \qquad \text{and} \qquad 2$$

$$y = 3 \mp 1 = 2 \qquad \text{and} \qquad 4$$

Consequently, if $x = 4$, then $y = 2$, and if $x = 2$, then $y = 4$.

If we replace u by $-\frac{1}{5}$ in Eq. (4), we get

$$\tfrac{1}{25} + 2v^2 - \tfrac{2}{5} = 17$$

$$1 + 50v^2 - 10 = 425 \qquad \text{multiplying each member by 25}$$

$$50v^2 = 434$$

$$v^2 = \frac{434}{50} = \frac{217}{25}$$

$$v = \pm \frac{\sqrt{217}}{5}$$

Hence, if $u = -\frac{1}{5}$, then $v = \pm\sqrt{217}/5$.

In order to obtain the values of x and y, we replace u by $-\frac{1}{5}$ and v by $\pm\sqrt{217}/5$ in Eq. (11.8) and get

$$x = \frac{-1 \pm \sqrt{217}}{5}$$

$$y = \frac{-1 \mp \sqrt{217}}{5}$$

Therefore, if

$$x = \frac{-1 + \sqrt{217}}{5}$$

then

$$y = \frac{-1 - \sqrt{217}}{5}$$

and if

$$x = \frac{-1 - \sqrt{217}}{5}$$

then

$$y = \frac{-1 + \sqrt{217}}{5}$$

Consequently, the solution set is

$$\left\{ (4, 2), (2, 4), \left(\frac{-1 + \sqrt{217}}{5}, \frac{-1 - \sqrt{217}}{5} \right), \left(\frac{-1 - \sqrt{217}}{5}, \frac{-1 + \sqrt{217}}{5} \right) \right\}$$

EXERCISE 11.4

Elimination by Addition or Subtraction and Substitution; Symmetric Equations

Find the solution set of the equations in each of the following problems.

1 $2x^2 - xy - 3y^2 = 0$
$x^2 - 2xy + 2y^2 = 5$

2 $3x^2 - 4xy + y^2 = 0$
$4x^2 - 3xy - 3y^2 = -32$

3 $x^2 - xy - 2y^2 = 0$
$5x^2 + 3xy + y^2 = 27$

4 $2x^2 - 3xy - 5y^2 = 0$
$6x^2 - 3xy - 2y^2 = 112$

5 $2x^2 + xy + y^2 = 16$
$5x^2 + 4xy - y^2 = 32$

6 $3x^2 - 3xy + 4y^2 = 40$
$5x^2 - 3xy + 12y^2 = 80$

7 $x^2 - 2xy + y^2 = 36$
$3x^2 - 2xy + 5y^2 = 90$

8 $4x^2 - 4xy + y^2 = 16$
$6x^2 - 11xy + 9y^2 = 64$

9 $2x^2 + 2xy + y^2 = 5$
$4x^2 + xy + 11y^2 = 25$

10 $x^2 - xy + 3y^2 = 15$
$6x^2 - 5xy + 9y^2 = 60$

11 $6x^2 + xy - 12y^2 = 112$
$2x^2 + 17xy + y^2 = 224$

12 $x^2 + 3xy - 2y^2 = 16$
$6x^2 + 13xy - 13y^2 = 80$

13 $2x^2 + 11xy - 9y^2 = -36$
$4x^2 + 17xy - 14y^2 = -54$

14 $8x^2 + 9xy + 7y^2 = -54$
$9x^2 + 17xy + 23y^2 = -135$

15 $5x^2 - 11xy + 14y^2 = 240$
$3x^2 - 18xy + 24y^2 = 360$

16 $2x^2 + 26xy - 24y^2 = -36$
$19x^2 + 23xy - 32y^2 = -90$

17 $x^2 + y^2 + 2x = 19$
$x^2 + y^2 - 2y = 9$

18 $x^2 + y^2 - 3x = 7$
$2x^2 + 2y^2 + y = 23$

19 $x^2 + y^2 + 6x = 52$
$x^2 + y^2 - 6y = 4$

20 $x^2 + y^2 + 6x = 65$
$x^2 + y^2 - 6y = 11$

21 $x^2 + y^2 + 5x + y = 26$
$x^2 + y^2 + 6x + 2y = 31$

22 $x^2 + y^2 - 5x - 11y = -24$
$x^2 + y^2 - 8x - 2y = -12$

23 $x^2 + y^2 - 3x + y = 4$
$2x^2 + 2y^2 - 5x + y = 13$

24 $3x^2 + 3y^2 + x - y = 124$
$x^2 + y^2 + 5x + 2y = 74$

25 $5x^2 + 5y^2 - 16x + 4y = 25$
$3x^2 + 3y^2 - 13x - y = -2$

26 $2x^2 + 2y^2 - 15x + 4y = 23$
$3x^2 + 3y^2 + x + 53y = 340$

27 $2x^2 + 2y^2 + 5x + 4y = 81$
$7x^2 + 7y^2 + x + 3y = 190$

28 $2x^2 + 2y^2 - 5x + y = 52$
$5x^2 + 5y^2 - 11x + 5y = 144$

29 $2x^2 - xy + 2y^2 - 3x - 3y = 5$
$x^2 - 2xy + y^2 + 3x + 3y = 16$

30 $x^2 - 3xy + y^2 + 3x + 3y = 20$
$3x^2 - 4xy + 3y^2 - 6x - 6y = 5$

31 $2x^2 + 2xy + 2y^2 - x - y = 22$
$3x^2 + 2xy + 3y^2 + 4x + 4y = 52$

32 $x^2 + 2xy + y^2 + x + y = 2$
$5x^2 + 6xy + 5y^2 + x + y = 30$

33 $4x^2 - 5xy + 4y^2 - 6x - 6y = -8$
$2x^2 - 3xy + 2y^2 - 2x - 2y = -2$

34 $4x^2 + 2xy + 4y^2 - 6x - 6y = 118$
$5x^2 + 2xy + 5y^2 - 3x - 3y = 176$

35 $2x^2 + xy + 2y^2 - 5x - 5y = 57$
$x^2 - xy + y^2 + 7x + 7y = 84$

36 $5x^2 + 3xy + 5y^2 + 4x + 4y = 211$
$2x^2 + 3xy + 2y^2 + 7x + 7y = 25$

37 $7x^2 + 9xy + 7y^2 - 15x - 15y = 73$
$5x^2 + 8xy + 5y^2 - 6x - 6y = 38$

38 $8x^2 + 6xy + 8y^2 - 17x - 17y = 256$
    $3x^2 + 2xy + 3y^2 - 4x - 4y = 112$

39 $22x^2 - 5xy + 22y^2 + 9x + 9y = 601$
    $9x^2 - 3xy + 9y^2 + 6x + 6y = 255$

40 $17x^2 - 6xy + 17y^2 + 8x + 8y = 489$
    $5x^2 - 2xy + 5y^2 + 3x + 3y = 146$

EXERCISE 11.5
Problems Solvable by Simultaneous Quadratics

1 Two numbers differ by 4, and the difference of their squares is 6 times the larger number. Find the numbers.

2 The difference of the squares of two positive numbers is twice their sum, and their difference is one-third of the smaller number. Find the numbers.

3 The sum of the squares of two positive numbers is 10 times the larger number, and the difference of their squares is 8 times the larger. Find the numbers.

4 The product of two positive numbers is 45, and the sum of their squares is 106. Find the numbers.

5 A ladder 20 ft long leans against the wall of a house and just touches the top of a fence that is parallel to the wall and is 6 ft high. If the foot of the ladder is $4\frac{1}{2}$ ft from the fence, find the distance from the ground to the top of the ladder and from the wall to the foot of the ladder.

6 A man 6 ft tall and a boy 4 ft tall are standing on opposite sides of a lamp post, and each is 6 ft from it. Find the height of the post and the lengths of their shadows if that of the man is twice that of the boy.

7 The cost for building a rectangular vat with a square base was $115.20. The materials for the base and the sides cost 25 cents per sq ft and 20 cents per sq ft, respectively. Find the dimensions of the vat if the sum of the areas of the base and the sides is 512 sq ft.

8 Airfield B is 495 miles north of field A. A pilot flew from A to B and returned in 5 hr. On the same day a second plane whose airspeed was 5 miles per hour more than the first flew from A to B in $2\frac{1}{5}$ hr. If the wind blew from the south at a constant velocity during the day, find the airspeed of the first plane and the velocity of the wind.

9 A cotton buyer declined an offer of $17,500 for a consignment of baled cotton. Two months later, when the price had increased by $5 per bale, he sold the consignment for $17,640. If in the meantime, 2 bales had been destroyed by fire, find the number of bales in the original consignment and the price per bale of the original offer.

10 What are the dimensions of a rectangle if its area is 540 sq ft and its diagonal is 39 ft long?

11 A small rectangular park that has an area of 15,000 sq ft is surrounded by a concrete walk 4 ft wide. Find the dimensions of the park if the walk has an area of 2064 sq ft.

12 The combined areas of two externally tangent circles is 25π sq in. Find the radius of each circle if the centers are 7 in. apart.

13 A rectangle of area 60 sq ft is inscribed in a circle of area $169\pi/4$ sq ft. Find the dimensions of the rectangle.

14 Mrs. Simmons had a square carpet and a rectangular carpet of which the length was $\frac{3}{2}$ the width. The areas of the two carpets combined were 375 sq ft. The square carpet cost $10 a sq yd, and the rectangular one $12 a sq yd. If she paid $50 more for the square carpet than for the rectangular one, what were the dimensions of each?

15 A rectangle is constructed in a right triangle by selecting a point on the hypotenuse and drawing perpendiculars to the legs. Find the dimensions of the rectangle if the area is 36 sq in. and the legs of the triangle are 16 and 12 in.

16 A florist expected to earn $1950 from the sale of corsages at a football game. At half time he had sold 500 corsages. He then reduced the price by 50 cents apiece and sold the rest by the end of the game. If his earnings during the second half were $375, how many corsages did he have, and what was his first price?

17 A block of stock was bought for $7800. After holding it for a year, the buyer received a cash dividend of $1.50 per share and a stock dividend of 10 shares. The stock was then sold for $3 per share more than it cost. Find the number of shares bought and the cost per share if the profit on the transaction was $1220.

18 Two circles are tangent to each other, and the smaller circle is inside the larger. The lengths of the radii differ by 3 in., and the area between the circles is 66 sq in. Find the lengths of the radii. (Use $\frac{22}{7}$ for π.)

19 A piece of wire 76 in. in length is cut into two pieces. One piece is bent into a square and the other into a circle. If the sum of the areas of the two figures is 218 sq in., find the length of the side of the square and the radius of the circle.

20 A man made two investments, the first being $500 more than the second. If, during the first year, the rate of the second investment was 2 percent more than that of the first and the income from each was $120, find the amount of the first investment and the rate that it earned.

21 A farmer owned a square field and a rectangular field of which the length was twice the width, and the sum of the areas of the fields was 60 acres. He sold the square field for $200 per acre, and the other for $150 per acre, and received $5000 more for the former than the latter. Find the dimensions of the two fields. (One acre contains 160 sq rods.)

22 A farmer and his son can paint a barn in $2\frac{2}{5}$ days. How long would it take each to paint the barn if the son requires 2 more days than his father?

23 The sides of a race track are parallel straight lines and the ends are semicircles. The length of the outer boundary of the track is 1760 yd, and the area enclosed by it is 107,800 sq yd. Find the length of each side and the radius of each semicircle.

24 The lengths of the northern and western boundaries of a rectangular field are 120 rods and 180 rods, respectively. An irrigation well is 50 rods from the northwest corner and 170 rods from the southeast corner. Find the distances from the well to the northern boundary and to the western boundary.

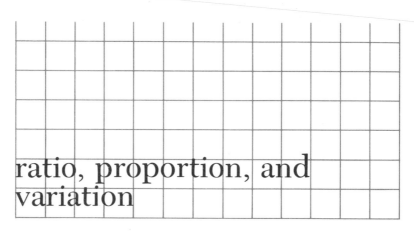

ratio, proportion, and variation

Chapter

12

In education circles one frequently hears the term *IQ*, or *intelligence quotient*. In a machine shop one may hear references to the *gear ratio*. Highway engineers are interested in the *grades* of highways, and carpenters continually deal with the *pitch* of a roof. Each of the italicized words is the quotient of two numbers and is the measure of some particular thing. For example, a person's IQ is a measure of his mental ability, and both the grade of a highway and the pitch of a roof are measures of a slope. In this chapter we discuss ratios, and we shall be especially interested in pairs of numbers that vary in such a way that their ratios or products never change.

12.1
RATIO

The *ratio* of the number a to the number b, $b \neq 0$, is defined as the quotient a/b which can also be written as

$$a \div b, \qquad \frac{a}{b}, \qquad \text{and} \qquad a{:}b$$

and represents the portion of a that corresponds to 1 unit of b. Thus, 60 cents/12 eggs indicates that one egg costs 60/12 cents $= 5$ cents, and $5700/10 months discloses the rate of $570 per month.

If a and b are measurements of the same kind, they are ordinarily expressed in terms of the same unit before a ratio is formed. Thus, if we want the ratio of 3 feet to 9 inches we must express 3 feet in terms of inches or 9 inches in terms of feet before forming their ratio. If we do the former, we have 36 inches/9 inches $= 4$.

12.2
PROPORTION

In applying ratios to problems, we often encounter situations in which two ratios are equal. The statement that two ratios are equal is called a *proportion*. Thus if a/b and c/d are equal, we write

$$\frac{a}{b} = \frac{c}{d} \tag{12.1}$$

or

$$a{:}b = c{:}d \tag{12.1'}$$

and read the proportion as a divided by b is equal to c divided by d, or a is to b as c is to d. The former is more generally used, and in this form a proportion is in reality a fractional equation or an equation that involves fractions.

Extremes and Means

Regardless of which of these forms is used, we say that a and d are the *extremes* and b and c are the *means*.

We shall now present several properties of proportions that facilitate work with proportions and their applications.

Product of Means and Extremes

If we multiply each member of (12.1) by bd, we get

$$\frac{abd}{b} = \frac{cbd}{d}$$

and, by dividing the members of each fraction by their common factor, we see that

$$\blacksquare \quad ad = bc \tag{12.2}$$

This can be expressed in words as follows:

In any proportion the product of the extremes is equal to the product of the means.

EXAMPLE 1

Find x if $x/15 = 2/5$

SOLUTION   If we apply (12.2) to the given equation, we have

$5x = (15)(2) = 30$

$\quad\quad x = 6$ **dividing each member by 5**

Alternation and Inversion

If we divide each member of (12.2) by cd, we get

$$\frac{ad}{cd} = \frac{bc}{cd}$$

Now, reducing the fraction in each member to lowest terms, we see that

■ $\dfrac{a}{c} = \dfrac{b}{d}$ $\qquad\qquad\qquad\qquad\qquad\qquad$ (12.3)

If we divide each member of (12.2) by ac and then reduce each fraction to lowest terms, we get

■ $\dfrac{d}{c} = \dfrac{b}{a}$ $\qquad\qquad\qquad\qquad\qquad\qquad$ (12.4)

We now can state the property

■   If $\dfrac{a}{b} = \dfrac{c}{d}$, then $\dfrac{a}{c} = \dfrac{b}{d}$ and $\dfrac{b}{a} = \dfrac{d}{c}$

The second proportion in this statement is said to be derived from the first by *alternation*, and the third to be derived from the first by *inversion*.

We shall derive two other proportions from (12.1). If we add 1 to each member of (12.1), we get

$$\frac{a}{b} + 1 = \frac{c}{d} + 1$$

Then since $\dfrac{a}{b} + 1 = \dfrac{a + b}{b}$ and $\dfrac{c}{d} + 1 = \dfrac{c + d}{d}$, we have

■ $\dfrac{a + b}{b} = \dfrac{c + d}{d}$ $\qquad\qquad\qquad\qquad\qquad$ (12.5)

Similarly, by subtracting 1 from each member of (12.1) and simplifying, we get

■ $$\frac{a-b}{b} = \frac{c-d}{d} \qquad (12.6)$$

If we now equate the quotients of the corresponding members of (12.5) and (12.6), we see that

■ $$\frac{a+b}{a-b} = \frac{c+d}{c-d} \qquad (12.7)$$

We can now state the following property:

■ If $\dfrac{a}{b} = \dfrac{c}{d}$, then $\dfrac{a+b}{b} = \dfrac{c+d}{d}, \dfrac{a-b}{b} = \dfrac{c-d}{d}$, and $\dfrac{a+b}{a-b} = \dfrac{c+d}{c-d}$ (12.8)

EXAMPLE 2
Find a and b if

$$\frac{a}{b} = \frac{c}{d}, \qquad a - b = 12, \qquad c = 3, \qquad \text{and} \qquad d = 2$$

SOLUTION If $a/b = c/d$, then by (12.6) we have

$$\frac{a-b}{b} = \frac{c-d}{d}$$

Consequently, substituting the given values for $a - b$, c, and d, we have

$$\frac{12}{b} = \frac{3-2}{2}$$

$$= \frac{1}{2}$$

Consequently,

$b = 24$ by (12.2)

Finally, since

$a - b = 12$

and

$b = 24$

it follows that

$$a - 24 = 12$$

$$a = 36$$

Mean, Third, and Fourth Proportional

If $c = b$ in $a/b = c/d$, then b is called the *mean proportional* to or between a and d, and d is the *third proportional* to a and b. If $c \neq b$ in $a/b = c/d$, then d is the *fourth proportional* to a, b, and c.

EXAMPLE 3

Find the mean proportional between 2 and 12.5.

SOLUTION   If we represent the desired mean proportional by x and make use of the definition, we have

$$\frac{2}{x} = \frac{x}{12.5}$$

$$x^2 = 25 \qquad \text{by (12.2)}$$

$$x = \pm 5 \qquad \text{solving for } x$$

EXAMPLE 4

Find the third proportional to 2 and 6.

SOLUTION   If we represent the third proportional by x, we have

$$\frac{2}{6} = \frac{6}{x}$$

$$2x = 36 \qquad \text{by (12.2)}$$

$$x = 18 \qquad \text{solving for } x$$

EXAMPLE 5

Find the fourth proportional to 3, 4, and 5.

SOLUTION   If we let x represent the fourth proportional, then

$$\frac{3}{4} = \frac{5}{x}$$

$$3x = 20 \qquad \text{by (12.2)}$$

$$x = \frac{20}{3} \qquad \text{solving for } x$$

Ratio and Proportion

In each of Probs. 1 to 8, express the indicated ratio as a fraction and simplify.

1 8 months to 2 years

2 3.5 quarts to 4 pints

3 $1.65 to 6 nickels

4 5 ft to 18 in.

5 8 weeks to 16 days

6 1.2 miles to 840 ft

7 90 cents to $12

8 5 qt to $2\frac{1}{2}$ gal

In each of Probs. 9 to 12, find the value of the indicated ratio, and interpret the result.

9 $10,320 to 12 months

10 384 miles to 8 hr

11 27 doughnuts to 9 boys

12 $10.20 to 12 lb of candy

13 The grade of a highway is the distance it rises per unit length along the surface. Find the average grade of a highway that rises 176 ft in a mile.

14 The specific gravity of a body is the ratio of the weight of the body to the weight of an equal volume of water. If a cubic foot of water weighs 62.50 lb and a cubic foot of platinum weighs 1343.75 lb, what is the specific gravity of platinum?

15 The specific heat of a substance is the number of calories required to increase the temperature of 1 g of it by 1°C. What is the specific heat of aluminum if 6.3 cal increase the temperature of a 30-g block by 1°C?

16 The pitch of a roof is the distance the roof rises per unit of horizontal distance covered. Find the pitch of a roof if a rafter is 13 ft in length and one of its ends is 5 ft higher than the other.

Find the value of x in each of Probs. 17 to 24.

17 $3:5 = x:8$

18 $x:4 = 6:15$

19 $8:4x = 32:48$

20 $6:7 = 5:x$

21 $(5 - x):(1 + x) = 6:2$

22 $(2 + x):(2 - x) = 2:3$

23 $(x + 1):1 = 4:(x - 2)$

24 $3:(x - 1) = (2x - 3):2$

Find the mean proportional to the pair of numbers in each of Probs. 25 to 28.

25 6, 24   **26** 7, 28   **27** 12, 27   **28** 8, 72

Find the third proportional to the pair of numbers in each of Probs. 29 to 32.

29 5, 8   **30** 3, 8   **31** 4, 12   **32** 6, 9

Find the fourth proportional to the pair of numbers in each of Probs. 33 to 36.

33 3, 8, 6   **34** 5, 2, 15   **35** 3, 1, 9   **36** 6, 9, 4

37 If $x:y = 7:2$ and $x - y = 15$, find x and y.

38 If $x:8 = y:2$ and $x - y = 9$, find x and y.

39 If $x{:}y = 5{:}6$ and $x + y = 22$, find x and y.

40 If $x{:}y = 2{:}3$ and $x + y = 20$, find x and y.

41 If a clap of thunder was heard 5445 ft from the flash of lightning that produced it and 5 sec after the flash, how far from the flash was a person who heard the thunder 8 sec after the flash occurred?

42 If 2152 cal are required to vaporize 4 g of water, how many grams of water can be vaporized by 4304 cal?

43 If a map is drawn to the scale of $\frac{1}{2}$ in. to 40 miles, what distance is represented by 1.75 in.

44 If 6 lb of ground meat are necessary to make a meat loaf for 25 people, how much meat should be bought to prepare meat loaf for 80 people?

12.3
VARIATION

There are many situations in science and elsewhere that deal with variables that are so related that their quotient or product is a constant. For example $V/T = k$ is a symbolic statement of Charles' law. In words, this law is: If the pressure remains constant, the volume of a confined mass of gas varies as the absolute temperature. The symbolic statement can be and often is expressed in the form $V = kT$. This illustrates the following definition.

Direct Variation

If two variables x and y are so related that $y = kx$, we say that y *varies directly* as x or merely that y *varies* as x or that y is *proportional* to x or y is *directly proportional* to x.

Other types of variation will be defined, and then all types will be discussed.

Inverse Variation

If two variables x and y are so related that $y = k/x$, we say that y *varies inversely* as x.

Joint Variation

If a variable varies directly as the product of several others, we say that it *varies jointly* as the others. Thus, if $y = kxwz$, we say that y varies jointly as x, w, and z.

Constant of Variation

The constant k in each type of variation is called the *constant of variation*. It can be determined if a set of corresponding values of the variables is given.

EXAMPLE 1

Find the constant of variation if y varies directly as x and is 36 for $x = 9$.

SOLUTION   Since y varies directly as x, we know that

$y = kx$ by definition of direct variation

Furthermore,

$36 = k9$ substituting the given pair of corresponding values

$k = 4$ solving for k

Therefore, substituting this for k in $y = kx$, we find that

$y = 4x$

is the equation of variation.

If, in addition to the data that enable us to determine k, we know a set of corresponding values of all the variables except one, we can determine that one as seen in Example 2.

EXAMPLE 2

The amount of gas used by a car traveling at a uniform rate varies jointly as the distance traveled and the square of the speed. If a car uses 5 gallons in going 100 miles at 40 miles per hour, how much will it use in going 80 miles at 60 miles per hour?

SOLUTION   The first step is to write the equation of variation. In doing this, we shall represent the number of gallons of gas used by g, the number of miles traveled by s, and the speed by r; then

$g = ksr^2$ by definition of joint variation

Consequently,

$5 = k(100)(40)^2$ since $g = 5$ when $s = 100$ and $r = 40$

$k = \dfrac{5}{100(40)^2}$ solving for k

$= \dfrac{1}{32,000}$

Therefore,

$g = \dfrac{1}{32,000}sr^2$ since $k = 1/32,000$

Finally, we can find the desired value of g by substituting $s = 80$ and $r = 60$ in this equation since they are a pair of corresponding values. Thus,

$$g = \frac{1}{32{,}000}(80)(60)^2$$

$$= 9$$

EXAMPLE 3

The safe load of a rectangular beam of given width varies directly as the square of the depth of the beam and inversely as the length between supports. If the safe load for a beam of given width that is 10 feet long and 5 inches deep is 2400 pounds, determine the safe load for a beam of the same material and width if it is 8 inches deep and 16 feet between supports.

SOLUTION We shall let

d = the number of inches in the depth of the beam

s = the number of feet between supports

L = the number of pounds in the safe load

Then,

$$L = \frac{kd^2}{s} \qquad \text{by definition of the variations}$$

If we now make use of the set of corresponding values of d, s, and L, we obtain

$$2400 = \frac{k5^2}{10}$$

$$k = 960 \qquad \text{solving for } k$$

Therefore,

$$L = \frac{960d^2}{s}$$

and, substituting $d = 8$ and $s = 16$, we have

$$L = \frac{960(8^2)}{16}$$

$$= 3840$$

Therefore, the safe load is 3840 pounds.

Variation

1 Express the following statements as equations: (*a*) *p* varies directly as *q*; (*b*) *a* varies inversely as *b*; (*c*) *x* varies jointly as *y* and *z*; (*d*) *u* varies directly as the square of *w* and inversely as *v*.

2 If *y* varies directly as *x* and is 18 when $x = 6$, find the value of *y* if $x = 2$.

3 If *w* varies directly as *x* and is 24 when $x = 4$, find the value of *w* if $x = 5$.

4 Given that *y* varies inversely as *x* and $y = 6$ when $x = 3$, find the value of *y* when $x = 9$.

5 If *w* varies inversely as *y* and is equal to 12 when $y = 3$, find the value of *w* if $y = 18$.

6 If *y* varies jointly as *x* and *w* and is 36 when $x = 3$ and $w = 2$, find the value of *y* if $x = 5$ and $w = 4$.

7 Given that *x* varies jointly as *w* and *y* and also that $x = 60$ when $w = 3$ and $y = 5$, find the value of *x* if $w = 4$ and $y = 7$.

8 Given that *w* varies directly as the product of *x* and *y* and inversely as the square of *z* and that $w = 4$ when $x = 2$, $y = 6$, and $z = 3$, find the value of *w* when $x = 1$, $y = 4$, and $z = 2$.

9 The area of a rhombus varies jointly as the length of the diagonals. If the area of a rhombus with diagonals 3 and 4 in. is 6 sq in., find the area of another rhombus whose diagonals are 2 and 5 in.

10 When aluminum is added to an excess of hydrochloric acid, the amount of hydrogen produced varies directly as the amount of aluminum added. If 18 g of aluminum produce 2 g of hydrogen, how much hydrogen can be produced by adding 63 g of aluminum to an excess of hydrochloric acid?

11 The simple interest earned in a given time varies jointly as the principal and the interest rate. If $600 earned $108 at 6 percent, find the interest earned in the same time by $810 at 5 percent.

12 The air resistance to a moving object is approximately proportional to the square of the speed of the object. Compare the air resistance of an automobile traveling 30 miles per hour with that of an automobile traveling 60 miles per hour.

13 According to Boyle's law, the volume of a confined mass of gas varies inversely as the pressure, provided that the temperature is constant. If a mass of gas has a volume of 1 liter under a pressure of 76 cm of mercury, what is its volume under 38 cm of mercury?

14 The amount of paint needed to paint the sides of a cylindrically shaped fence post varies jointly as the radius and height of the post. Compare the amount of paint needed for a post 5 ft high of radius $\frac{1}{2}$ ft with that needed for a post 6 ft high of radius $\frac{1}{3}$ ft.

15 The volume of gas discharged in a given time from a horizontal pipe under constant pressure and specific gravity varies inversely as the square root of the length of the pipe. If 246 cu ft of gas is discharged in 1 hr from a pipe 2500 ft long, what volume would be discharged in 1 hr from a pipe 3600 ft long?

16 The weight of a body above the surface of the earth varies inversely as the square of the distance from the body to the center of the earth. If a boy weighs 121 lb on the surface, how much would he weigh 400 miles above the surface? Assume the radius of the earth to be 4000 miles.

17 The amount that a bar will bend under a given force varies inversely as the width of the bar when the thickness and length of the bar remain the same. If a bar 2.5 in. wide will bend 7° under a certain force, what is the width of a bar of the same material, length, and thickness that will bend 5° under the same force?

18 The horsepower that a rotating shaft can safely transmit varies jointly as the cube of the radius of the shaft and the number of revolutions through which the shaft turns per minute. Compare the safe load of a shaft of radius 2 in. which turns 600 revolutions per minute (rpm) with that of another shaft which has a radius of 3 in. and revolves 800 times a minute.

19 The weight of a clothesline varies jointly as the length and the square of the diameter. If the weight of a 36-ft line $\frac{1}{4}$ in. in diameter is 5.4 lb, find the weight of a 48-ft line $\frac{1}{8}$ in. in diameter.

20 On the ocean, the square of the distance in miles to the horizon varies as the height in feet that the observer is above the surface of the water. If a 6-ft man on a surfboard can see 3 miles, how far can he see if he is standing on the deck of a ship that is 48 ft above the water?

21 The power available in a jet of water varies jointly as the cube of the water's velocity and the cross-sectional area of the jet. Compare the power of a jet with that of another that is moving two times as fast through an opening one-half large.

22 The centrifugal force at a point of a revolving body varies as the radius of the circle in which the point is revolving. If the centrifugal force is 450 lb when the radius is 12 in., at what radius is the force 375 lb?

23 The force of a wind on a flat surface perpendicular to the direction of the wind varies as the area of the surface and the square of the wind velocity. When the wind is blowing 8 miles an hour, the force on a 4- by 6-ft signboard is 7.5 lb. What is the force on a window 3 by 4 ft when the wind is blowing at 16 miles per hour?

24 The crushing load of a pillar varies as the fourth power of its diameter and inversely as the square of its height. If 1953 tons will crush a pillar 1 ft in diameter and 15 ft high, find the load that will crush a pillar 4 in. in diameter and 5 ft high.

25 One of Kepler's laws states that the square of the time in days required by a planet to make 1 revolution about the sun varies directly as the

cube of the average distance of the planet from the sun. If Mars is $1\frac{1}{2}$ times as far from the sun, on the average, as the earth is, find the approximate number of days required for it to make a revolution about the sun.

26 The horsepower required to operate a fan varies as the cube of the speed. If 2 horsepower will drive a fan at 600 rpm, find the speed derived from $\frac{1}{4}$ horsepower.

27 The exposure time necessary to obtain a good photographic negative varies directly as the square of the f number of the camera lens. If an exposure of $\frac{1}{50}$ sec produces a good negative at $f16$, what exposure time would be necessary at a lens setting of $f8$?

28 The amount of oil used by a ship traveling at a uniform speed varies jointly as the distance traveled and the square of the velocity. If a ship used 1500 barrels of oil traveling 480 miles at 25 knots, at what uniform rate did the ship travel if it covered 540 miles and used 1080 barrels of oil?

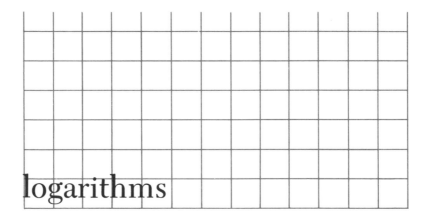

logarithms

Chapter

13

Logarithms, invented in the seventeenth century, are useful and efficient in arithmetical computation, important in the application of mathematics to chemistry, physics, and engineering, and indispensable for some parts of advanced mathematics. In this chapter we shall develop some properties of logarithms and show how they are used in numerical computation. We also use logarithms to obtain the solution sets of certain types of equations. The theory of logarithms is based on the laws of exponents, and the reader is advised to review these laws now.

13.1
APPROXIMATIONS

In this chapter we deal with computation of numbers that are approximations. Therefore we must consider the degree of accuracy in the result of such computation. This problem involves the concept of significant figures and the method of rounding off a number to the desired degree of accuracy.

Any number that is obtained by measurement is an approximation. For example, if the thickness of a sheet of metal is measured with a micrometer calibrated in thousandths of an inch and found to be 0.053, it is understood that the actual thickness is between 0.0525 and 0.0535 including the former number but not the latter, and we write $0.00525 \leq$ thickness < 0.0535. Similarly, if the length and width of a rectangle are obtained with a rule calibrated in tenths of a centimeter and are 10.6 centimeters and 5.2 centimeters, then

$$10.55 \leq \text{length} < 10.65 \qquad \text{and} \qquad 5.15 \leq \text{width} < 5.25 \qquad (1)$$

Now in connection with the statement that the area is $(10.6)(5.2) = 55.12$ square centimeters, the question arises: How many digits in 55.12 are accurate? We answer this question by referring to statement (1) and seeing that

$$(10.55)(5.15) \leq \text{area} < (10.65)(5.25)$$

or

■ $54.3325 \leq \text{area} < 55.9125 \qquad (2)$

Since the first number in statement (2) is only slightly more than 54 and the other is nearly 56, we cannot be certain of the area even to two digits. We shall say, however, that the area is probably nearer 55 square centimeters than to any other two-digit number. This discussion illustrates the following definition:

■ In a number that is an approximation, the digits known to be correct are called significant.

The digits 1, 2, 3, 4, 5, 6, 7, 8, and 9 are always significant if used in connection with a measurement, as are any zeros between any two of them. Zeros whose only function is to place the decimal point are never significant. Such zeros occur between the decimal point and the first non-zero digit and only in numbers between 1 and -1.

Rounding Off

In computation including approximate numbers, we do not use a digit if its accuracy is doubtful. Eliminating these digits is called *rounding off a number*. The procedure of rounding a number off to n significant digits consists of the following steps:

1 If the decimal point is to the right of the nth digit, we replace all digits between the nth and the decimal point by zeros and drop all digits to the right of the decimal point.
2 If the decimal point is to the left of the nth digit we drop all digits to the right of the nth.

3 If the digit following the nth is 5, 6, 7, 8, or 9, we increase the nth digit by 1.

4 If the digit following the nth digit is 0, 1, 2, 3, or 4, the nth digit is not changed.

We shall illustrate rounding off by observing that 48.257 rounded off to three digits is 48.3, since the fourth digit is 5 and the decimal point is to the left of the third; 3.842 rounded off to three digits is 3.84, since the fourth digit is 2 and the decimal point is to the left of the third; 3842 rounded off to three digits is 3840, since the fourth digit is 2 and the decimal point is to the right of the third; and .0020651 rounded off to three digits is .00207, since the first digit dropped is 5 and the decimal point is to the left of the third.

Computation with Approximate Numbers

As illustrated in the discussion following Eq. (2), the result of a computation depends upon the significant digits in the numbers involved. We shall now present three rules that are used in such cases.

Rule 1 If the number of significant figures in M and N are equal or differ by one, we round off the product MN and the quotient M/N to the number of significant figures in the less accurate one of M and N.

Rule 2 If the number of significant figures in M and N differ by two or more, we round off the more accurate so that it contains one more significant digit than the other and then proceed as in Rule 1.

For example, if $M = 27.3$ and $N = 145.9$, then the number of significant digits in M and N differ by one, and the least accurate number 27.3 has three significant digits. Hence, we round off the product MN $= (27.3)(145.9) = 3983.07$ to three significant digits and obtain $MN = 3980$, accurate to three digits. Note that the final zero is not significant. Furthermore, since the number of significant digits in $M = 583.27$ is two more than in $N = 34.7$, we round M off to one more than in N and have $M = 583.3$ to four figures. Then

$$\frac{M}{N} = \frac{583.3}{34.7} = 16.81 = 16.8 \qquad \textbf{to three digits}$$

When dealing with a sum, we are not primarily interested in the number of significant figures in the data but rather in the precision of it. That is, we are interested in whether the data measure the integral number of units, the tenth of a unit, the hundredth of a unit, or a smaller fraction of a unit. Consequently, we are interested in the number of significant figures in the *decimal portion* of the addends, since we add digits having the same place value; hence, we shall use the following rule in adding approximate numbers.

Rule 3 Round off each addend to one more decimal place than there is in the addend which has the least number of decimal places, then add the resulting numbers, and round off the last digit in the sum. A similar procedure is used in subtraction.

EXAMPLE
Find the sum of 28.72, 3.683, 7.2, and 23.7864.

SOLUTION The third number contains only one decimal place; hence the others should be rounded off to two decimal places before adding. If this is done, the addends and their sum are as shown on the right; hence, after rounding off to one decimal place, we find the sum to be 63.4.

$$
\begin{array}{r}
28.72 \\
3.68 \\
7.2 \\
\underline{23.79} \\
63.39
\end{array}
$$

13.2
SCIENTIFIC NOTATION

In Sec. 13.1 we stated the conditions under which some zeros in an approximate number are significant and others are not. We did not discuss the significance of a zero if it is the last digit in the number. The fact is that in some instances the final zero or zeros are significant, and in others they are not. For example, the diameter of the earth to the nearest multiple of 1000 miles is 8000, and to the nearest multiple of 100 miles, it is 7900. Since the diameter to the nearest mile is 7918, no one of the zeros in 8000 or 7900 is significant. Their only purpose is to fill the space between the last significant number and the decimal point. On the other hand, if we know that the length of a rectangular pasture to the nearest yard is 600, both zeros in 600 are significant.

The ambiguity with respect to the final zeros is removed if the number is expressed in scientific notation. A number is in scientific notation if it is expressed as the product of two factors, with the first factor a number equal to or greater than 1 but less than 10 and having all of its digits significant and with the other factor an integral power of 10. As examples we shall express the numbers mentioned in the preceding paragraph in scientific notation. Note that only the significant digits appear in the first factor.

8000 (zeros not significant) $= 8(10^3)$

7900 (zeros not significant) $= 7.9(10^3)$

600 (zeros significant) $= 6.00(10^2)$

7918 (all digits significant) $= 7.918(10^3)$

Reference Position of the Decimal Point

The procedure for expressing a number in scientific notation makes use of the reference position of the decimal point, which we now define.

■ The *reference position* for the decimal point in a number N is immediately to the right of the first nonzero digit in N.

For example, the reference position for the decimal point in each of the following numbers is indicated by a caret:

$$2_\wedge 16.34 \qquad 3_\wedge 021 \qquad 0.004_\wedge 623$$

The scientific notation for the number N is $N'(10^c)$, with all significant digits in N appearing in N' and with the decimal point in N' in reference position. The exponent c is numerically equal to the number of digits between the reference position and the decimal point in N and is positive or negative according as the decimal point is to the right or to the left of the reference position. Of course if a number is not an approximation or is not a measurement, all of its digits are significant. The following table provides the scientific notation for six numbers together with N' and c.

Number	N'	c	Scientific Notation
312	3.12	2	$3.12(10^2)$
1235	1.235	3	$1.235(10^3)$
0.0621	6.21	-2	$6.21(10^{-2})$
0.00004326	4.326	-5	$4.326(10^{-5})$
1250, zero significant	1.250	3	$1.250(10^3)$
3650, zero not significant	3.65	3	$3.65(10^3)$

EXAMPLE

Convert $M = 3270$ (zero not significant) and $N = 43$ to scientific notation, and then find the product MN. Express the result in scientific notation.

SOLUTION   In scientific notation

$$M = 3.27(10^3), \qquad N = 4.3(10^1),$$

and

$$
\begin{aligned}
MN &= (3.27)(10^3)(4.3)(10^1) \\
&= (3.27)(4.3)(10^4) \\
&= 14.061(10^4) \qquad \text{since (3.27)(4.3)} = 14.061 \\
&= 14(10^4) \qquad \text{rounding off to two digits} \\
&= 1.4(10)(10^4) \qquad \text{since } 14 = 1.4(10) \\
&= 1.4(10^5)
\end{aligned}
$$

EXERCISE 13.1
Approximations

Round off the number in each of Probs. 1 to 16 to four significant figures.

1   62.5612   **2**   723.925

3	3.15644	4	0.0298619
5	708,935.86	6	90,532,099
7	611,929.85	8	8,165,395
9	9077.916	10	318.7592
11	339,672.5	12	29,875.13
13	47.335	14	516.25
15	9061.5	16	49.825

Express the number in each of Probs. 17 to 32 in scientific notation.

17	5917	18	56.29
19	695,752	20	9173.15
21	0.0576	22	0.0004987
23	0.00000935	24	0.000075
25	0.008133	26	1280, zero significant
27	29,700, zeros significant	28	7000, zeros significant
29	7510, zero not significant	30	97,100, zeros not significant
31	18,300,000, zeros not significant		
32	596,700,000, zeros not significant		

Assume that the numbers in Probs. 33 to 52 are approximations. Perform the indicated operations; then round off the result to the correct number of digits and express it in scientific notation.

33	(4.6)(19)	34	(0.51)(1.5)
35	(6.9)(193)	36	(4.85)(0.065)
37	(55.3)(917.2)	38	(57.3)(9.8165)
39	(5.1)(71.23)	40	(91)(6.135)
41	$91.2 \div 3.6$	42	$7.01 \div 49$
43	$0.31972 \div 14.2$	44	$9.1324 \div 12.9$
45	$19.1 + 0.695 + 6.17$	46	$29.1 + 916 + 1.7$
47	$96.17 + 1.113 + 0.9175$	48	$6.75 + 93.114 + 817.1$
49	$813 + 6.97 - 39.5$	50	$6.965 - 2.5 + 5.76$
51	$5.31 + 9.913 - 11.5$	52	$0.6812 + 0.365 - 0.77$

13.3
DEFINITION OF A LOGARITHM

In the introduction to this chapter we stated that the theory of logarithms is based on the laws of exponents. This is due to the relationship between logarithms and exponents established by the following definition:

■ The *logarithm* L of a number N to the *base b* is the exponent that indicates the power to which the base b must be raised in order to be equal to N.

■ The abbreviation of the phrase "logarithm to the base b of N" is $log_b N$. In terms of this abbreviation the preceding definition can be stated as follows:

■ $log_b N = L$ if and only if $b^L = N$ $\qquad\qquad$ (13.1)

By use of this definition, we see that

$log_2 64 = 6$ $\qquad$ since $2^6 = 64$

$log_4 64 = 3$ $\qquad$ since $4^3 = 64$

$log_8 64 = 2$ $\qquad$ since $8^2 = 64$

$log_{16} 64 = \frac{3}{2}$ $\qquad$ since $16^{3/2} = (16^{1/2})^3 = 4^3 = 64$

$log_b 1 = 0$ $\qquad$ since $b^0 = 1$

By use of (13.1), we can also see that

If $log_3 N = 4$, then $N = 3^4 = 81$

If $log_b 125 = 3$, then $b^3 = 125 = 5^3$ and $b = 5$

If $log_{16} 4 = L$, then $16^L = 4 = 16^{1/2}$ and $L = \frac{1}{2}$

We see from these examples that a logarithm may be integral or fractional. In fact, it can be and often is irrational. Furthermore, in many cases, if two of the three letters in (13.1) are known, the third can be found by inspection.

EXAMPLE 1

Find the replacement for N in $log_7 N = 2$.

SOLUTION By use of (13.1), we have

$7^2 = N$

hence,

$N = 49$

EXAMPLE 2

Find the replacement for b if $log_b 125 = 3$.

SOLUTION By use of (13.1), we have

$b^3 = 125$

$b = \sqrt[3]{125} = 5$ $\qquad$ **solving for b**

EXAMPLE 3

319

13.4 Properties of
Logarithms

Find the replacement for a if $\log_{27} 3 = a$.

SOLUTION   Again using (13.1), we have

$$27^a = 3$$

and, since $27^{1/3} = 3$, it follows that

$$a = \tfrac{1}{3}$$

13.4
PROPERTIES OF LOGARITHMS

In this article we shall use the laws of exponents and the definition of a logarithm to derive three important properties of logarithms. In the next article we shall show how to use these properties in numerical computation.

We shall first show how to find the logarithm of a product of two numbers in terms of the logarithms of the two numbers.

If we are given

$$\log_b M = m \quad \text{and} \quad \log_b N = n \tag{1}$$

then, by (13.1), we have

$$M = b^m \quad \text{and} \quad N = b^n \tag{2}$$

Hence,

$$MN = (b^m)(b^n)$$
$$= b^{m+n}$$

and

$$\log_b MN = m + n \qquad \text{by (13.1)}$$
$$= \log_b M + \log_b N \qquad \text{by (1)}$$

Consequently, we have proved the following theorem:

■   The logarithm of the product of two positive numbers is equal to the sum of the logarithms of the numbers.

EXAMPLE 1   If

$$\log_{10} 3 = 0.4771 \quad \text{and} \quad \log_{10} 2 = 0.3010$$

then

$$\log 6 = \log (3 \times 2)$$
$$= \log 3 + \log 2$$
$$= 0.4771 + 0.3010$$
$$= 0.7781$$

This theorem can be extended to three or more numbers by the following process:

$$\log_b MNP = \log_b (MN)(P)$$
$$= \log_b MN + \log_b P$$
$$= \log_b M + \log_b N + \log_b P$$

Again, using relations (2), we have

$$\frac{M}{N} = \frac{b^m}{b^n}$$
$$= b^{m-n}$$

Hence,

$$\log_b \frac{M}{N} = m - n \qquad\qquad \text{by (13.1)}$$

$$= \log_b M - \log_b N \qquad \text{by (1)}$$

Therefore, we have proved the following theorem:

■ The logarithm of the quotient of two positive numbers is equal to the logarithm of the dividend minus the logarithm of the divisor.

EXAMPLE 2 If

$$\log_{10} 3 = 0.4771 \qquad \text{and} \qquad \log_{10} 2 = 0.3010$$

then

$$\log_{10} 1.5 = \log_{10} \frac{3}{2}$$
$$= \log_{10} 3 - \log_{10} 2$$
$$= 0.4771 - 0.3010$$
$$= 0.1761$$

Finally, if we raise both members of $M = b^m$ to the kth power, we have

$$M^k = (b^m)^k$$

$$= b^{km}$$

Therefore,

$$\log_b M^k = km \qquad \text{by (13.1)}$$

$$= k \log_b M \qquad \text{by (1)}$$

Thus we have proved the following theorem:

■ The logarithm of a power of a positive number is equal to the product of the exponent of the power and the logarithm of the number.

EXAMPLE 3   If

$$\log_{10} 3 = 0.4771$$

then

$$\log_{10} 3^2 = 2 \log_{10} 3$$

$$= 2(0.4771)$$

$$= 0.9542$$

Note   Since a root of a number can be expressed as a fractional power, the last theorem can be used to find the logarithm of a root of a number. Thus,

$$\log_b \sqrt[r]{M} = \log (M)^{1/r} = \frac{1}{r} \log M$$

EXAMPLE 4   If

$$\log_{10} 2 = 0.3010$$

then

$$\log_{10} \sqrt[3]{2} = \log_{10} 2^{1/3}$$

$$= \tfrac{1}{3} \log_{10} 2$$

$$= \tfrac{1}{3}(0.3010)$$

$$= 0.1003$$

Computational Theorems

For the convenience of the reader, we shall restate the preceding properties in symbolic form.

$$\log_b MN = \log_b M + \log_b N \qquad \begin{matrix} M > 0 \\ N > 0 \end{matrix} \qquad (13.2)$$

$$\log_b \frac{M}{N} = \log_b M - \log_b N \qquad \begin{matrix} M > 0 \\ N > 0 \end{matrix} \qquad (13.3)$$

$$\log_b M^k = k \log_b M \qquad \begin{matrix} M > 0 \\ M > 0 \end{matrix} \qquad (13.4)$$

EXERCISE 13.2
Conversion of Logarithms and Exponential Forms

By use of (13.1), change the statement in each of Probs. 1 to 16 to logarithmic form and the statement in each of Probs. 17 to 28 to exponential form.

1	$3^3 = 27$	**2**	$2^4 = 16$
3	$3^5 = 243$	**4**	$2^6 = 64$
5	$(\frac{1}{3})^{-2} = 9$	**6**	$(\frac{1}{4})^{-3} = 64$
7	$(\frac{1}{2})^{-5} = 32$	**8**	$(\frac{1}{10})^{-7} = 10,000,000$
9	$4^{-2} = \frac{1}{16}$	**10**	$3^{-4} = \frac{1}{81}$
11	$5^{-4} = \frac{1}{625}$	**12**	$2^{-8} = \frac{1}{256}$
13	$36^{1/2} = 6$	**14**	$125^{1/3} = 5$
15	$4^{3/2} = 8$	**16**	$27^{4/3} = 81$
17	$\log_3 9 = 2$	**18**	$\log_5 3125 = 5$
19	$\log_6 216 = 3$	**20**	$\log_2 128 = 7$
21	$\log_{10} \frac{1}{10} = -1$	**22**	$\log_4 \frac{1}{16} = -2$
23	$\log_5 \frac{1}{125} = -3$	**24**	$\log_2 \frac{1}{512} = -9$
25	$\log_{1/3} 9 = -2$	**26**	$\log_{1/8} 512 = -3$
27	$\log_{1/2} 256 = -8$	**28**	$\log_{1/5} 625 = -4$

In each of Probs. 29 to 52, find the replacement for b, n, or L so that the statement in the problem will be true.

29	$\log_2 64 = L$	**30**	$\log_5 625 = L$
31	$\log_3 243 = L$	**32**	$\log_4 256 = L$
33	$\log_9 27 = L$	**34**	$\log_{64} 16 = L$
35	$\log_{16} 8 = L$	**36**	$\log_{125} 25 = L$
37	$\log_5 n = 3$	**38**	$\log_2 n = 5$

39 $\log_{1/3} n = 4$

40 $\log_{3/4} n = 3$

41 $\log_{64} n = \frac{2}{3}$

42 $\log_{36} n = \frac{3}{2}$

43 $\log_{16/9} n = \frac{1}{2}$

44 $\log_{8/27} n = \frac{4}{3}$

45 $\log_b 25 = 2$

46 $\log_b 216 = 3$

47 $\log_b 512 = 9$

48 $\log_b 729 = 3$

49 $\log_b \frac{27}{125} = 3$

50 $\log_b \frac{64}{125} = \frac{3}{2}$

51 $\log_b \frac{25}{36} = \frac{2}{3}$

52 $\log_b \frac{8}{27} = \frac{3}{5}$

Use (13.2), (13.3), and (13.4) to obtain the logarithm required in each of Probs. 53 to 68. Given: $\log_2 8 = 3$, $\log_2 16 = 4$, $\log_2 64 = 6$, $\log_2 256 = 8$, $\log_2 1024 = 10$, $\log_2 4096 = 12$, $\log_2 16{,}384 = 14$, $\log_2 65{,}536 = 16$.

53 $\log_2 (256)(1024)$

54 $\log_2 (64)(16{,}384)$

55 $\log_2 (4096)(65{,}536)$

56 $\log_2 (1024)(4096)$

57 $\log_2 \dfrac{16{,}384}{4096}$

58 $\log_2 \dfrac{1024}{64}$

59 $\log_2 \dfrac{65{,}536}{256}$

60 $\log_2 \dfrac{4096}{1024}$

61 $\log_2 (1024)^3$

62 $\log_2 (4096)^4$

63 $\log_2 \sqrt{16{,}384}$

64 $\log_2 \sqrt[4]{65{,}536}$

65 $\log_2 \dfrac{(8)(16{,}384)}{16}$

66 $\log_2 \dfrac{(1024)(65{,}536)}{16{,}384}$

67 $\log_2 \dfrac{(256)(64)}{4096}$

68 $\log_2 \dfrac{(16)(16{,}384)}{65{,}536}$

13.5
COMMON, OR BRIGGS, LOGARITHMS

We stated previously that logarithms can be used efficiently in numerical computation, and we shall use logarithms to the base 10 for this purpose. Logarithms to the base 10 are called *common*, or *Briggs*, logarithms.

The common logarithm of a number that is not an integral power or root of 10 cannot be computed by elementary methods, but tables have been prepared that enable us to obtain a decimal approximation to the common logarithm of any positive number. The use of the tables will be explained in the next three sections.

It is customary to omit the subscript indicating the base in the notation for a common logarithm. Consequently, in the statement $\log N = L$, it is understood that the base is 10.

If c is a real number, then 10^c is positive. Hence the common logarithm of zero or of a negative number does not exist as a real number.

13.6
CHARACTERISTIC AND MANTISSA

If we express a positive number $N \neq 1$ in scientific notation, we have

$$N = N'(10^c) \tag{1}$$

where

$$1 < N' < 10$$

and c is an integer.

We now consider the following three situations:

1 If $N > 10$, then in Eq. (1), $c \geq 1$. For example, $231 = 2.31(10^2)$.
2 If $N < 1$, then $c < 0$. For example, $0.0231 = 2.31(10^{-2})$.
3 If $1 < N < 10$, then in Eq. (1), $N = N'$ and $c = 0$. For example, $2.31 = 2.31(10^0)$.

Now, if we equate the common logarithms of the members of Eq. (1), we have

$$\log N = \log N' + \log 10^c \qquad \text{by (13.2)}$$

$$= \log N' + c \log 10 \qquad \text{by (13.4)}$$

$$= \log N' + c \qquad \text{since log 10} = 1$$

Thus, we have

$$\log N = c + \log N' \qquad \text{by commutative axiom of addition} \tag{2}$$

Since $1 < N' < 10$, it follows that $10^0 < N' < 10^1$, and hence, that $0 < \log N' < 1$. From Table I in the Appendix, we can obtain a decimal approximation to the common logarithm of any number between 1 and 10, correct to four decimal places. Consequently, by referring to Eq. (2), we see that the common logarithm of any positive number not equal to 1 can be expressed approximately as an integer plus a nonnegative decimal fraction. Since $1 = 10^0$, then $\log 1 = 0$. Thus, for log 1, the integer is zero and the decimal fraction is zero. We are now in position to state the following definition:

■ If the common logarithm of a positive number is expressed as an integer plus a nonnegative decimal fraction, the integer is called the *characteristic* of the logarithm and the decimal fraction is the *mantissa*.

In the expression for log N in Eq. (2), the characteristic is the integer c and the mantissa is $\log N'$. In Sec. 13.2 we demonstrated that c is numerically equal to the number of digits between the reference position and the decimal point in N and that it is positive or negative according as the decimal is to the right or to the left of the reference position. Therefore, we have the following rule for determining the characteristic of the common logarithm of a positive number:

■ The characteristic of the common logarithm of a positive number N is numerically equal to the number of digits between the reference position and the decimal point in N and is positive or negative according as the decimal point is to the right or to the left of the reference position.

EXAMPLE 1

The reference position in 236.78 is between 2 and 3. Hence there are two digits, 3 and 6, between the reference position and the decimal point. Since the decimal point is to the right of the reference position, the characteristic of the common logarithm of 236.78 is 2.

EXAMPLE 2

The characteristic of the common logarithm of 2.3678 is zero since the decimal point is in reference position.

EXAMPLE 3

The decimal point in 0.0023678 is three places to the left of the reference position. Therefore, the characteristic of log 0.0023678 is -3.

Since the position of the decimal point in the number N affects only the integer c in the scientific notation for N, and since the mantissa of log N is log N', the mantissa of log N depends only upon the sequence of digits in N.

If $N \geq 1$, the characteristic of log N is zero or a positive integer. Therefore, log N can be written as a single number. For example, if the mantissa of log 23,678 is .3743, then log $236.78 = 2 + .3743 = 2.3743$, and log $2.3678 = 0 + .3743 = 0.3743$. In finding the logarithm of a positive number less than 1, however, we have a different situation. For example, log $0.0023678 = -3 + 0.3743 = -2.6257 = -2 - 0.6257$. Now the decimal fraction $-.6257$ is negative and consequently is not a mantissa. We use the following device for dealing with such situations. If the characteristic of log N is $-c$, where $c > 0$, we express $-c$ in the form $(10 - c) - 10$; then we write the mantissa to the right of $10 - c$. For example, we express log 0.0023678 in the form $7.3743 - 10$ since the characteristic is -3, and $(10 - 3) - 10 = 7 - 10$. Similarly, we have log $0.23678 = 9.3743 - 10$, log $0.023678 = 8.3743 - 10$, and log $0.0000023678 = 4.3743 - 10$.

13.7
GIVEN N, TO FIND LOG N

The mantissa of the logarithm of a number does not depend upon the position of the decimal point since, as is pointed out earlier in this chapter, shifting the position of the decimal point is equivalent to multiplying by an integral power of 10 and, therefore, only the characteristic is affected.

Use of Tables

The mantissa is determined by the sequence of digits and can be found by use of a table such as Table I in the Appendix. From this table we can find to four figures the mantissa of the logarithm of any three-digit number.

To get the mantissa of the logarithm of such a number, we find the first two digits of the number in the column headed by N at the left of the page and then move across the page to the column headed by the third digit. The desired mantissa is in the column headed by the third digit and in the line containing the first two digits.

In keeping with this procedure, to find log 327, we look in Table I across from 32 and under 7 and find 5145; hence log 327 = 2.5145, since the characteristic is 2 and the decimal point that is a part of each mantissa is not printed. Furthermore, to get log 0.914, we look across from 91 and under 4 and find 9609; hence log 0.914 = 9.9609 − 10, since the characteristic is −1 = 9 − 10.

13.8
GIVEN LOG N, TO FIND N

If log N is given and the mantissa is in the table, we find the first two digits of N to the left of the mantissa and the third digit above the mantissa. The position of the decimal point is determined by the value of the characteristic. Thus, if log N = 1.9212, we first find 9212 in the body of the table. It is in the column headed by 4 and in line with 83 in the N column. Hence, the sequence of digits in N is 834. The characteristic of log N is 1. Hence, N = 83.4.

If log N is given and the mantissa is not in the table and we want N to only three digits, we find the mantissa that is in the table and nearest to the mantissa of log N than any other entry and then use the corresponding value of N. If the mantissa of log N is midway between two entries, we use the value of N which corresponds to the larger entry. Accordingly, we see that if log N = 0.4822, then the entry in the table that is nearest it is 4829 and the corresponding sequence of idgits is 304. Furthermore, N = 3.04 since the characteristic is zero. If log N = 1.1804, the mantissa is midway between 1790 and 1818; hence the sequence of digits used is 152 and N = 15.2 because the characteristic is 1.

13.9
INTERPOLATION

Quite often we need the value of the logarithm of a number that is not listed in the table or want the sequence of digits that corresponds to a mantissa value that is not in the table. Under such circumstances we resort to a procedure that is known as *linear interpolation*.

We shall illustrate the procedure by using it to get the mantissa of log 2537, and in the discussion we use *ml N* to stand for the phrase, "the mantissa of log N."

Since

$$2570 < 2573 < 2580$$

then

ml 2570 < ml 2573 < ml 2580

Furthermore,

ml 2580 = ml 258   and   ml 2570 = ml 257

We now let

ml 2573 = ml 2570 + c

and shall find the value of c by the following procedure:

$$10 \left[\begin{array}{c} \text{ml } 2580 = .4116 \\ 3 \left[\begin{array}{c} \text{ml } 2573 = .4099 + c \\ \text{ml } 2570 = .4099 \end{array} \right] c \end{array} \right] .0017$$

The numbers to the right and to the left of the brackets in this diagram are the differences of the numbers connected by the brackets. Now these differences are approximately proportional. Therefore,

$$\frac{c}{.0017} = \frac{3}{10}$$

and it follows that

$$c = \frac{3}{10}(.0017) = .0005$$

Therefore,

ml 2573 = .4099 + .0005 = .4104

Furthermore, since the decimal point in 2573 is three places to the right of the reference position, the characteristic of the logarithm is 3. Hence,

log 2573 = 3.4104

We use the same principle to obtain N if log N is given and ml N is not listed in the table. We illustrate the method by finding the value of N if log $N = 1.2869$. The mantissa .2869 is not in the table, but the two mantissas nearest it are .2856 and .2878. Now, .2856 = ml 1930 and .2878 = ml 1940. We let n stand for the number composed of the four digits of N, then we shall find n, and finally we shall place the decimal point in the position indicated by the characteristic and thus obtain N.

Since

.2856 < .2869 < .2878

then,

$$1930 < n < 1940$$

Now we let $n = 1930 + c$ and determine c as follows:

$$.0022 \left[.0013 \left[\begin{array}{l} .2878 = \text{ml } 1940 \\ .2869 = \text{ml } 1930 + c \\ .2856 = \text{ml } 1930 \end{array} \right] c \right] 10$$

Since the ratios of the differences to the right and left of the brackets are proportional, we have

$$\frac{c}{10} = \frac{.0013}{.0022}$$

$$c = \frac{.0013}{.0022}(10) = \frac{13}{22}(10) = 6 \qquad \textbf{to one digit}$$

Hence,

$$n = 1930 + 6 = 1936$$

Since the characteristic of $\log N = 1.2869$ is 1, we place the decimal point to the right of 9 in 1936 and then get

$$N = 19.36$$

EXERCISE 13.3
Numbers and Logarithms of Numbers

Find the common logarithm of the number in each of Probs.1 to 8.

1	27.6	**2**	3.17	**3**	504	**4**	72.1
5	.0726	**6**	.849	**7**	.993	**8**	.00277

If the number given in each of Probs. 9 to 24 is log N, find N to three digits.

9	1.4624	**10**	2.5977	**11**	0.6345	**12**	1.7300
13	0.8089	**14**	1.8733	**15**	2.9460	**16**	0.9939
17	9.8848 − 10	**18**	8.7723 − 10	**19**	8.6928 − 10	**20**	7.3118 − 10
21	2.4497	**22**	1.6698	**23**	0.8951	**24**	2.9760

Use interpolation to find the logarithm of the number in each of Probs. 25 to 32.

25	692.6	**26**	71.48	**27**	3027	**28**	4793
29	.2509	**30**	.02584	**31**	.007102	**32**	.5315

Round off the number in each of Probs. 33 to 40 to four digits, and then find its common logarithm.

33	35.916	**34**	5.9345	**35**	472.75	**36**	6.0253
37	.61114	**38**	.076537	**39**	.0023478	**40**	.49932

If the number given in each of Probs. 41 to 52 is log N, find N to four digits.

41	1.6678	**42**	0.4170	**43**	2.5398	**44**	1.7394
45	2.7975	**46**	3.8761	**47**	1.9880	**48**	0.9931
49	$9.8105 - 10$	**50**	$8.6792 - 10$	**51**	$7.3650 - 10$	**52**	$9.0729 - 10$

13.10
LOGARITHMIC COMPUTATION

As we stated previously, one of the most useful applications of logarithms is in the field of numerical computation. We shall explain presently the methods involved by means of several examples. Before considering special problems, however, we wish to call attention again to the fact that results obtained by the use of four-place tables are correct at most to four places. If the numbers in any computation problem contain only three places, the result is dependable to only three places. If a problem contains a mixture of three-place and four-place numbers, we cannot expect more than three places of the result to be correct, and so we round it off to three places. Hence, in the problems that follow, we shall not obtain any answer to more than four nonzero places, and sometimes not that far. Tables exist from which logarithms may be obtained to five, six, seven, and even more places. If results that are correct to more than four places are desired, longer tables should be used. The methods which we have presented may be applied to a table of any length.

We shall now present examples with explanations that illustrate the methods for using logarithms to obtain (1) products and quotients, (2) powers and roots, and (3) miscellaneous computation problems.

In all computation problems we use the properties of logarithms developed in Sec. 13.4 to find the logarithm of the result. Then the value of the result can be obtained from the table.

Products and Quotients

EXAMPLE 1
Find the value of $R = (8.56)(3.47)(198)$.

SOLUTION Since R is equal to the product of three numbers, log R is equal to the sum of the logarithms of the three factors. Hence, we shall obtain the logarithm of each of the factors, add them together, and thus have log R. Then we may use the table to get R. Before we turn to the table, it is advisable to make an outline, leaving blanks in which to enter the logarithms as they are found. It is also advisable to arrange the outline so that the logarithms to be added are in a column. We suggest the following plan:

$$\log R = \log (8.56)(3.47)(198)$$
$$= \log 8.56 + \log 3.47 + \log 198$$

$$\log 8.56 =$$
$$\log 3.47 =$$
$$\underline{\log 198 =}$$
$$\log R =$$
$$R = \qquad \underline{\qquad}\text{ enter sum here}$$

Next we enter the characteristics in the blanks and have

$$\log 8.56 = 0.$$
$$\log 3.47 = 0.$$
$$\underline{\log 198 = 2.}$$
$$\log R =$$
$$R =$$

Now we turn to the tables, get the mantissas, and, as each is found, enter it in the proper place in the outline. Then we perform the addition and finally determine R by the method of Sec. 13.8. The completed solution then appears as

$$\log 8.56 = 0.9325$$
$$\log 3.47 = 0.5403$$
$$\underline{\log 198 = 2.2967}$$
$$\log R = 3.7695$$
$$R = 5.88(10^3)$$

Note   Each of the numbers in the problem contains only three digits. Hence, we can determine only three digits in R. Since the mantissa 7695 is between the two entries 7694 and 7701 and nearer the former than the latter, the first three digits in R are 588, the number corresponding to the mantissa .7694. The characteristic of $\log R$ is 3. Hence, the decimal point is three places to the right of the reference position. Therefore, we use scientific notation and multiply by 10^3 to place the decimal point.

We have written the outline of the solution three times in order to show how it appears at the conclusion of each step. In practice, it is necessary to write the outline only once, since only the original blanks are required for computing all the operations.

EXAMPLE 2   Use logarithms to find R, where

$$R = \frac{(337)(2.68)}{(521)(0.763)}$$

SOLUTION   In this problem R is a quotient in which the dividend and divisor are each the product of two numbers. Hence, we shall add the logarithms of the two numbers in the dividend and also add the logarithms of the two in the divisor, then subtract the latter sum from the former, and so obtain $\log R$. We suggest the following outline for the solution:

$$\log R = \log \frac{(337)(2.68)}{(521)(0.763)}$$

$$= \log (337)(2.68) - \log (521)(0.763)$$

$$= \log 337 + \log 2.68 - (\log 521 + \log 0.763)$$

$$\begin{array}{l} \log 337 = \\ \underline{\log 2.68 =} \\ \log \text{dividend} = \end{array} \quad \underline{}\text{enter sum here}$$

$$\begin{array}{l} \log 521 = \\ \underline{\log 0.763 =} \\ \log \text{divisor} = \end{array} \quad \underline{}\text{enter sum here}$$

$$\log R = \quad \underline{}\text{enter difference of the two sums here}$$

$$R =$$

After the characteristics are entered and the mantissas are found and listed in the proper places, the problem is completed as follows:

$$\begin{array}{ll} \log 337 = 2.5276 & \\ \underline{\log 2.68 = 0.4281} & \\ \log \text{dividend} = & 2.9557 \\ \log 521 = 2.7168 & \\ \underline{\log 0.763 = 9.8825 - 10} & \\ \log \text{divisor} = & \cancel{1}2.5993 - \cancel{10} \\ \log R = & 0.3564 \\ R = 2.27 & \end{array}$$

Note   The logarithm of the divisor turned out to be $12.5993 - 10$. Hence, the characteristic is 2. Therefore, in the above outline, we strike out the 10 and the first digit in 12 before completing the solution.

EXAMPLE 3   Use logarithms to evaluate

$$R = \frac{2.68}{33.2}$$

SOLUTION

$$\begin{array}{l} \log 2.68 = 0.4281 \\ \underline{\log 33.2 = 1.5211} \\ \log R = \end{array}$$

where $\log R$ is obtained by subtracting the second logarithm from the first. If we perform this subtraction, we get $\log R = -1.0930$. This is a correct value of $\log R$, but since the fractional part .0930 is negative, it is not a mantissa, and the value of R cannot be obtained from the table. We avoid this type of difficulty by adding $10 - 10$ to $\log 2.68$ before performing the subtraction. Then we have

$$
\begin{array}{rl}
\log 2.68 = & 10.4281 - 10 \\
\log 33.2 = & 1.5211 \\
\hline
\log R = & 8.9070 - 10 \\
R = & 0.0807
\end{array}
$$

We use the device shown in Example 3 whenever a necessary subtraction of one logarithm from another leads to a negative remainder. Thus, in order to subtract $9.2368 - 10$ from 2.6841, we add $10 - 10$ to the latter and have

$$
\begin{array}{r}
12.6841 - 10 \\
9.2368 - 10 \\
\hline
3.4473
\end{array}
$$

Similarly, in performing the indicated subtraction in $(7.3264 - 10) - (9.4631 - 10)$, we should again obtain a negative number, -2.1367. Hence, we add $10 - 10$ to $7.3264 - 10$ and proceed with the subtraction as follows:

$$
\begin{array}{r}
17.3264 - 20 \\
9.4631 - 10 \\
\hline
7.8633 - 10
\end{array}
$$

Powers and Roots

EXAMPLE 4

By use of logarithms, obtain the value of $R = (3.74)^5$.

SOLUTION

$$
\begin{aligned}
\log R &= \log (3.74)^5 \\
&= 5(\log 3.74) \\
&= 5(0.5729) \\
&= 2.8645
\end{aligned}
$$

Hence,

$$
R = 732
$$

We may also obtain the root of a number by means of logarithms. The method is illustrated in the following example:

EXAMPLE 5 Evaluate

$$R = \sqrt[3]{62.3}$$

SOLUTION   If we rewrite the problem in the exponential form, we get

$$R = (62.3)^{1/3}$$

Hence,

$$\log R = \tfrac{1}{3} \log 62.3$$
$$= \tfrac{1}{3}(1.7945)$$
$$= 0.5982$$

Therefore,

$$R = 3.96$$

In the application of logarithms to the problem of extracting a root of a decimal fraction, we use a device similar to that described in Example 3 in order to avoid a troublesome situation.

EXAMPLE 6
By use of logarithms evaluate $R = \sqrt[6]{0.0628}$.

SOLUTION

$$\log R = \tfrac{1}{6} \log 0.0628$$
$$= \frac{8.7980 - 10}{6}$$

If we perform the division indicated, we get

$$\log R = 1.4663 - 1.6667$$
$$= -0.2004$$

Thus, we have a negative logarithm and cannot obtain R from the table. We avoid a situation of this sort by adding $50 - 50$ to $\log 0.0628$ and obtaining $\log 0.0628 = 58.7980 - 60$. Now we have

$$\log R = \frac{58.7980 - 60}{6}$$
$$= 9.7997 - 10$$

The last logarithm is in the customary form, and by referring to the table, we find that

$$R = 0.631$$

MISCELLANEOUS PROBLEMS Many computation problems require a combination of the processes of multiplication, division, raising to powers, and the extraction of roots. We shall illustrate the general procedure for solving such problems by Example 7.

EXAMPLE 7 Use logarithms to find R if

$$R = \sqrt{\frac{\sqrt{2.689}\,(3.478)}{(52.18)^2(51.67)}}$$

SOLUTION Since all the numbers in this problem contain four digits, we must obtain the value of R to four places. Furthermore, we must use interpolation to obtain the mantissas. The steps in the solution are indicated in the following suggested outline:

$\log \sqrt{2.689} = \tfrac{1}{2} \log 2.689 = \tfrac{1}{2}(\qquad) =$
$\log 3.478 \qquad\qquad\qquad\qquad\qquad =$

$\qquad\qquad\qquad\qquad\qquad\qquad \log \text{dividend} = \qquad\qquad \underline{\qquad\qquad}\text{ sum}$

$\log (52.18)^2 = 2 \log 52.18 = 2(\qquad) =$
$\log 5167 \qquad\qquad\qquad\qquad\qquad =$

$\qquad\qquad\qquad\qquad\qquad\qquad \log \text{divisor} = \qquad\qquad \underline{\qquad\qquad}\text{ sum}$

$\qquad\qquad\qquad\qquad\qquad\qquad\qquad \log R = \qquad 5\underline{\qquad\qquad}\text{ difference}$

$\qquad\qquad\qquad\qquad\qquad\qquad\qquad\qquad R =$

We now enter the characteristics in the proper places, then turn to the table, get the mantissas, enter each in the space left for it, and complete the solution. Then the outline appears as here.

$\log \sqrt{2.689} = \tfrac{1}{2} \log 2.689 = \tfrac{1}{2}(0.4296) = 0.2148$
$\log 3.478 \qquad\qquad\qquad\qquad\qquad\qquad = 0.5413$

$\qquad\qquad\qquad\qquad \log \text{dividend} = \qquad\qquad 10.7561\dagger - 10$

$\log (52.18)^2 = 2 \log 52.18 = 2(1.7175) = 3.4350$
$\log 51.67 \qquad\qquad\qquad\qquad\qquad = 1.7132$

$\qquad\qquad\qquad\qquad \log \text{divisor} = \qquad\qquad\qquad 5.1482$
$\qquad\qquad\qquad\qquad\qquad\qquad\qquad\qquad 5\,|\,\overline{45.6079 - 50\ddagger}$
$\qquad\qquad\qquad\qquad\qquad \log R = \qquad\qquad\qquad 9.1216 - 10$
$\qquad\qquad\qquad\qquad\qquad\qquad R = 0.1323$

$\dagger$ Note that we add $10 - 10$ here so that we can subtract 5.1482.
$\ddagger$ We add $40 - 40$ here so that we can divide by 5.

Logarithmic Computation

Use logarithms to perform the computations indicated in Probs. 1 to 48. In Probs. 1 to 32, obtain the answers to three digits. In Probs. 33 to 48, obtain the answers to four digits.

1 $(76.5)(0.112)(11.9)$

2 $(37.5)(1.39)(8.16)$

3 $(54.3)(1.76)(0.796)$

4 $(9.13)(26.1)(0.843)$

5 $(0.0984)(0.0413)(0.875)$

6 $(6.95)(0.0907)(0.00595)$

7 $(0.583)(0.0312)(0.0827)$

8 $(0.392)(0.0752)(0.436)$

9 $\dfrac{(7.02)(0.0541)}{83.2}$

10 $\dfrac{(5.46)(62.3)}{97.6}$

11 $\dfrac{(58.3)(2.77)}{16.5}$

12 $\dfrac{(7.33)(6.26)}{5.05}$

13 $\dfrac{9.03}{(38.4)(4.29)}$

14 $\dfrac{932}{(27.6)(39.6)}$

15 $\dfrac{8.94}{(3.82)(26.3)}$

16 $\dfrac{6.87}{(3.25)(4.17)}$

17 $\sqrt[3]{(4.02)^2}$

18 $\sqrt[4]{0.695}$

19 $\sqrt[3]{0.571}$

20 $\sqrt[4]{0.587}$

21 $\sqrt{\dfrac{(215)(52.3)}{(255)(2.12)}}$

22 $\sqrt[3]{\dfrac{(6.53)(30.1)}{(0.612)(2.07)}}$

23 $\left[\dfrac{(0.0522)(4.51)}{(0.702)(822)}\right]^2$

24 $\sqrt{\left[\dfrac{(8.02)(0.786)}{(84.3)(7.26)}\right]^3}$

25 $\sqrt{80.5}\ \sqrt[3]{(5.68)^2}$

26 $\sqrt[3]{7.09}\ \sqrt{55.3}$

27 $\sqrt[4]{0.824}\ \sqrt{0.231}$

28 $\sqrt{87.3}\ \sqrt[3]{4.06}$

29 $\dfrac{\sqrt{673}}{\sqrt[3]{812}}$

30 $\dfrac{\sqrt[3]{93.6}}{\sqrt{38.4}}$

31 $\dfrac{366}{\sqrt[3]{139}\ \sqrt{27.4}}$

32 $\dfrac{(293)^{2/3}}{(35.6)\ \sqrt{44.8}}$

33 $\sqrt[3]{87.25}$

34 $\sqrt[4]{493.6}$

35 $\sqrt[4]{0.5762}$

36 $\sqrt[4]{0.6285}$

37 $\dfrac{\sqrt{848.5}}{\sqrt[3]{75.42}}$

38 $\dfrac{\sqrt[3]{351.6}}{\sqrt{28.35}}$

39 $\dfrac{\sqrt{263.7}}{33.71}$

40 $\dfrac{(773.3)^{2/3}}{(370.5)^{3/4}}$

41 $\dfrac{(55.82)(2.816)}{(482.7)(0.5715)}$

42 $\dfrac{(729.6)(8145)}{(3.526)(6.984)}$

43 $\dfrac{(564.2)(292.1)}{(0.4675)(0.1673)}$

44 $\dfrac{(22.33)(3.745)}{(285.2)(14.76)}$

45 $\sqrt{\dfrac{(40.476)(0.1751)^2}{3674}}$

46 $\sqrt[3]{\dfrac{(41.63)^2\sqrt{0.32513}}{376.3}}$

47 $\sqrt[3]{\dfrac{\sqrt{0.5134}\,(70.347)^3}{443.1}}$

48 $\left[\dfrac{(27.15)^2(3.054)}{335.24}\right]^4$

In Probs. 49 to 56, express the right member as the sum or difference of the logarithms of the first power of a number.

49 $y = \log \dfrac{cx}{x+4}$

50 $y = \log \dfrac{c^2x}{x-5}$

51 $y = \log cx^4\sqrt{x+1}$

52 $y = \log \dfrac{\sqrt{x+3}}{(x-1)^3}$

53 $z = \log \dfrac{c(x+y)^3}{\sqrt{x-y}}$

54 $z = \log \dfrac{cx^2y}{(x+y)^3}$

55 $z = \log \dfrac{c\sqrt[4]{x}}{x(x+y)^2}$

56 $z = \log \dfrac{x^3c}{y^3\sqrt{x+2y}}$

13.11
LOGARITHMS TO BASES OTHER THAN 10

There are only two bases ordinarily used for determining a system of logarithms. One of them is the base 10 (the system of logarithms to that base has been discussed at some length in this chapter) and the other is the irrational number $e = 2.718\ldots$.

Natural Logarithms

If e is used as the base, we have a system of *natural* logarithms. Such a system is often called the *Napierian* system after John Napier, who lived from 1550 to 1617. This system is used in calculus and for most noncomputational purposes, and tables of natural logarithms are included in many handbooks. However, we shall see how to find $\log N$ to any base if $\log N$ to some other base is given and shall make use of this relation to express $\log_e N$, usually written as $\ln N$, in terms of $\log_{10} N$. The relation is given by the following statement: *If a and b are any two acceptable bases, then*

■ $\log_a N = \dfrac{\log_b N}{\log_b a}$ (13.5)

PROOF We shall let $N = a^y$; then

$$\log_a N = \log_a a^y$$

$$= y$$

Furthermore,

$$\log_b N = \log_b a^y$$

$$= y \log_b a \qquad \text{by (13.4)}$$

$$= \log_a N \log_b a \qquad \text{since } y = \log_a N$$

Now, solving this equation for $\log_a N$, we have

■ $$\log_a N = \frac{\log_b N}{\log_b a}$$

Corollary

The relation between $\log_{10} N$ and $\log_e N$ is given by

$$\log_e N = \frac{\log_{10} N}{\log_{10} e} \tag{13.6}$$

PROOF If in (13.5), we let $a = e$ and $b = 10$, we have the equation as stated in the corollary.
    Since

$$\log_{10} e = 0.4343 \qquad \text{and} \qquad \frac{1}{0.4343} = 2.3026$$

Eq. (13.6) can be expressed in the form

■ $$\log_e N = 2.3026 \log_{10} N \tag{13.7}$$

EXAMPLE
By means of Eq. (13.5), find the value of $\log_7 236$.

SOLUTION If we substitute in (13.5), we get

$$\log_7 236 = \frac{\log_{10} 236}{\log_{10} 7}$$

$$= \frac{2.3729}{0.8451} = 2.808$$

13.12
LOGARITHMIC AND EXPONENTIAL EQUATIONS

An equation in which the variable appears in one or more exponents is an *exponential equation*. A *logarithmic equation* is one in which the variable appears in the logarithm of a function value. For example,

$$3^x = 7 \qquad \text{and} \qquad 2^{x-1} = 3^{x+y}$$

are exponential equations and

$$\log_2 x + \log_2(y - 1) = 2$$

is a logarithmic equation.

In general, exponential and logarithmic equations cannot be solved by the methods previously discussed. Many such equations, however, can be solved by use of the properties of logarithms. The following examples illustrate the procedure.

EXAMPLE 1 Solve the equation

$$3^{x+4} = 5^{x+2}$$

SOLUTION We first equate the logarithms of the members of the equation and then proceed as indicated.

$$\log 3^{x+4} = \log 5^{x+2}$$

$$(x + 4) \log 3 = (x + 2) \log 5 \qquad \text{by (13.4)}$$

$$x \log 3 + 4 \log 3 = x \log 5 + 2 \log 5$$

$$x(\log 3 - \log 5) = 2 \log 5 - 4 \log 3 \qquad \text{adding } -x \log 5 - 4 \log 3 \text{ to each member}$$

$$= \log 25 - \log 81 \qquad \text{by (13.4)}$$

$$x = \frac{\log 25 - \log 81}{\log 3 - \log 5} \qquad \text{solving for } x$$

$$= \frac{1.3979 - 1.9085}{0.4771 - 0.6990}$$

$$= \frac{-0.5106}{-0.2219}$$

$$= 2.301$$

EXAMPLE 2 Solve $\log_6 (x + 3) + \log_6 (x - 2) = 1$.

SOLUTION By applying the theorem for the logarithm of a product to the left member of the given equation, we have

$$\log_6 (x + 3)(x - 2) = 1$$

Hence,

$$(x + 3)(x - 2) = 6^1 \qquad \text{by (13.1)}$$

$$x^2 + x - 6 = 6 \qquad \text{performing indicated operations}$$

$$x^2 + x - 12 = 0$$

$$(x + 4)(x - 3) = 0$$

Solving this equation, we get $x = -4$ and $x = 3$, but we cannot admit -4 as a root since then $x + 3$ is negative, and we have not defined the logarithm of a negative number. Hence, the solution set is $\{3\}$.

EXAMPLE 3 Solve the exponential equations for x and y:

$$5^{x-2y} = 100 \tag{1}$$

and

$$3^{2x-y} = 10 \tag{2}$$

SOLUTION If we equate the logarithms of each member of Eq. (1) and of Eq. (2), we get

$$(x - 2y) \log 5 = 2 \tag{3}$$

$$\text{by (13.4)}$$

$$(2x - y) \log 3 = 1 \tag{4}$$

Therefore,

$$x - 2y = \frac{2}{\log 5} \tag{3'}$$

and

$$2x - y = \frac{1}{\log 3} \tag{4'}$$

If we perform the computation indicated in the right member, we have

$$x - 2y = 2.861 \tag{3''}$$

$$2x - y = 2.096 \tag{4''}$$

Multiplying each member of (4″) by 2 and subtracting from the corresponding member of (3″), we obtain

$$-3x = -1.331$$

$$x = 0.4437$$

Substituting this value for x in (3″) and solving for y, we get

$$-2y = 2.861 - 0.4437$$

$$= 2.4173$$

Therefore,

$$y = -1.209$$

Hence, the solution set of the system is

$\{0.4437, -1.209\}$

EXERCISE 13.5
Equations, $\log_b N,\ b \neq 10$

By use of a table of common logarithms, find the logarithm to three digits in each of Probs. 1 to 20.

1 $\log_e 36.2$	**2** $\log_e 8.17$	**3** $\log_e 1350$	**4** $\log_e 427$
5 $\log_5 18.3$	**6** $\log_5 9.31$	**7** $\log_5 139$	**8** $\log_5 1270$
9 $\log_4 63.2$	**10** $\log_4 9.13$	**11** $\log_4 6240$	**12** $\log_4 437$
13 $\log_2 862$	**14** $\log_2 13.7$	**15** $\log_2 5.43$	**16** $\log_2 4130$
17 $\log_6 9312$	**18** $\log_6 64.31$	**19** $\log_6 372.8$	**20** $\log_6 5.124$

In Probs. 21 to 44, find the value of x exactly or to four digits.

21 $3^{x+1} = 5^{x-1}$ **22** $2^{2x+1} = 3^{x+1}$ **23** $2^{2x+1} = 5^x$ **24** $7^{x-1} = 2^{2x}$

25 $5^{2x+1} = 7^{x+1}$ **26** $7^{x+2} = 5^{2x+1}$ **27** $3^{2x+1} = 13^{x-1}$ **28** $5^{2x+1} = 11^{x+1}$

29 $\log_6 (x + 2) + \log_6 (x + 3) = 1$ **30** $\log_8 (x + 2) + \log_8 (x + 9) = 1$

31 $\log_9 (x - 2) + \log_9 (2x + 3) = 1$ **32** $\log_{13} (x - 4) + \log_{13} (3x - 2) = 1$

33 $\log_2 (x + 1) + \log_2 (x + 3) = 3$ **34** $\log_5 (x + 3) + \log_5 (2x + 1) = 2$

35 $\log_7 (x + 3) + \log_7 (2x - 1) = 2$ **36** $\log_2 (x - 2) + \log_2 (2x + 3) = 2$

37 $\log_2 (3x + 1) - \log_2 (x + 1) = 1$ **38** $\log_3 (1 - 4x) - \log_3 (5 + x) = 1$

39 $\log_7 (5x + 9) - \log_7 (x - 1) = 1$ **40** $\log_2 (3x - 1) - \log_2 (x - 1) = 2$

41 $\log_2 (x^2 + x - 2) - \log_2 (x + 2) = 1$

42 $\log_7 (x^2 + 7x + 19) - \log_7 (x + 4) = 1$

43 $\log_2 (x^2 + 5x + 2) - \log_2 x = 3$

44 $\log_3 (x^2 - 5x + 3) - \log_3 (x + 4) = 3$

Solve the following pairs of equations simultaneously.

45 $3^{x+y} = 10$ $\ 3^{x-y} = 2$	**46** $2^{x+y} = 9$ $\ 2^{x-y} = 3$	**47** $5^{x+y} = 120$ $\ 5^{2x-y} = 113$	**48** $7^{2x-y} = 407$ $\ 7^{x-y} = 14$
49 $2^{x+2y} = 35$ $\ 3^{2x-y} = 237$	**50** $3^{2x-y} = 12$ $\ 5^{x+y} = 601$	**51** $3^{x+2y} = 222$ $\ 5^{x-2y} = 4$	**52** $5^{2x-y} = 136$ $\ 7^{x-2y} = 2374$

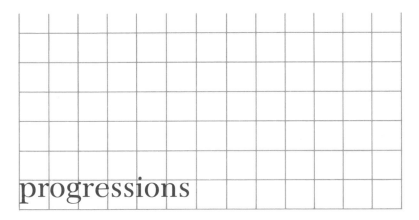

progressions

14

Sequences

Any collection of numbers is called a set, and an ordered set is called a *sequence* of numbers. Many sequences have properties that are intriguing to anyone who is interested in numbers merely as a plaything, other sequences are interesting because of their use in mathematics and its application, and others are both intriguing and useful. This chapter will be devoted to two types of sequences.

14.1
ARITHMETIC PROGRESSIONS

A sequence of numbers so related that each after the first can be obtained from the immediately preceding one by adding a fixed number is called an *arithmetic progression.*

Common Difference

The fixed quantity that is added is known as the *common difference.* Thus the sequence 2, 5, 8, 11, 14, and 17 is an arithmetic progression since each, after the first, can be obtained from the one just before it by adding the number 3; furthermore the sequence 7, 5, 3, 1, −1, and −3 is an arithmetic progression with −2 as the common difference.

Most problems in arithmetic progressions deal with three or more of the following five quantities: the first term, the last term, the number of terms, the common difference, and the sum of all the terms. Hence we shall derive formulas which enable us to determine any one of these five quantities if we know the values of three of the others.

We shall let

a = the first term in the progression

l = the last term

d = the common difference

n = the number of terms

s = the sum of all the terms

14.2
THE LAST TERM OF AN ARITHMETIC PROGRESSION

In terms of the notation just given, the first four terms of an arithmetic progression are

$$a \qquad a + d \qquad a + 2d \qquad a + 3d$$

We notice that d enters with the coefficient 1 in the second term and that this coefficient increases by 1 as we move from each term to the next. Hence, the coefficient of d in any term is 1 less than the number of that term in the progression. Therefore, the sixth term is $a + 5d$, the ninth is $a + 8d$, and finally the last, or nth, term is $a + (n − 1)d$. Hence, we have the formula

■ $\quad l = a + (n − 1)d$ $\qquad\qquad\qquad\qquad\qquad\qquad\qquad$ (14.1)

EXAMPLE 1

If the first three terms of an arithmetic progression are 2, 6, and 10, find the eighth term.

SOLUTION Since the first and second terms, as well as the second and third, differ by 4, it follows that $d = 4$. Furthermore, $a = 2$ and $n = 8$. Hence, if we substitute these values in (14.1), we have

$$l = 2 + (8 − 1)4$$

$$= 2 + 28$$

$$= 30$$

EXAMPLE 2

343

14.3 The Sum of
an Arithmetic
Progression

If the first term of an arithmetic progression is -3 and the eighth term is 11, find d and write the eight terms of the progression.

SOLUTION In this problem, $a = -3$, $n = 8$, and $l = 11$. If these values are substituted in (14.1), we have

$$11 = -3 + (8 - 1)d \quad \text{or} \quad 11 = -3 + 7d$$

Hence,

$$-7d = -14 \quad \text{and} \quad d = 2$$

Therefore, since $a = -3$, the first eight terms of the desired progression are $-3, -1, 1, 3, 5, 7, 9, 11$.

14.3
THE SUM OF AN ARITHMETIC PROGRESSION

In order to obtain the formula for the sum s of the n terms of an arithmetic progression in which the first term is a and the common difference is d, we note that the terms in the progression are a, $a + d$, $a + 2d$, and so on, until we reach the last term, which by Formula (14.1) is $l = a + (n - 1)d$. Hence,

$$s = a + (a + d) + (a + 2d) + \cdots + [a + (n - 1)d] \tag{1}$$

Since there are n terms in Eq. (1) and each term contains a, we may rearrange the terms and write s as

$$s = na + [1 + 2 + \cdots + (n - 1)]d \tag{2}$$

Now, if we reverse the order of the terms in the progression by writing l as the first term, then the second term is $l - d$, the third $l - 2d$, and so on to the nth term, which by (14.1) is $l + (n - 1)(-d)$. Hence we can write the sum as

$$s = l + (l - d) + (l - 2d) + \cdots + [l + (n - 1)(-d)]$$

Next, combining the l's and the d's, we get

$$s = nl - [1 + 2 + \cdots + (n - 1)]d \tag{3}$$

Finally, if we add the corresponding members of Eqs. (2) and (3), we have

$$2s = na + nl + 0 \quad \text{since coefficients of } d \text{ in Eqs. (2) and (3) are equal}$$
$$\text{in absolute value but have opposite signs}$$

$$= n(a + l)$$

Hence, dividing by 2, we obtain the formula

■ $s = \dfrac{n}{2}(a + l)$ (14.2)

This can be expressed in the form

■ $s = \dfrac{n}{2}[2a + (n - 1)d]$ (14.3)

by replacing l by $a + (n - 1)d$.

EXAMPLE 1

Find the sum of all the even integers from 2 to 1000 inclusive.

SOLUTION Since the even integers 2, 4, 6, etc., taken in order, form an arithmetic progression with $d = 2$, we may use (14.2) with $a = 2$, $n = 500$, and $l = 1000$ to obtain the desired sum. The substitution of these values in (14.2) yields

$s = \frac{500}{2}(2 + 1000)$

$ = 250(1002) = 250,500$

EXAMPLE 2

A man buys a used car for $600 and agrees to pay $100 down and $100 per month plus interest at 6 percent on the outstanding indebtedness until the car is paid for. How much will the car cost him?

SOLUTION The rate of 6 percent per year is 0.5 percent per month. Hence, when he makes his first payment, he will owe 1 month's interest on $500, or $(0.005)(\$500) = \$2.50$. Since he pays $100 on the principal, his interest from month to month is reduced by 0.5 per cent of $100 or by $.50 per month. The final payment will be $100 plus interest on $100 for 1 month, which is $100.50. Hence, his payments constitute an arithmetic progression with $a = \$102.50$, $l = \$100.50$, and $n = 5$. Therefore, by (14.2), the sum of his payments is

$s = \frac{5}{2}(\$102.50 + \$100.50)$

$ = \frac{5}{2}(\$203) = \$507.50$

Thus, the total cost of the car will be $607.50.

14.4

SIMULTANEOUS USE OF THE FORMULAS FOR l AND s

If any three of the quantities l, a, n, d, and s are known, the other two can be found by use of (14.1) and (14.2) or (14.3). If all three known quantities appear in either of the two formulas, the two unknowns can be found by use of the formulas separately. If, however, only two of the three known quantities appear in each of the formulas, we get the other two by solving (14.1) and (14.2) simultaneously.

345

14.4 Simultaneous
Use of the
Formulas for
l and *s*

EXAMPLE 1 Find d and s if

$$a = 4, \qquad n = 10, \qquad \text{and} \qquad l = 49$$

SOLUTION Since each of (14.1) and (14.2) contains a, n, and l, we may find d and s by using the formulas separately. If we substitute the given values for a, n, and l in (14.1) we get

$$49 = 4 + (10 - 1)d \qquad \text{or} \qquad 49 = 4 + 9d$$

Hence,

$$9d = 45 \qquad \text{and} \qquad d = 5$$

Similarly, substituting in (14.2), we have

$$s = \tfrac{10}{2}(4 + 49)$$
$$= 5(53)$$
$$= 265$$

EXAMPLE 2 Find a and n if

$$l = 23, \qquad d = 3, \qquad \text{and} \qquad s = 98$$

SOLUTION If we substitute these values in (14.1) and (14.2), we obtain

$$23 = a + (n - 1)3 \tag{1}$$

from the former and

$$98 = \frac{n}{2}(a + 23) \tag{2}$$

from the latter. Each of these equations contains the two desired unknowns a and n. Hence, we may complete the solution by solving Eqs. (1) and (2) simultaneously. If we solve Eq. (1) for a, we get

$$a = 23 - (n - 1)3 \qquad \text{or} \qquad a = 26 - 3n \tag{3}$$

Substituting this value for a in Eq. (2), we obtain

$$98 = \frac{n}{2}(26 - 3n + 23)$$

which we may solve for n as follows:

$196 = n(49 - 3n)$ multiplying each member by 2 and combining terms

$196 = 49n - 3n^2$ performing indicated operations

$3n^2 - 49n + 196 = 0$

$n = \dfrac{49 \pm \sqrt{(49)^2 - (4)(3)(196)}}{6}$ by quadratic formula

$ = \dfrac{49 \pm \sqrt{2401 - 2352}}{6}$

$ = \dfrac{49 \pm \sqrt{49}}{6}$

$ = \dfrac{49 \pm 7}{6}$

$ = \frac{28}{3}$ and 7

Since n cannot be a fraction, we discard $\frac{28}{3}$ and have

$n = 7$

If we substitute 7 for n in Eq. (3), we obtain

$a = 26 - 3(7)$

$ = 5$

Hence, the progression consists of the seven terms 5, 8, 11, 14, 17, 20, and 23.

EXERCISE 14.1
Arithmetic Progressions

Write the terms of the arithmetic progression that satisfy the conditions in each of Probs. 1 to 4.

1 $a = 5, d = 3, n = 6$
2 $a = 8, d = -2, n = 5$
3 Second term, 7; third term, 9; $n = 7$
4 Third term, -1; fourth term, 2; $n = 6$

In each of Probs. 5 to 20, find the values of the terms l, a, n, d, and s that are not given.

5 $a = 3, d = 2, n = 6$ 6 $a = -7, d = 3, n = 5$
7 $a = 17, d = -3, n = 4$ 8 $a = -5, d = -4, n = 5$

9 $a = 3, n = 5, l = 11$ 10 $a = 11, n = 6, l = 1$

11 $n = 5, l = 13, s = 35$ 12 $a = 4, l = 14, s = 54$

13 $a = -2, d = 3, l = 7$ 14 $a = 5, n = 5, s = 55$

15 $n = 8, d = 1, l = 4$ 16 $n = 6, d = 3, s = 21$

17 $n = 7, d = -5, s = 42$ 18 $a = 13, d = -4, s = 18$

19 $l = 6, d = 3, s = -21$ 20 $l = -4, d = -7, s = 81$

21 Find the sum of the first 61 positive odd integers.

22 If a compact body falls 16 ft during the first second, 48 ft during the next, 80 ft during the third, and so on, how far will it fall during the eighth second? During the first 8 sec?

23 A man invests $1000 at the beginning of each year for 6 years at 4 percent simple interest. How much will be in his account at the end of the 6 years?

24 A machine costs $3700 and depreciates 27 percent of the cost during the first year, 22 percent during the second year, 17 percent during the third year, and so on, for 6 years. Find its value at the end of the 6 years.

25 Mr. Tompkins earned $5100 in 1959 and had an increase of $350 in each of the next 9 years. How much did he earn in the 10-year period?

26 The promoters of a bazaar bought 14 prizes for $23.13 and sold them for the first 14 integral multiples of 25 cents. How much did they make?

27 An ill-prepared but capable and industrious student made 35 on the first of seven algebra tests. What was his average grade if he made 10 more on each quiz than he did on the immediately preceding one?

28 The first of two runners travels at 395 yd per min, and the second runs at a uniform rate for each minute but goes 50 yd further each minute than in the immediately preceding one. Which will win a mile race if the second one runs 320 yd in the first minute?

29 Show that ka, kb, kc is an arithmetic progression (AP) if a, b, c is an AP.

30 Find the replacement for a so that $a, 2a^2, 3a^3$ is an AP.

31 Find x and y so that the sum of the AP $x, y, 5x$ is 81.

32 If $x, y, x + y + 1$ is an AP with sum $7x$, find x and y.

14.5
GEOMETRIC PROGRESSIONS

A sequence of numbers so related that each, after the first, can be obtained from the immediately preceding one by multiplying it by a fixed quantity is called a *geometric progression*. The fixed quantity is known as the *common ratio*. Thus, the sequence 2, 6, 18, 54, 162 is a geometric progression since each, after the first, can be obtained from the one just before it by multiplying by 3; furthermore the sequence 8, -4, 2, -1, $\frac{1}{2}$ is a geometric progression with $-\frac{1}{2}$ as common ratio.

In order to obtain formulas for dealing with a geometric progression, we shall let

a = first term

l = last term

r = common ratio

n = number of terms

s = sum of the terms

14.6
THE LAST TERM OF A GEOMETRIC PROGRESSION

In terms of the above notation, the first six terms of a geometric progression in which the first term is a and the common ratio is r are

$$a \qquad ar \qquad ar^2 \qquad ar^3 \qquad ar^4 \qquad ar^5$$

We notice here that the exponent of r in the second term is 1, and that this exponent increases by 1 as we proceed from each term to the next. Hence, the exponent of r in any term is 1 less than the number of that term in the progression. Therefore, the nth term is ar^{n-1}, and we have the formula

■ $\qquad l = ar^{n-1}$ (14.4)

EXAMPLE

Find the seventh term of the geometric progression $36, -12, 4, \ldots$.

SOLUTION In this progression each term, after the first, is obtained by multiplying the preceding term by $-\frac{1}{3}$. Hence, $r = -\frac{1}{3}$. Obviously, $a = 36$, $n = 7$, and the seventh term is l. Hence, if we substitute these values in (14.4), we have

$$l = 36(-\tfrac{1}{3})^{7-1}$$

$$= \frac{36}{(-3)^6}$$

$$= \tfrac{36}{729}$$

$$= \tfrac{4}{81}$$

14.7
THE SUM OF A GEOMETRIC PROGRESSION

If we add the terms of the geometric progression $a, ar, ar^2, \ldots, ar^{n-2}, ar^{n-1}$, we have

$$s = a + ar + ar^2 + \cdots + ar^{n-2} + ar^{n-1}$$ (1)

By use of an algebraic device, however, we can obtain a more compact formula for s. First, we multiply each member of (1) by r and get

$$rs = ar + ar^2 + ar^3 + \cdots + ar^{n-1} + ar^n \tag{2}$$

Now, we subtract the corresponding members of (1) and (2), arrange terms in the difference, apply the distributive axiom, and get

$$s - rs = a + (a - a)r + (a - a)r^2 + \cdots + (a - a)r^{n-1} - ar^n$$

Hence, we have

$$s - rs = a - ar^n \qquad \text{or} \qquad s(1 - r) = a - ar^n$$

Solving the last equation for s, we obtain

$$\blacksquare \qquad s = \frac{a - ar^n}{1 - r} \qquad r \neq 1 \tag{14.5}$$

This can be expressed in the form

$$\blacksquare \qquad s = \frac{a - rl}{1 - r} \qquad r \neq 1 \tag{14.6}$$

by replacing ar^{n-1} by l.

EXAMPLE 1

Find the sum of the first six terms of the progression $2, -6, 18, \ldots$.

SOLUTION   In this progression $a = 2$, $r = -3$, and $n = 6$. Hence, if we substitute these values in (14.5), we have

$$s = \frac{2 - 2(-3)^6}{1 - (-3)}$$

$$= \frac{2 - 2(729)}{1 + 3}$$

$$= \frac{2 - 1458}{4}$$

$$= \frac{-1456}{4}$$

$$= -364$$

EXAMPLE 2

The first term of a geometric progression is 3; the fourth term is 24. Find the tenth term and the sum of the first 10 terms.

SOLUTION   In order to find either the tenth term or the sum, we must have the value of r. We may obtain this value by considering the progression made up of the first four terms. Then we have $a = 3$, $n = 4$, and $l = 24$. If we substitute these values in Formula (14.4), we get

$$24 = 3r^{4-1} \quad \text{or} \quad 3r^3 = 24$$

Hence,

$$r^3 = 8 \quad \text{and} \quad r = 2$$

Now, again using (14.4) with $a = 3$, $r = 2$, and $n = 10$, we get

$$l = 3(2^{10-1})$$
$$= 3(2^9)$$
$$= 3(512)$$
$$= 1536$$

Hence, the tenth term is 1536.

In order to obtain s, we shall use (14.5), with $a = 3$, $r = 2$, and $n = 10$, and get

$$s = \frac{3 - 3(2)^{10}}{1 - 2} = \frac{3 - 3(1024)}{-1} = \frac{3 - 3072}{-1} = 3069$$

14.8
SIMULTANEOUS USE OF THE FORMULAS FOR l AND s

If any three of the numbers s, n, a, r, and l are given, the other two can be found by use of Formulas (14.4) and (14.5) or (14.6). For example, if a, r, and n are given, we can find l and s by separate use of (14.4) and (14.5), but if l, s, and n are given, the two equations must be solved simultaneously to obtain a and r.

EXAMPLE 1 Find r and n if

$$a = 2, \quad l = 162, \quad \text{and} \quad s = 242$$

SOLUTION If we substitute the given values in (14.6), r is the only unknown and (14.6) becomes

$$242 = \frac{2 - r162}{1 - r}$$

$$242 - 242r = 2 - 162r \qquad \text{multiplying by } 1 - r$$

$$-80r = -240 \qquad \text{combining similar terms}$$

$$r = 3$$

If we now substitute $a = 2$ and $l - 162$ as given, along with $r = 3$, in (14.4), we have

351
14.8 Simultaneous
Use of the
Formulas for
l and s

$$162 = (2)3^{n-1}$$

$$81 = 3^{n-1} \qquad \text{dividing by 2}$$

$$3^4 = 3^{n-1} \qquad \text{since } 81 = 3^4$$

$$4 = n - 1$$

$$n = 5$$

EXAMPLE 2 Find a and l if

$$r = 2, \qquad s = 127, \qquad \text{and} \qquad n = 7$$

SOLUTION Since both unknowns enter in Eqs. (14.4) and (14.6) and only one in (14.5), we shall use (14.5) to determine a and then use (14.4) to find l. If we substitute in (14.5), we get

$$127 = \frac{a - a(2^7)}{1 - 2}$$

$$127 = -a + 128a$$

$$a = 1$$

Now substituting $r = 2$ and $n = 7$ as given, along with $a = 1$, in (14.4) yields

$$l = 1(2^6)$$

$$= 64$$

EXERCISE 14.2
Geometric Progressions

Write the n terms of the geometric progressions that have the elements given in Probs. 1 to 12.

1 $a = 3, r = 2, n = 6$ 2 $a = 2, r = 3, n = 5$

3 $a = 1, r = -3, n = 5$ 4 $a = -2, r = -2, n = 7$

5 $a = 5$; second term, 15; $n = 6$ 6 $a = 4$; second term, -8; $n = 5$

7 $a = -2$; third term, -8; $n = 7$ 8 $a = 3$; fourth term, -24; $n = 6$

9 Second term, 4; fifth term, -108; $n = 5$

10 Third term, 27; sixth term, 729; $n = 6$

11 Second term, -8; fourth term, -32; $n = 7$

12 Fourth term, 243; sixth term, 2187; $n = 7$

Find the nth term of the geometric progression described in each of Probs. 13 to 20.

13 $a = 4, r = 2, n = 5$

14 $a = 3, r = -3, n = 6$

15 $a = -1, r = 4, n = 4$

16 $a = 5, r = -2, n = 7$

17 $a = 243, r = \frac{1}{3}, n = 6$

18 $a = 256, r = \frac{1}{4}, n = 6$

19 $a = 1296, r = -\frac{2}{3}, n = 5$

20 $a = 15{,}625, r = \frac{1}{5}, n = 7$

Find the sum of each geometric progression described in Probs. 21 to 28.

21 $a = 32, r = \frac{1}{2}, n = 6$

22 $a = 243, r = \frac{1}{3}, n = 5$

23 $a = 2, r = -3, n = 7$

24 $a = -5, r = -2, n = 6$

25 $a = 1, r = 3, l = 243$

26 $a = 2, r = -2, l = 128$

27 $a = 1458, r = -\frac{1}{3}, l = 2$

28 $a = -3072, r = -\frac{1}{2}, l = -48$

Find the two of s, n, a, r, and l that are missing in each of Probs. 29 to 44.

29 $a = 2, r = 3, l = 162$

30 $a = 343, r = \frac{1}{7}, l = 1$

31 $n = 4, r = 3, l = 3$

32 $n = 6, r = 3, l = 27$

33 $n = 9, a = 256, l = 1$

34 $n = 7, a = \frac{1}{8}, l = 8$

35 $s = 400, r = \frac{1}{7}, l = 1$

36 $s = 126, r = \frac{1}{2}, l = 2$

37 $s = 242, a = 2, r = 3$

38 $s = \frac{3906}{125}, a = \frac{1}{125}, r = 5$

39 $s = -\frac{5}{8}, a = \frac{1}{8}, l = -1$

40 $s = 126, a = 64, l = 2$

41 $s = 781, n = 5, r = \frac{1}{5}$

42 $s = \frac{127}{8}, n = 7, r = 2$

43 $s = 448, l = 64, r = \frac{1}{2}$

44 $s = -\frac{104}{125}, n = 4, r = -5$

45 Show that $a^{1/2}, (ab)^{1/2}, (bc)^{1/2}$ is a geometric progression (GP) if $1, a, b, c$ is a GP.

46 Show that the products of corresponding terms of two GPs of n terms form a GP.

47 Show that the quotients of corresponding terms of two GPs of n terms form a GP.

48 Show that the squares of the terms of a GP form a GP.

49 Ten men are fishing from a pier. The first is worth $2000, the second $4000, the third $8000, and so on. Is there a millionaire among them? If so, how many?

50 If there are no duplicates, how many ancestors does a set of triplets have in the 11 generations just before them?

51 A man willed one-third of his estate to one friend, one-third of the remainder to another, and so on until the fifth friend received $1600. What was the value of the estate?

52 A lady started a chain letter by writing to three of her friends and asking that each write to three persons. If the chain was unbroken when the sixth set of letters was mailed, how much was spent for postage at 6 cents per letter?

53 The number of bacteria in a culture doubles every 2 hr. If there were n bacteria present at 5 a.m. one day, how many were there at 5 a.m. the next day?

54 The first stroke of a pump removes one-fourth of the air from a bell jar, and each stroke thereafter removes one-quarter of the remaining air. What part of the original amount is left after six strokes?

55 For what values of k are $2k$, $5k + 2$, and $20k - 4$ consecutive terms of a GP?

56 If it were possible for a person to save 1 cent on the first day of November, 2 cents on the second day, 4 cents on the third day, and so on for the entire month, find the amount he would save.

57 If 1, 4, and 19 are added to the first, second, and third terms, respectively, of an AP, with $d = 3$, a GP is obtained. Find the AP and the common ratio of the GP.

58 If a man deposits $200 at the beginning of each year in a bank that pays 4 percent compounded annually, how much will be to his credit at the end of 6 years?

59 If $1/(y - x)$, $1/2y$, and $1/(y - z)$ form an AP, prove that x, y, and z form a GP.

60 If $p \neq q$ and both are positive numbers, show that $\sqrt{pq}$ is less than $(p + q)/2$.

14.9
INFINITE GEOMETRIC PROGRESSIONS

In this section we discuss the sum of a geometric progression in which the common ratio is between -1 and 1 and the number of terms increases indefinitely. As an example, we shall consider the progression

$$1, \frac{1}{2}, \frac{1}{4}, \ldots, \frac{1}{2^{n-1}} \tag{1}$$

Here $a = 1$ and $r = \frac{1}{2}$. We shall let $s(n)$ stand for the sum of the first n terms in progression (1) and shall tabulate the corresponding values of n and $s(n)$ for $n = 1, 2, 3, \ldots, 7$.

n	1	2	3	4	5	6	7
s	1	$1\frac{1}{2}$	$1\frac{3}{4}$	$1\frac{7}{8}$	$1\frac{15}{16}$	$1\frac{31}{32}$	$1\frac{63}{64}$

The tabulated values show that as n increases from 1 to 7, $s(n)$ approaches nearer and nearer to 2. We now prove that this is true as n increases beyond

7 by replacing a by 1 and r by $\frac{1}{2}$ in (14.5). This replacement yields

$$s(n) = \frac{1 - (\frac{1}{2})^n}{1 - \frac{1}{2}}$$

$$= 2 - \frac{1}{2^{n-1}}$$

Now by choosing n sufficiently large, we can make $1/2^{n-1}$ less than any number ϵ selected in advance. Hence, for the chosen value of n, $s(n)$ differs from 2 by a number less than ϵ. This situation is described in mathematical language by the statement, "the limit of $s(n)$ as n approaches infinity is 2," and the statement is abbreviated thus:

$$\lim_{n \to \infty} s(n) = 2$$

We now consider a progression with $|r| < 1$, and shall prove that the limit of the sum s as n approaches infinity is given by the formula

$$\blacksquare \quad s = \frac{a}{1 - r} \qquad |r| < 1 \tag{14.7}$$

Note that in progression (1), $a = 1$ and $r = \frac{1}{2}$, and if we substitute these values in (14.7), we get $s = 2$.

The proof of (14.7) follows. We let $s(n)$ stand for the sum of the first n terms of a progression with $|r| < 1$. Then by (14.5),

$$s(n) = \frac{a - ar^n}{1 - r}$$

and this can be expressed in the form

$$s(n) = \frac{a}{1 - r}(1 - r^n) \tag{2}$$

We next notice that the binomial $1 - r^n$ in (2) differs from 1 by r^n. Furthermore, if $|r| < 1$, then $|r| > |r^2| > |r^3| > \cdots > |r^n|$, and it can be proved† that by choosing n sufficiently large, $|r^n|$ can be made less than any preassigned positive number. Hence, for this value and larger values of n, $1 - r^n$ differs from 1 by an amount that is less than the preassigned number. Hence,

$$\lim_{n \to \infty} (1 - r^n) = 1$$

Now we return to Formula (2) and complete the proof as follows:

---

†P. K. Rees and F. W. Sparks, "Algebra and Trigonometry," 2d ed. p. 420, McGraw-Hill Book Company, New York, 1969.

$$\lim_{n \to \infty} s_n = \lim_{n \to \infty} \frac{a}{1-r}(1 - r^n)$$

$$= \frac{a}{1-r} \lim_{n \to \infty} (1 - r^n)$$

$$= \frac{a}{1-r} \qquad \text{since } \lim_{n \to \infty} (1 - r^n) = 1$$

EXAMPLE 1

Find the sum of the infinite geometric progression with first term 2 and ratio $\frac{1}{3}$.

SOLUTION In this progression $a = 2$ and $r = \frac{1}{3}$ hence,

$$s = \frac{2}{1 - \frac{1}{3}} \qquad \text{substituting the given values in (14.7)}$$

$$= 3$$

Any nonterminating decimal fraction with a sequence of digits repeated indefinitely is an infinite geometric progression with r equal to a negative integral power of 10. For example,

$$.3333 \cdots = .3 + .03 + .003 + .0003 + \cdots$$

Here

$$a = .3 \qquad \text{and} \qquad r = .1$$

EXAMPLE 2

Find the rational number represented by $.351351351. \ldots$

SOLUTION This nonterminating, repeating decimal is equal to

$$.351 + .000351 + .000000351 + \cdots$$

It is now clear that $a = .351$ and $r = .001$; hence,

$$s = \frac{.351}{1 - .001} = \frac{.351}{.999} = \frac{13}{37}$$

EXAMPLE 3

A share in a mine was given to a college endowment fund. The dividend the first year was \$2500 and each year thereafter was 80 percent of the amount received the previous year. How much did the college receive?

SOLUTION In this problem $a = \$2500$ and $r = 0.8$. Consequently,

$$s = \frac{\$2500}{1 - 0.8} = \$12,500$$

14.10
ARITHMETIC MEANS

The terms between the first and last terms of an arithmetic progression are called *arithmetic means*. If the progression contains only three terms, the middle term is called *the arithmetic mean* of the first and last term. We may obtain the arithmetic means between two numbers by first using (14.1) to find d, and then the means can be computed. If the progression consists of the three terms a, m, and l, then by (14.1),

$$l = a + (3 - 1)d = a + 2d$$

Hence,

$$d = \frac{l - a}{2} \quad \text{and} \quad m = a + \frac{l - a}{2} = \frac{a + l}{2}$$

Therefore, the following is proved:

■ The arithmetic mean of two numbers is equal to one-half their sum.

EXAMPLE
Insert five arithmetic means between 6 and -10.

SOLUTION   Since we are to find five means between 6 and -10, we shall have seven terms in all. Hence, $n = 7$, $a = 6$, and $l = -10$.

Thus, by (14.1), we have

$$-10 = 6 + (7 - 1)d$$

Hence,

$$6d = -16$$
$$d = -\frac{16}{6} = -\frac{8}{3}$$

and the progression consists of the terms

$$6, \quad \frac{10}{3}, \quad \frac{2}{3}, \quad -\frac{6}{3}, \quad -\frac{14}{3}, \quad -\frac{22}{3}, \quad -\frac{30}{3}$$

14.11
GEOMETRIC MEANS

The terms between the first and last terms of a geometric progression are called the *geometric means*. If the progression contains only three terms, the middle term is called *the geometric mean* of the other two. In order to obtain the geometric means between a and l, we use (14.4) to find the value of r, and then the means can be computed. If there are only three terms in the progression, then by (14.4),

$$l = ar^2$$

Hence,

$$r = \pm \sqrt{\frac{l}{a}}$$

Thus the second term, or the geometric mean between a and l, is

$$a\left(\pm \sqrt{\frac{l}{a}}\right) = \pm \sqrt{\frac{a^2 l}{a}} = \pm \sqrt{al}$$

Hence, the following is proved:

■ The geometric means between two quantities are plus and minus the square root of their product.

EXAMPLE 1

Find two sets of five geometric means between 3 and 192.

SOLUTION A geometric progression starting with 3 and ending with 192 with five intermediate terms contains seven terms. Hence,

$n = 7$, $a = 3$, and $l = 192$. Therefore by (14.4),

$$192 = 3r^{7-1}$$

Hence,

$$r^6 = \frac{192}{3} = 64 \qquad \text{and} \qquad r = \pm \sqrt[6]{64} = \pm 2$$

Consequently, the two sets of geometric means of five terms each between 3 and 192 are

$$6, \quad 12, \quad 24, \quad 48, \quad 96 \qquad \text{and} \qquad -6, \quad 12, \quad -24, \quad 48, \quad -96$$

EXAMPLE 2

Find the geometric means of $\frac{1}{2}$ and $\frac{1}{8}$.

SOLUTION By the statement just before Example 1, the geometric means of $\frac{1}{2}$ and $\frac{1}{8}$ are

$$\pm \sqrt{\left(\tfrac{1}{2}\right)\left(\tfrac{1}{8}\right)} = \pm \sqrt{\tfrac{1}{16}} = \pm \tfrac{1}{4}$$

EXERCISE 14.3

Infinite Geometric Progressions, Arithmetic and Geometric Means

Find the sum of each geometric progression described in Probs. 1 to 12.

1 $a = 4$, $r = \frac{1}{2}$ **2** $a = 4$, $r = -\frac{1}{2}$ **3** $a = 5$, $r = -\frac{1}{4}$

4 $a = 5$, $r = \frac{4}{5}$ **5** $a = 10$; second term, 6

6 $a = 12$; second term, -8 **7** Second term, 12; third, 9

8 Second term, 20; third, -16 **9** Second term, 3; fifth, $\frac{1}{9}$

10 First term, 8; fourth, 1   **11** Second term, 16; fourth, 1

12 Third term, 1; fifth, $\frac{1}{9}$

Express each of the following repeating decimals as a rational number.

13 .888...   **14** .555...   **15** .606060...   **16** 8.38383...

17 .757575...   **18** 2.12121...   **19** 2.1060606...   **20** 21.351351...

Find the sum of all numbers of the form indicated in Probs. 21 to 24 if *n* is a positive integer.

21 3^{-n} **22** 5^{-n} **23** $\left(\frac{3}{5}\right)^n$ **24** $\left(\frac{3}{7}\right)^n$

25 If a ball rebounds two-thirds as far as it falls, how far will it travel before coming to rest if dropped 17 ft?

26 In rolling up an inclined plane a ball rolls three-fourths as far each second as in the immediately preceding one. If it rolls 6 ft the first second, what total distance will it roll?

27 If the midpoints of the sides of a square are joined in order, another square is formed. If one begins with an 8-in. square and forms a sequence of squares in the manner described, what is the limit of the sum of their areas as the number of squares increases beyond limit?

28 Find the limit of the sum of the perimeters of the squares described in Prob. 27.

29 The tip of a pendulum swings through an arc of 7 in., and thereafter, each arc is six-sevenths as long as the immediately preceding one. How far does the tip move before the pendulum comes to rest?

30 The day of his birth a child receives $55, and on each birthday thereafter ten-elevenths as much as on the preceding one is credited to his account. About how much will he have from this source when the young folks in the neighborhood think he is an old man?

31 A college was given the royalty from some oil wells. It paid $8000 the first year and increased by 50 percent of the amount received the previous year until six payments had been made. Each year thereafter it paid 80 percent of the amount of the previous year. About how much was received in 52 payments?

32 If stored potatoes shrink 5 percent the first week and one-half as much each week thereafter as during the previous one, can a dealer afford to hold 1000 lb of potatoes for 6.7 cents per lb that could have been sold for 6 cents per lb at the time they were stored? How much is gained by the proper choice?

33 If $x > 1$, find the sum of $1/x, 1/x^2, 1/x^3, \ldots$.

34 If $x > 0$, find the sum of $\dfrac{1}{2x+1}, \dfrac{1}{(2x+1)^2}, \dfrac{1}{(2x+1)^3}, \ldots$.

35 If $x > 2$, find the sum of $\dfrac{2}{3x-4}, \dfrac{4}{(3x-4)^2}, \dfrac{8}{(3x-4)^3}, \ldots$.

36 If $x > \frac{1}{2}$, find the sum of $\dfrac{4}{2x+3}, \dfrac{16}{(2x+3)^2}, \dfrac{64}{(2x+3)^3}, \ldots$.

Insert the specified number of arithmetic means between the given terms in Probs. 37 to 44.

37 One between 6 and 10

38 One between 4 and 11

39 Two between -2 and 7

40 Three between 5 and -3

41 Three between 3 and -9

42 Four between 41 and 76

43 Four between -17 and 18

44 Five between 6 and 18

Insert the specified number of geometric means between the given terms in Probs. 45 to 52.

45 One between 2 and 8

46 One between 1 and 81

47 Two between 4 and $\frac{1}{2}$

48 Two between 27 and 1

49 Three between -2 and $-\frac{1}{8}$

50 Three between $\frac{25}{36}$ and $\frac{36}{25}$

51 Four between 1 and 32

52 Four between $\frac{81}{16}$ and $\frac{2}{3}$

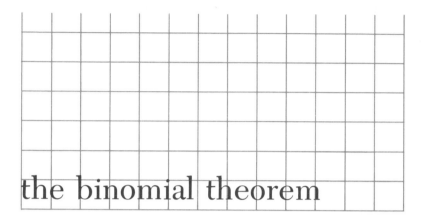

the binomial theorem

In earlier chapters we developed formulas for obtaining the square and the cube of a binomial. In this chapter we shall develop the binomial formula, which expresses the nth power of a binomial as a polynomial. This formula is known as the *binomial formula,* and it has many applications in more advanced mathematics and in all fields in which mathematics is applied. The polynomial yielded by the formula is called the *expansion of a power of a binomial.*

15.1
THE BINOMIAL FORMULA

By actual multiplication we may obtain the following expansions of the first, second, third, fourth, and fifth powers of $x + y$:

$$(x + y)^1 = x + y$$

$$(x + y)^2 = x^2 + 2xy + y^2$$

$$(x + y)^3 = x^3 + 3x^2y + 3xy^2 + y^3$$

$$(x + y)^4 = x^4 + 4x^3y + 6x^2y^2 + 4xy^3 + y^4$$

$$(x + y)^5 = x^5 + 5x^4y + 10x^3y^2 + 10x^2y^3 + 5xy^4 + y^5$$

By referring to these expansions, we may readily verify the fact that the following properties of $(x + y)^n$ exist when $n = 1, 2, 3, 4$, and 5:

1   The first term in the expansion is x^n.
2   The second term is $nx^{n-1}y$.
3   The exponent of x decreases by 1 and the exponent of y increases by 1 as we procced from term to term.
4   There are $n + 1$ terms in the expansion.
5   The $(n + 1)$st term, or the last term, is y^n.
6   The nth, or the next to the last, term of the expansion is nxy^{n-1}.
7   If we multiply the coefficient of any term by the exponent of x in that term and then divide the product by the number of the term in the expansion, we obtain the coefficient of the next term.
8   The sum of the exponents of x and y in any term is n.

If we assume that these properties hold for all integral values of n, we may write the first five terms in the expansion of $(x + y)^n$ as follows:

First term: x^n by property 1

Second term: $nx^{n-1}y$ by property 2

Third term: $\dfrac{n(n-1)}{2}x^{n-2}y^2$ by properties 7 and 3

Fourth term: $\dfrac{n(n-1)(n-2)}{(3)(2)}x^{n-3}y^3$ by properties 7 and 3

Fifth term: $\dfrac{n(n-1)(n-2)(n-3)}{(4)(3)(2)}x^{n-4}y^4$ by properties 7 and 3

We continue this process until we reach the nth term, which is

nth term: nxy^{n-1} by property 6

and finally we reach the last, or the

$(n + 1)$st term: y^n by property 5

Factorial n

The sum of the above terms is the expansion of $(x + y)^n$. If, however, we introduce a new notation, we can write the expansion in a slightly more compact form.

■   The product of any positive integer n and all the positive integers less than n is called *factorial n* and is designated by the symbol $n!$.

In keeping with this, we have

$$3! = 3 \times 2 \times 1 = 6 \qquad \text{and} \qquad 5! = 5 \times 4 \times 3 \times 2 \times 1 = 120$$

Now, if we notice that $4 \times 3 \times 2 = 4 \times 3 \times 2 \times 1 = 4!$, $3 \times 2 = 3 \times 2 \times 1 = 3!$, and $2 = 2 \times 1 = 2!$, we may write the expansion of $(x + y)^n$ as follows:

$$\blacksquare \quad (x + y)^n = x^n + nx^{n-1}y + \frac{n(n - 1)}{2!}x^{n-2}y^2 + \frac{n(n - 1)(n - 2)}{3!}x^{n-3}y^3$$
$$+ \frac{n(n - 1)(n - 2)(n - 3)}{4!}x^{n-4}y^4 + \cdots + nxy^{n-1} + y^n \quad (15.1)$$

Formula (15.1) is called the *binomial formula*, and the statement that it is true is called the *binomial theorem*. The proof of the binomial formula is not given in this book.

EXAMPLE 1

Use the binomial formula to obtain the expansion of $(2a + b)^6$.

SOLUTION   We shall first apply (15.1) with $x = 2a$, $y = b$, and $n = 6$. Then we shall simplify each term in the expansion. By (15.1),

$$(2a + b)^6 = (2a)^6 + 6(2a)^5b + \frac{(6)(5)}{2!}(2a)^4b^2 + \frac{(6)(5)(4)}{3!}(2a)^3b^3$$
$$+ \frac{(6)(5)(4)(3)}{4!}(2a)^2b^4 + \frac{(6)(5)(4)(3)(2)}{5!}(2a)b^5$$
$$+ \frac{(6)(5)(4)(3)(3)(2)(1)}{6!}b^6$$

Now we shall compute the coefficients and raise $2a$ to the indicated powers and obtain

$$(2a + b)^6 = 64a^6 + 6(32a^5)b + 15(16a^4)b^2 + 20(8a^3)b^3$$
$$+ 15(4a^2)b^4 + 6(2a)b^5 + b^6$$

Finally, we perform the indicated multiplication in each term above and get

$$(2a + b)^6 = 64a^6 + 192a^5b + 240a^4b^2 + 160a^3b^3 + 60a^2b^4 + 12ab^5 + b^6$$

The computation of the coefficients can, in most cases, be performed mentally by use of property 7, and thus we can avoid writing the first step in the expansion in this example.

EXAMPLE 2

Expand $(a - 3b)^5$.

SOLUTION   The first term in the expansion is a^5, and the second is $5a^4(-3b)$. To get the coefficient of the third, we multiply 5 by 4 and divide the product by 2, obtaining 10. Hence, the third term is $10a^3(-3b)^2$. Similarly, the fourth terms is $\frac{30}{3}a^2(-3b)^3 = 10a^2(-3b)^3$. Continuing this process, we obtain the following expansion:

$$(a - 3b)^5 = a^5 + 5a^4(-3b) + 10a^3(-3b)^2 + 10a^2(-3b)^3 + 5a(-3b)^4 + (-3b)^5$$

$$= a^5 - 15a^4b + 90a^3b^2 - 270a^2b^3 + 405ab^4 - 243b^5$$

Note that we carry the second term of the binomial $-3b$ through the first step of the expansion as a single term. Then we raise $-3b$ to the indicated power and simplify the result.

EXAMPLE 3

Expand $(2x - 5y)^4$.

SOLUTION   We shall carry through the expansion with $2x$ as the first term and $-5y$ as the second and get

$$(2x - 5y)^4 = (2x)^4 + 4(2x)^3(-5y) + 6(2x)^2(-5y)^2 + 4(2x)(-5y)^3 + (-5y)^4$$

$$= 16x^4 + 4(8x^3)(-5y) + 6(4x^2)(25y^2) + 4(2x)(-125y^3) + 625y^4$$

$$= 16x^4 - 160x^3y + 600x^2y^2 - 1000xy^3 + 625y^4$$

15.2
THE rth TERM OF THE BINOMIAL FORMULA

In the preceding examples we explained the method for obtaining any term of a binomial expansion from the term just before it. By use of this method, however, it is impossible to obtain any specific term of the expansion without first computing all the terms which precede it. We shall next develop a formula for finding the general, or rth term without reference to the other terms. Our development is an example of the method of inductive reasoning—a method that is very important in all scientific investigations.

We shall consider the fifth term in (15.1) and note the following properties:

1   The exponent of y in the fifth term is 1 less than the number 5 of the term in the expansion.
2   The exponent of x is n minus the exponent of y.
3   The denominator of the coefficient is the exponent of y followed by the exclamation point, or the factorial of the exponent of y.
4   The first factor in the numerator is n, and the last factor is n minus a number that is 2 less than the number of the term, or $n - (5 - 2)$, and the intervening factors are the consecutive integers between the first and the last factors.

We may also verify the fact that these properties are true for the other terms of the expansion.

We now assume that the properties hold for each term in the expansion, and hence, for the rth term we have:

1 The exponent of y is $r - 1$.
2 The exponent of x is $n - (r - 1) = n - r + 1$.
3 The denominator of the coefficient is $(r - 1)!$
4 The last factor in the numerator is $n - (r - 2) = n - r + 2$, and hence, the numerator is $n(n - 1)(n - 2) \cdots (n - r + 2)$.

Therefore we have the formula:

■ The rth term in the expansion of $(x + y)^n$ is

$$\frac{n(n - 1)(n - 2) \cdots (n - r + 2)}{(r - 1)!} x^{n-r+1} y^{r-1} \tag{15.2}$$

It should be noted that this formula is based on the *assumption* that the four properties following the second paragraph of this section hold for *all* terms in the expansion. The proof of this fact depends upon the method known as *mathematical induction*, which is beyond the scope of this book.

EXAMPLE
Find the sixth term in the expansion of $(2a - b)^9$.

SOLUTION In this problem,

$$x = 2a, \quad y = -b, \quad n = 9, \quad \text{and} \quad r = 6$$

Hence,

$$r - 1 = 5, \quad n - r + 1 = 9 - 6 + 1 = 4, \quad \text{and} \quad n - r + 2 = 5$$

Hence, if we substitute these values in (15.2), we get

$$\text{6th term} = \frac{(9)(8)(7)(6)(5)}{(5)(4)(3)(2)(1)} (2a)^4 (-b)^5$$

$$= 126(16a^4)(-b^5)$$

$$= -2016a^4 b^5$$

EXERCISE 15.1
Use of the Binomial Formula

Expand the binomials in Probs. 1 to 36.

1 $(x + y)^4$	2 $(c + d)^5$	3 $(a + y)^7$	4 $(b + c)^6$
5 $(m - t)^6$	6 $(s - a)^5$	7 $(a - y)^7$	8 $(x - s)^4$

9 $(a + 2x)^6$ **10** $(x + 3y)^3$ **11** $(4x - a)^7$ **12** $(5x - b)^4$

13 $(5x + y)^5$ **14** $(3a + b)^6$ **15** $(s - 5t)^5$ **16** $(h - 3y)^6$

17 $(2a + 3b)^4$ **18** $(3x + 4y)^5$ **19** $(5t - 3b)^6$ **20** $(4h - 3s)^5$

21 $(x^2 + 2y)^4$ **22** $(s^3 + 2h)^5$ **23** $(t^4 - 2s)^6$ **24** $(p^5 - 3q)^4$

25 $(2a + 3y^2)^4$ **26** $(3x + 5y^2)^5$ **27** $(2x - 7y^3)^3$ **28** $(4b - 3y^3)^5$

29 $(2x^2 + 3y^4)^3$ **30** $(3x^2 + 2y^3)^5$ **31** $(5x^3 - 3y^2)^6$ **32** $(2x^4 - 5y^3)^6$

33 $(\frac{1}{2}x + 2y^{1/3})^4$ **34** $\left(\dfrac{a}{3} + 3y^{2/3}\right)^3$ **35** $\left(\dfrac{2x}{5} - \dfrac{5y^{1/2}}{2}\right)^4$ **36** $\left(\dfrac{3b}{4} - \dfrac{4a^{1/3}}{3}\right)^5$

Find the first four terms in the expansion of the binomial in each of Probs. 37 to 44.

37 $(x + 3y)^{11}$ **38** $(a + 2b)^{17}$ **39** $(x - 3a)^{19}$ **40** $(s - 5t)^{13}$

41 $(x^2 + y)^{21}$ **42** $(x^3 + y^2)^{18}$ **43** $(a^2 - a^{-1})^{16}$ **44** $(b^3 - b^{-3})^{22}$

Evaluate the given power of the number in each of Probs. 45 to 48 to four decimal places.

45 $1.02^5 = (1 + 0.02)^5$ **46** 1.03^4 **47** 1.04^3 **48** 1.05^4

Find the specified term of the expansion in each of Probs. 49 to 64.

49 Fifth term of $(a + 3y)^{12}$ **50** Seventh term of $(2x - y)^{11}$

51 Sixth term of $(3x - b)^7$ **52** Eighth term of $(s + 2t)^9$

53 Fourth term of $(x^2 + 2y)^9$ **54** Sixth term of $(x^3 - y)^8$

55 Seventh term of $(3a - b^3)^{13}$ **56** Sixth term of $(2x + a^2)^{10}$

57 Middle term of $(x^{1/2} - y)^6$ **58** Middle term of $(a + 2b^{1/3})^8$

59 Middle term of $(c + 2b^{1/2})^{10}$ **60** Middle term of $(x^{1/2} - x^{1/3})^6$

61 The term that involves y^5 in $(x - 2y)^7$

62 The term that involves x^4 in $(x + 3y)^6$

63 The term that involves x^2 in $(2x^{1/3} + a)^7$

64 The term that involves d in $(c^2 - d^{1/4})^9$

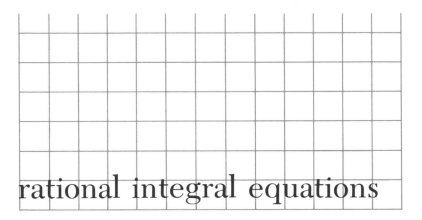

rational integral equations

An equation of the type

$$a_0 x^n + a_1 x^{n-1} + \cdots + a_{n-1} x + a_n = 0 \tag{16.1}$$

if n is a positive integer, $a_0 \neq 0$, and a_i, $i = 1, 2, 3, \ldots, n$ are constants, is called a *rational integral equation* of *degree n*. Such an equation is also called a *polynomial equation*. In previous chapters we presented methods for solving polynomial equations of degrees 1 and 2. The direct methods for solving such equations of degrees 3 and 4 are very tedious, and no direct method exists for solving a general polynomial equation of degree greater than 4. Fortunately, however, there are trial-and-error methods that enable us to obtain the rational roots, if such roots exist, and to compute the irrational roots to any degree of accuracy desired. We shall discuss these methods in this chapter.

16.1

THE REMAINDER THEOREM; THE FACTOR THEOREM
AND ITS CONVERSE

367
16.1 The
Remainder
Theorem; the
Factor Theorem
and Its Converse

The left member of (16.1)

$$a_0 x^n + a_1 x^{n-1} + \cdots + a_{n-1} x + a_n \qquad (16.2)$$

where n is a positive integer, $a_0 \neq 0$, and a_i, $i = 1, 2, 3, \ldots, n$ are constants, is called *rational integral* function, or a *polynomial*. We shall see that there is a direct connection between the factors of (16.2) and the roots of (16.1). For this reason we shall next present three theorems that are very useful in factoring (16.2).

The Remainder Theorem

The first of these three theorems is known as the *remainder theorem* and is stated as follows:

■   If a polynomial $f(x)$ is divided by $x - r$ until a remainder independent of x is obtained, then the remainder is equal to $f(r)$.

For example, if we divide $f(x) = x^3 - 2x^2 + 3x - 5$ by $x - 3$, we obtain the quotient $x^2 + x + 6$ and the remainder 13. Now, since in this problem $x - r = x - 3$, it follows that $r = 3$. Furthermore, $f(3) = 3^3 - 2(3^2) + 3(3) - 5 = 27 - 18 + 9 - 5 = 13$. Consequently, the remainder is equal to $f(3)$.

To prove the remainder theorem, we shall use the following identity which states the relation between the dividend, the divisor, the quotient, and the remainder:

$$\text{dividend} = (\text{quotient})(\text{divisor}) + \text{remainder}$$

Now if we replace the dividend by $f(x)$, the quotient by $q(x)$, the divisor by $x - r$, and the remainder by R, we obtain

■ $$f(x) = q(x)(x - r) + R \qquad (16.3)$$

Equation (16.3) is true for every value of x including $x = r$, and if we replace x by r, we obtain

$$f(r) = q(r)(r - r) + R$$

Hence,

$$f(r) = R$$

since

$$r - r = 0$$

Factor Theorem

The second theorem is known as the *factor theorem*; it is stated here and then proved.

■ If r is a root of the polynomial equation $f(x) = 0$, then $x - r$ is a factor of $f(x)$.

PROOF Since r is a root of $f(x) = 0$, it follows that $f(r) = 0$. Consequently, we know by the remainder theorem that if $f(x)$ is divided by $x - r$, the remainder is zero. Therefore by (16.3), we have

$$f(x) = q(x)(x - r)$$

Hence, $x - r$ is a factor of $f(x)$.

EXAMPLE 1

Use the factor theorem to show that $x + 2$ is a factor of

$$f(x) = x^3 - x^2 - 5x + 2$$

SOLUTION In order to apply the factor theorem to this problem, we first notice that $x - r = x + 2$. Consequently, $r = -2$. Now the factor theorem states that if $f(r) = 0$, then $x - r$ is a factor of $f(x)$. Since in this problem $r = -2$, we replace x by -2 in $f(x)$ and get

$$f(-2) = (-2)^3 - (-2)^2 - 5(-2) + 2$$

$$= -8 - 4 + 10 + 2$$

$$= -12 + 12$$

$$= 0$$

Therefore $x + 2$ is a factor of $f(x)$.

Converse of the Factor Theorem

The converse of the factor theorem follows:

■ If $x - r$ is a factor of the polynomial $f(x)$, then r is a root of $f(x) = 0$.

PROOF If $x - r$ is a factor of $f(x)$, then the remainder when $f(x)$ is divided by $x - r$ is zero. Hence, by the remainder theorem, $f(r) = 0$. Consequently $x = r$ is a root of $f(x) = 0$.

EXAMPLE 2

Use the converse of the factor theorem to show that 2 is a root of

$$x^3 - 2x^2 - 3x + 6 = 0$$

SOLUTION In order to apply the converse of the factor theorem to this problem, we shall first factor the left member of the equation as follows:

16.2 Synthetic
Division

$$x^3 - 2x^2 - 3x + 6 = x^2(x - 2) - 3(x - 2) \qquad \text{by distributive axiom}$$

$$= (x - 2)(x^2 - 3) \qquad \text{by distributive axiom}$$

Hence, since $x - 2$ is a factor of $x^3 - 2x^2 - 3x + 6$, then 2 is a root of

$$x^3 - 2x^2 - 3x + 6 = 0$$

16.2
SYNTHETIC DIVISION

In the application of the theorems in the preceding section, it will frequently be necessary to divide a polynomial by $x - r$. When the dividend is a polynomial and the divisor is of the form $x - r$, most of the steps in long division can be eliminated and the process can be reduced to a very short form called *synthetic division*. We shall illustrate the process by first dividing $2x^3 - 10x^2 + 7x - 9$ by $x - 3$ by long division and then indicating the steps that can be eliminated.

$$
\begin{array}{l}
2x^2 - 4x -5 \\
\overline{2x^3 - 10x^2 + 7x - 9}\,(x - 3 \\
(2x^3) - 6x^2 \\
\overline{} \\
-4x^2 + [7x] \\
(-4x^2) + 12x \\
\overline{} \\
-5x - [9] \\
(-5x) + 15 \\
\overline{} \\
-24
\end{array}
$$

We shall now examine the above process to ascertain the portions that can be eliminated without interfering with the essential steps. The division process requires that each term in the above problem that is enclosed in parentheses be the same as the term above it. Furthermore, the terms enclosed in brackets are terms in the dividend written in a new position. If these two sets of terms are eliminated, we have

$$
\begin{array}{l}
2x^2 - 4x -5 \\
\overline{2x^3 - 10x^2 + 7x - 9}\,(x - 3 \\
 - 6x^2 \\
\overline{} \\
 - 4x^2 \\
 + 12x \\
\overline{} \\
 - 5x \\
 + 15 \\
\overline{} \\
 - 24
\end{array}
$$

We can save space by writing $12x$ and 15 on the same line with $-6x^2$. Also it is not necessary to write the various powers of x, because the problem tells us what they should be. Then a shorter form for the work is

$$\frac{2-\ 4-\ 5}{2-10+\ 7-\ 9}\underline{(1-3}$$
$$\frac{-\ 6+12+15}{-\ 4-\ 5-24}$$

This method will be applied only when the divisor is $x - r$. Therefore, it is not necessary for the coefficient of x to appear in the divisor. Furthermore, in subtraction, we change the signs in the subtrahend and add. This change becomes automatic if we replace -3 in the divisor by 3. When these changes are made, the problem becomes

$$\frac{2-\ 4-\ 5}{2-10+\ 7-\ 9}\underline{(3}$$
$$\frac{6-12-15}{-\ 4-\ 5-24}$$

Finally, we write the 2 in the divisor as the first term in the last line. Then the numbers in the quotient are the same as the first three numbers in the last line. Therefore, the quotient can be omitted, and we have

$$\frac{2-10+\ 7-\ 9}{}\underline{(3}$$
$$\frac{6-12-15}{2-\ 4-\ 5-24}$$

Steps in Synthetic Division

Now we see that the essential steps in the process can be carried out mechanically as follows: Write the first 2 in the third line, multiply by 3 and place the product 6 under -10, add and obtain -4, multiply -4 by 3 and obtain -12, write -12 under 7, add and obtain -5, multiply -5 by 3, place the product under -9, and add and obtain -24. Then the coefficients in the quotient are 2, -4, and -5, and so the quotient is $2x^2 - 4x - 5$ and the remainder is -24.

Rule for Synthetic Division

This problem illustrates the following rule for synthetic division. In order to divide $f(x)$ by $x - r$ synthetically:

1   Arrange the coefficients of $f(x)$ in order of descending powers of x, supplying zero as the coefficient of each missing power.
2   Replace the divisor $x - r$ by r.
3   Bring down the coefficient of the largest power of x, multiply it by r, place the product beneath the coefficient of the second largest power of x, and add the product to that coefficient. Multiply the sum just obtained by r, place the product beneath the coefficient of the next larg-

est power of x, and add the product to the coefficient. Continue this process until there is a product added to the constant term.

4   The last number in the third row is the remainder, and the others, reading from left to right, are the coefficients of the quotient, which is of degree 1 less than $f(x)$.

For example, to divide $2x^4 + x^3 - 16x^2 + 18$ by $x + 2$ synthetically, we first notice that $x - r = x + 2$, so that $r = -2$. We write the coefficients of the dividend in a line, supplying zero as the coefficient of the missing power of x, and carry out the steps in the synthetic division and get

$$
\begin{array}{r}
2 + 1 - 16 \quad\ \ 0 + 18 \,\underline{(-2} \\
-4 + \ \ 6 + 20 - 40 \\
\hline
2 - 3 - 10 + 20 - 22
\end{array}
$$

Hence, the quotient is $2x^3 - 3x^2 - 10x + 20$, and the remainder is -22.

EXERCISE 16.1
Applications of the Factor and Remainder Theorems

By use of the remainder theorem, find the remainder when the polynomial in each of Probs. 1 to 8 is divided by the binomial.

1   $x^3 + 5x^2 - 7x + 4$; $x - 1$

2   $x^3 - 8x^2 + 6x + 3$; $x - 3$

3   $3x^3 - 2x^2 - 5x + 16$; $x + 2$

4   $3x^3 + 5x^2 - 3x + 2$; $x + 3$

5   $3x^4 + 7x^3 - 15x^2 + 16x - 15$; $x + 4$

6   $2x^4 + 7x^2 - 6x + 6$; $x + 3$

7   $2x^4 - 8x^3 - 9x^2 - 3x - 8$; $x - 5$

8   $3x^4 + x^3 - 5x^2 - 2x - 10$; $x - 2$

Using the factor theorem, show that the binomial in each of Probs. 9 to 16 is a factor of the polynomial.

9   $x^3 + 4x^2 - 8x + 3$; $x - 1$

10   $2x^3 + x^2 - 7x - 2$; $x + 2$

11   $2x^3 - 5x^2 - 4x + 12$; $x - 2$

12   $2x^3 + 6x^2 - 2x + 24$; $x + 4$

13   $5x^4 + 8x^3 + x^2 + 2x + 4$; $x + 1$

14   $3x^4 - 9x^3 - 4x^2 + 9x + 9$; $x - 3$

15   $2x^4 - 8x^3 - 9x^2 - 3x - 10$; $x - 5$

16   $6x^4 - 5x^3 + 10x^2 - 15x + 4$; $x - \frac{1}{3}$

By use of the converse of the factor theorem, find the solution set of the equation in each of Probs. 17 to 24.

17 $(x + 1)(x - 3)(x + 5) = 0$

18 $(x - 5)(x + 2)(x + 7) = 0$

19 $(x + \frac{1}{2})(x - \frac{1}{3})(x + \frac{2}{3}) = 0$

20 $(x - a)(x + b)(x - c) = 0$

21 $(x^2 + 2x - 8)(x - 3) = 0$

22 $(x^2 - x - 2)(x + 5) = 0$

23 $(2x^2 - x - 3)(3x^2 + 4x - 4) = 0$

24 $(3x^2 + 5ax - 2a^2)(2x^2 + bx - b^2) = 0$

Using synthetic division, find the quotient and remainder if the polynomial in each of the following problems is divided by the binomial.

25 $x^3 - 5x^2 + 2x + 3; \; x - 1$

26 $x^3 + 3x^2 + 4x + 2; \; x + 2$

27 $2x^3 + 6x^2 - 5x + 8; \; x + 4$

28 $-3x^3 + 11x^2 + 5x - 4; \; x - 4$

29 $x^5 - 3x^4 + 2x^3 - 5x^2 - 3x - 1; \; x - 3$

30 $-2x^4 - 6x^3 + 5x^2 - 7x + 8; \; x + 4$

31 $x^5 - 3x^4 + 9x^3 - 3x^2 - 16x - 10; \; x - 2$

32 $x^5 - 2x^3 + 2x; \; x + 2$

33 $-x^5 + 3x^4 + 2x^3 - 4x^2 - 5x - 5; \; x + 1$

34 $x^6 + 3x^2 - 3; \; x - 1$

35 $x^5 - 5x^4 + 3x^3 + 4x^2 - 2x + 8; \; x - 2$

36 $x^5 + 3x^4 - 2x^3 - 4x^2 + x + 8; \; x + 3$

16.3
THE NUMBER OF ROOTS OF A POLYNOMIAL EQUATION

In previous chapters we proved that an equation of the first degree has exactly one root and that an equation of the second degree has exactly two roots. In this section we shall prove that a polynomial equation of degree n has exactly n roots.

Fundamental Theorem of Algebra

The method for determining the number of roots of a polynomial equation depends upon the fundamental theorem of algebra which states that every polynomial equation has at least one root†

†For proof see L. E. Dickson, "First Course in Theory of Equations," p. 155, John Wiley & Sons, Inc., New York, 1922.

By this theorem, the equation

∎ $f(x) = a_0 x^n + a_1 x^{n-1} + \cdots + a_{n-1}x + a_n = 0$ (16.4)

has at least one root. If r_1 is this root, then $x - r_1$ is a factor of $f(x)$, and we can write $f(x) = (x - r_1)Q_1(x)$, where $Q_1(x)$ is a polynomial of degree $n - 1$. Again by this above theorem, $Q_1(x) = 0$ has at least one root r_2, and therefore $Q_1(x) = (x - r_2)Q_2(x)$, where $Q_2(x)$ is a polynomial of degree $n - 2$. Consequently,

∎ $f(x) = (x - r_1)(x - r_2)Q_2(x)$

By continuing this process, we obtain

∎ $f(x) = (x - r_1)(x - r_2) \cdots (x - r_n)Q_n(x)$ (16.5)

where Q_n is of degree $n - n = 0$ and is therefore a constant. From (16.5) we see that Q_n is the coefficient of x^n in $f(x)$ and, consequently, is equal to a_0. Hence, the factored form of $f(x)$ is

∎ $f(x) = a_0(x - r_1)(x - r_2)(x - r_3) \cdots (x - r_n)$ (16.6)

Multiplicity of a Root

By the converse of the factor theorem, r_1, r_2, r_3, $\ldots$, r_n are roots of $f(x) = 0$. Furthermore, $f(x)$ has no other roots since no one of the factors in (16.6) is zero for $x = r_{n+1}$ unless r_{n+1} is one of r_1, r_2, r_3, $\ldots$, r_n. If $r_1 = r_2 = r_3 = \cdots = r_s$, then

$f(x) = (x - r_s)^s(x - r_{s+1}) \cdots (x - r_n)$

and r_s is called a root of *multiplicity s*.

Theorem on the Number of Roots of a Polynomial Equation

Therefore we have the following theorem:

∎ A polynomial equation of degree n has exactly n roots, where a root of multiplicity s is counted as s roots.

16.4
UPPER AND LOWER BOUNDS OF THE REAL ROOTS

We stated earlier in this chapter that we shall use the method of trial and error for obtaining the rational roots or an approximation to the irrational roots of a polynomial equation of degree greater than 2. By this statement we mean that we select a number that we have reason to believe may be a root, then we replace x in the polynomial in the equation by this number, and thus determine whether or not the chosen number is a root. In this and in the next section we shall develop theorems that serve as guides for selecting the numbers to be tried. The following theorem gives us a criterion for deciding whether or not a given number is greater than the largest real root or less than the smallest real root.

Definition of the Upper and Lower Bounds of the Roots

Since each real root of a polynomial equation is a finite number, there is a number that is greater than, or equal to, the largest real root and another number that is less than, or equal to, the least real root. Any number that is greater than, or equal to, the greatest real root of an equation is called an *upper bound* of the real roots. Any number that is smaller than, or equal to, the least real root of an equation is called a *lower bound* of the real roots.

Theorem on the Upper and Lower Bounds of the Roots

We shall now state, illustrate, and then prove a theorem that enables us to determine these bounds.

■ If the coefficient of x^n in the polynomial equation $f(x) = 0$ is positive and if there are no negative terms in the third line of the synthetic division of $f(x)$ by $x - k$, $k > 0$, then k is an upper bound of the real roots of $f(x) = 0$. Furthermore, if the signs in the third line of the synthetic division of $f(x)$ by $x - (-k) = x + k$ are alternately plus and minus,† then $-k$ is a lower bound of the real roots.

As an example of this theorem, we shall find an upper bound and a lower bound of the roots of $f(x) = x^3 - 4x^2 - 2x - 13 = 0$. In order to obtain an upper bound, we must find a postive number k such that all signs in the third line of the synthetic division of $f(x)$ by $x - k$ are positive. It can readily be verified that if $f(x)$ is divided synthetically by $x - 1$, $x - 2$, or $x - 3$, the second term in the third line is negative, and if $f(x)$ is divided by $x - 4$, the third term in the third line is negative. Hence, neither 1, 2, 3, nor 4 is an upper bound. If, however, $f(x)$ is divided synthetically by $x - 5$, we have

$$\underline{\begin{array}{l} 1 - 4 - 2 - 13 \,\underline{(5} \\ 5 + 5 + 15 \end{array}} \\ 1 + 1 + 3 + 2$$

Therefore, since no negative terms occur in the third line, 5 is an upper bound of the roots. Furthermore, since 4 is not an upper bound, 5 is the least *integral* upper bound.

To determine a lower bound of the roots, we find a number $-k$, $k > 0$, such that the signs in the third line of the synthetic division of $f(x)$ by $x -$

---

† If one or more zeros occur in the third line of the synthetic division of $f(x)$ by $x + k$, $-k$ is a lower bound of the real roots if, after each zero is replaced by either a plus or a minus sign, the signs in the third line are alternately plus and minus.

$(-k) = x + k$ are alternately plus and minus. If we try $x + 1$, we get

$$1 - 4 - 2 - 13 \,\underline{(-1}$$
$$\underline{-1 + 5 - 3}$$
$$1 - 5 + 3 - 16$$

Hence, since the signs in the third line are alternately plus and minus, -1 is a lower bound of the roots. Consequently, the real roots of the given equation are between -1 and 5.

To prove the first part of the theorem, we shall divide the polynomial $f(x)$ by $x - k$, $k > 0$, and get the quotient $Q(x)$ and the remainder R. Then

$$f(x) = Q(x)(x - k) + R$$

The last term in the third line of the synthetic division of $f(x)$ by $x - k$ is R, and the other terms in this line are the coefficients in $Q(x)$. The first of these terms is positive, since it is the coefficient of x^n in $f(x)$. If the other terms in $Q(x)$ are positive or zero, then all terms in $Q(x)$ are positive or zero. Hence, if $s > k$, we have $Q(s)(s - k) > 0$ and $R > 0$. Therefore $f(s) > 0$, and s is not a root of $f(x) = 0$. Therefore k is an upper bound.

To prove the second part of the theorem, we shall assume that the signs in the third line of the synthetic division of $f(x)$ by $x - (-h) = x + h$ are alternately plus and minus. We shall further assume that the quotient is $q(x)$ and the remainder is R'. Thus, we have

$$f(x) = q(x)(x + h) + R'$$

Again the coefficients in $Q(x)$ and R' are the numbers in the third line of the synthetic division of $f(x)$ by $x + h$ and are, therefore, by our assumption, alternately plus and minus.

We shall first consider the case in which the degree of $f(x)$ is n, where n is a positive odd integer. Then the degree of $q(x)$ is the even integer $n - 1$. By referring to Eq. (16.1), we see that a polynomial of degree n has $n + 1$ terms. Hence $q(x)$ has n terms. Furthermore, the coefficient of the first term in $q(x)$ is positive, since it is the coefficient of the first term in $f(x)$, and the other terms are alternately minus and plus. Therefore, since there is an odd number of terms in $q(x)$, the final term is positive and R' is negative. Also, since the exponent of x in the first term of $q(x)$ is the even number $n - 1$ and the term is positive, all terms in $q(x)$ involving even exponents are positive and those involving odd exponents are negative. We may therefore write $q(x)$ in the form

$$q(x) = b_1 x^{n-1} - b_2 x^{n-2} + b_3 x^{n-3} - \cdots + b_{n-1}$$

where b_i, $i = 1, 2, 3, \ldots, n - 1$, is positive. Now if $-t < -h$, we have

$$f(-t) = [b_1(-t)^{n-1} - b_2(-t)^{n-2} + b_3(-t)^{n-3} - \cdots + b_{n-1}](-t + h) - R''$$

where R'' is positive. Since the exponents in the positive terms of $q(x)$ are even and the exponents in the negative terms are odd, $q(-t)$ is positive. Furthermore, $-t + h$ is negative and $-R''$ is negative. Therefore $f(-t)$ is negative and $-t$ is not a root of $f(x) = 0$. Hence, $-h$ is a lower bound of the roots.

By a similar argument we can prove that if $-t < -h$, if n is even, and if the signs in the third line of the synthetic division of $f(x)$ by $x + h$ are alternately plus and minus, then $f(-t) = q(-t)(-t + h) + R'$ is positive, and again $-t$ is not a root of $f(x) = 0$.

16.5
RATIONAL ROOTS OF A POLYNOMIAL EQUATION

In solving a polynomial equation, it is advantageous to determine the rational roots first, provided rational roots exist. In this section we discuss a theorem that enables us to select a set of rational numbers that contains the rational roots of the equation as a subset, provided that all the coefficients in the equation are integers. For example, the theorem enables us to say that if $3x^3 - 4x^2 + x + 2 = 0$ has a rational root, it must be one of the set of numbers $\pm 1, \pm 2, \pm\frac{1}{3}, \pm\frac{2}{3}$. We shall next state and prove the theorem and shall then illustrate its use.

Theorem on Rational Roots

■ If p and q are integers that have no common factor greater than 1, and if q/p is a root of the polynomial equation

$$a_0 x^n + a_1 x^{n-1} + \cdots + a_{n-1} x + a_n = 0 \tag{16.7}$$

where a_i, $i = 0, 1, 2, 3, \ldots, n$, are integers, then q is a factor of a_n and p is a factor of a_0.

PROOF If we replace x by q/p in Eq. (16.7), we have

$$a_0\left(\frac{q}{p}\right)^n + a_1\left(\frac{q}{p}\right)^{n-1} + \cdots + a_{n-1}\left(\frac{q}{p}\right) + a_n = 0 \tag{16.8}$$

since q/p is a root of the equation. We next multiply each member of (16.8) by p^n and get

$$a_0 q^n + a_1 p q^{n-1} + \cdots + a_{n-1} p^{n-1} q + a_n p^n = 0 \tag{16.9}$$

Finally, we add $-a_n p^n$ to each member of (16.9), divide by q, and have

$$a_0 q^{n-1} + a_1 p q^{n-2} + \cdots + a_{n-1} p^{n-1} = -\frac{a_n p^n}{q} \tag{16.10}$$

The left member of Eq. (16.10) is an integer since each of its terms is the product of an integer and the power of an integer. Hence, the right member

$a_n p^n / q$ is an integer. Consequently, since p and q have no common factor greater than 1, it follows that q is a factor of a_n.

To prove that p is a factor of a_0, we add $-a_0 q^n$ to each member of (16.9), then divide by p, and obtain

$$a_1 q^{n-1} + \cdots + a_{n-1} p^{n-2} q + a_n p^{n-1} = -\frac{a_0 q^n}{p} \tag{16.11}$$

Consequently, since the left member of (16.11) is an integer, $a_0 q^n / p$ is an integer, and since q and p have no common factor greater than 1, it follows that p is a factor of a_0.

EXAMPLE 1

Find the set of numbers that contains all the rational roots of

$$3x^3 + 2x^2 - 19x + 6 = 0 \tag{1}$$

as a subset, and then find all the rational roots.

SOLUTION By comparing Eq. (1) with Eq. (16.7), we see that $a_0 = 3$ and $a_n = a_3 = 6$. Hence, by the theorem on rational roots, the numerator of each root is a factor of 6 and the denominator is a factor of 3. Consequently, the possibilities for the numerators of the rational roots are $\pm 6, \pm 3, \pm 2$, and ± 1, and for the denominators, ± 3 and ± 1. Therefore, the set of numbers that contains the rational roots of (1) as a subset is

$$\{\pm \tfrac{6}{3}, \ \pm \tfrac{3}{3}, \ \pm \tfrac{2}{3}, \ \pm \tfrac{1}{3}, \ \pm 6, \ \pm 3, \ \pm 2, \ \pm 1\}$$

If we eliminate duplications and arrange the numbers in order of magnitude, we have

$$\{-6, \ -3, \ -2, \ -1, \ -\tfrac{2}{3}, \ -\tfrac{1}{3}, \ \tfrac{1}{3}, \ \tfrac{2}{3}, \ 1, \ 2, \ 3, \ 6\}$$

In order to select the rational roots from the above set, we can test them one by one in the left member of Eq. (1) by synthetic division. It is usually advisable to test the positive integers first, starting with the smallest. The test applied to the number 1 reveals that it is neither a root nor a bound of the roots. However, when we test 2, we get

$$
\begin{array}{r}
3 + 2 - 19 + 6\ \underline{(2} \\
\underline{+6 + 16 - 6} \\
3 + 8 - 3 \quad 0
\end{array}
$$

Hence 2 is a root, and the quotient obtained by dividing the left member of Eq. (1) by $x - 2$ is

$$3x^2 + 8x - 3$$

Consequently, we can express (1) in the form

$$(x - 2)(3x^2 + 8x - 3) = 0$$

from which we can see that any root of

$$3x^2 + 8x - 3 = 0 \tag{2}$$

is also a root of Eq. (1). Hence, we complete the process of solving Eq. (1) by finding the roots of Eq. (2). We solve this equation by the factoring method as follows:

$(3x - 1)(x + 3) = 0$ factoring left member of Eq. (2)

$x = \frac{1}{3}$ setting first factor equal to zero and solving resulting equation

$x = -3$ setting second factor equal to zero and solving

Hence, the solution set of Eq. (1) is $\{2, \frac{1}{3}, -3.\}$

The Depressed Equation

Example 1 illustrates the meaning and use of the depressed equation, which we next define.

■ If the polynomial equation $f(x) = 0$ is expressed in the form $f(x) = (x - r)q(x) = 0$, then the equation $q(x) = 0$ is called the *depressed equation* of $f(x) = 0$ that corresponds to the root r.

Since $f(x) = (x - r)q(x)$, it follows that any root of $q(x) = 0$ is also a root of $f(x) = 0$. Furthermore, if $f(x)$ is of degree n, then $q(x)$ is of degree $n - 1$. Consequently, the computation involved in finding the roots of $f(x) = 0$ is simplified if after one root is found, the corresponding depressed equation is used in searching for the remaining roots. The use of the depressed equation is further illustrated in Example 2.

EXAMPLE 2

Find all roots of

$$6x^4 + 5x^3 - 27x^2 + 2x + 8 = 0 \tag{1}$$

SOLUTION By the theorem on rational roots, the numerator of each root is a factor of 8 and the denominator is a factor of 6. Consequently, the set of numbers that contains all the rational roots of Eq. (1) as a subset is

$$\{-8, \quad -4, \quad -\tfrac{8}{3}, \quad -2, \quad -\tfrac{4}{3}, \quad -1, \quad -\tfrac{2}{3}, \quad -\tfrac{1}{2}, \quad -\tfrac{1}{3},$$
$$-\tfrac{1}{6}, \quad \tfrac{1}{6}, \quad \tfrac{1}{3}, \quad \tfrac{1}{2}, \quad \tfrac{2}{3}, \quad 1, \quad \tfrac{4}{3}, \quad 2, \quad \tfrac{8}{3}, \quad 4, \quad 8\}. \tag{2}$$

We shall now test each of these numbers one at a time by synthetic division, using the left member of Eq. (1) as the dividend. We first test the positive integers, starting with 1. The test reveals that 1 is neither a root nor an upper bound and that 2 is not a root but is an upper bound. Hence, we discard all numbers in the set (2) that are greater than 2 and start testing the negative integers. This procedure reveals that neither -1 nor -2 is a root or a lower bound. Furthermore, -4 is not a root but is a lower bound. Consequently, we discard -8 in the set (2) and start testing the fractions, beginning with $\frac{1}{6}$. We find that $\frac{1}{6}$, $\frac{1}{3}$, and $\frac{1}{2}$ are neither roots nor upper bounds. When we test $\frac{2}{3}$, however, we obtain

$$\begin{array}{r} 6 + 5 - 27 + 2 + 8 (\tfrac{2}{3} \\ \underline{4 + 6 - 14 - 8} \\ 6 + 9 - 21 - 12 0 \end{array}$$

Hence $\frac{2}{3}$ is a root, and the corresponding depressed equation is

$$6x^3 + 9x^2 - 21x - 12 = 3(2x^3 + 3x^2 - 7x - 4) = 0$$

We can now write Eq. (1) in the form

$$3(x - \tfrac{2}{3})(2x^3 + 3x^2 - 7x - 4) = 0$$

and see that any root of the equation

$$2x^3 + 3x^2 - 7x - 4 = 0$$

is a root of Eq. (1) and that any root of Eq. (1) other than $\frac{2}{3}$ is a root of Eq. (3). Consequently, we complete the process of solving Eq. (1) by solving Eq. (3).

The set of numbers that contains the roots of Eq. (3) as a subset is

$$\{-4, \quad -2, \quad -1, \quad -\tfrac{1}{2}, \quad \tfrac{1}{2}, \quad 1, \quad 2, \quad 4\}$$

We discard 4, however, since it was found to be an upper bound. We also discard all numbers previously tested, and thus have only $-\frac{1}{2}$ remaining. When we test $-\frac{1}{2}$, we get

$$\begin{array}{r} 2 + 3 - 7 - 4 (-\tfrac{1}{2} \\ \underline{-1 - 1 + 4} \\ 2 + 2 - 8 0 \end{array}$$

Therefore, $-\frac{1}{2}$ is a root, and the depressed equation is

$$2x^2 + 2x - 8 = 0$$

or, dividing by 2,

$$x^2 + x - 4 = 0 \tag{4}$$

Equation (4) is quadratic; we solve it by the quadratic formula and get

$$x = \frac{-1 \pm \sqrt{1 + 16}}{2} = \frac{-1 \pm \sqrt{17}}{2}$$

Consequently, the solution set of Eq. (1) is

$$\{-\tfrac{1}{2}, \quad \tfrac{2}{3}, \quad \tfrac{1}{2}(-1 + \sqrt{17}), \quad \tfrac{1}{2}(-1 - \sqrt{17})\}$$

We suggest the following procedure as a systematic and efficient method for investigating the rational roots of a polynomial equation.

Steps in the Process of Obtaining All Rational Roots

1 List the set of numbers that contains the set of rational roots as a sub-set.
2 Test the positive integers in the set, starting with the smallest. If an upper bound that is not a root is found, discard this bound and all larger numbers in the set. If a root is found, use the depressed equation for the remainder of the investigation.
3 Repeat the procedure for negative integers, using the depressed equation if a positive root has been found.
4 Test the fractions that remain after considering any bound and any root that has been found.

EXERCISE 16.2
Bounds, Rational Roots

In each of Probs. 1 to 8, state the degree of the equation. Find all roots and the multi-plicity of each. Verify the fact that the sum of the multiplicities is equal to the degree of the equation.

1 $(x + 3)^2(x - 2)^3 = 0$
2 $(x - 1)^3(x + 2)(x - 4)^2 = 0$
3 $(x + 5)^4(x + 1)^2(x - 3) = 0$
4 $(x - 6)^3(x + 2)^2(x - 4)^4 = 0$
5 $(3x + 1)^2(2x - 1)^3 = 0$
6 $(2x - 5)^3(x + 2)^2(5x + 1) = 0$
7 $(2x - 7)(3x + 8)^4(4x + 5)^5 = 0$
8 $(5x + 1)^2(3x - 1)^7(2x + 5)^2 = 0$

In each of Probs. 9 to 20, determine the integral lower and upper bounds of the roots by use of the theorem in Sec. 16.4.

9 $2x^3 - 9x^2 + 6x + 2 = 0$
10 $3x^3 + 10x^2 - 2x - 4 = 0$
11 $3x^3 - 7x^2 - 10x + 4 = 0$
12 $2x^3 - 3x^2 - 3x + 3 = 0$

13 $3x^3 + x^2 - 6x + 1 = 0$ **14** $x^3 - 3x^2 - x + 2 = 0$

15 $x^4 - 2x^3 - 11x^2 + 14x + 28 = 0$

16 $x^4 - 2x^3 - 16x^2 + 22x + 7 = 0$

17 $x^4 + 2x^3 - 16x^2 - 12x + 12 = 0$

18 $x^4 - 10x^2 + 4x + 8 = 0$

19 $x^4 - 23x^2 - 6x + 42 = 0$

20 $x^4 - 2x^3 - 20x^2 - 8x + 4 = 0$

Find the solution set of each of the following equations.

21 $x^3 - 4x^2 + x + 6 = 0$ **22** $x^3 - 2x^2 - 11x + 12 = 0$

23 $x^3 + 3x^2 - 10x - 24 = 0$ **24** $x^3 + 4x^2 - 11x - 30 = 0$

25 $2x^3 + x^2 - 13x + 6 = 0$ **26** $3x^3 - 8x^2 - 5x + 6 = 0$

27 $2x^4 - 9x^3 - 3x^2 + 34x - 24 = 0$

28 $2x^4 + 3x^3 - 12x^2 - 7x + 6 = 0$

29 $4x^3 + 20x^2 + 19x - 15 = 0$ **30** $6x^3 - 11x^2 - 24x + 9 = 0$

31 $6x^4 - x^3 - 46x^2 + 39x + 18 = 0$

32 $8x^4 - 6x^3 - 47x^2 - 24x + 9 = 0$

33 $32x^3 + 16x^2 - 42x + 9 = 0$ **34** $12x^3 + 16x^2 - 7x - 6 = 0$

35 $36x^4 + 36x^3 - x^2 - 9x - 2 = 0$

36 $36x^4 + 72x^3 - 31x^2 - 67x + 30 = 0$

37 $x^3 - 4x^2 + 3x + 2 = 0$ **38** $x^3 - 6x - 4 = 0$

39 $x^3 + 5x^2 + x - 7 = 0$ **40** $x^3 - x^2 - 10x + 12 = 0$

16.6
THE GRAPH OF A POLYNOMIAL

Since we shall use a process for obtaining an approximation to the irrational roots of the polynomial equation $f(x) = 0$ that depends upon the graph of $y = f(x)$, an understanding of this graph is essential. The procedure for constructing the graph of $y = f(x)$ is the same as that used in the preceding chapters except that synthetic division may be used for obtaining the corresponding values of x and y. Except for the values of x in a comparatively short interval, the absolute value of y increases much more rapidly than the value of x. For this reason it is usually advisable to use different scales on the X and Y axes. For example, in Figs. 16.1 and 16.2, the unit length on the X axis is four times the unit length on the Y axis. We shall illustrate the method with the following example.

EXAMPLE

Construct the graph of $y = 2x^3 - 3x^2 - 12x + 16$.

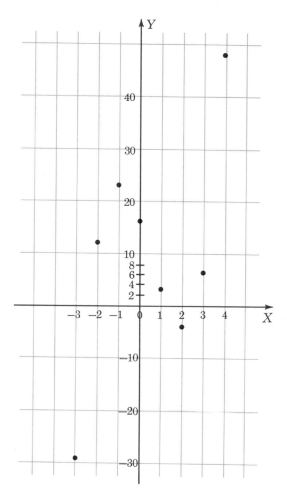

FIGURE 16.1

SOLUTION　In this problem

$$y = f(x) = 2x^3 - 3x^2 - 12x + 16$$

We first assign the integral values $-3, -2, -1, 0, 1, 2, 3,$ and 4 to x and compute the corresponding value of y either by direct substitution or by synthetic division. For example, if $x = 4$, $f(4)$ is the remainder when $2x^3 - 3x^2 - 12x + 16$ is divided by $x - 4$. By the synthetic division below, we see that $f(4) = 48$.

$$
\begin{array}{r}
2 - 3 - 12 + 16\ \underline{(4} \\
8 + 20 + 32 \\
\hline
2 + 5 +\ \ 8 + 48
\end{array}
$$

382

The following table shows the value of y that corresponds to each of the above values of x:

x	-3	-2	-1	0	1	2	3	4
y	-29	12	23	16	3	-4	7	48

The next problem is to choose a scale so that the points determined by the table will lie in a convenient region. Here we shall let the unit length on the X axis be four times the unit length on the Y axis, as indicated in Figs. 16.1 and 16.2. We now plot the points determined by the pairs of numbers in the table and obtain Fig. 16.1. As seen by viewing the figure, these points do

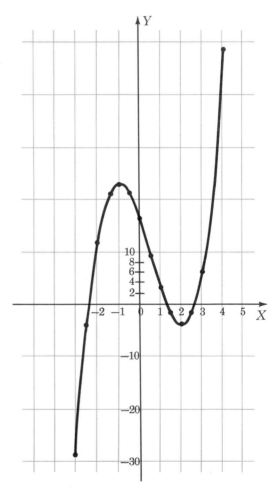

FIGURE 16.2

not indicate clearly the shape of the curve. Consequently, we assign the fractions $-\frac{5}{2}, -\frac{3}{2}, -\frac{1}{2}, \frac{1}{2}, \frac{5}{2}$, and $\frac{7}{2}$ to x, and comute each corresponding value of y, and thus tabulate the numbers as shown below.

x	$-\frac{5}{2}$	$-\frac{3}{2}$	$-\frac{1}{2}$	$\frac{1}{2}$	$\frac{3}{2}$	$\frac{5}{2}$	$\frac{7}{2}$
y	-4	$20\frac{1}{2}$	21	$9\frac{1}{2}$	-2	$-1\frac{1}{2}$	23

We now plot the points determined by the ordered pairs in the two tables, join them with a smooth curve, and obtain the graph in Fig. 16.2.

16.7
LOCATING THE REAL ROOTS OF A POLYNOMIAL EQUATION

Since $y = 0$ at all points where the graph of $y = f(x)$ crosses the X axis, the abscissa of each of these points is a root of the equation $f(x) = 0$. If these abscissas are rational numbers, the exact value of each of them can be determined by the method in Sec. 16.5. If, however, the abscissas are irrational, only an approximation to the value can be obtained. The method for computing this approximation is explained in the next section. The first step in the process is to locate bounds for the roots. By this statement we mean that we determine two numbers, usually two successive integers, between which one or more roots of the equation lie. The purpose of this section to explain how this is done.

Location Theorem

We can prove that if $f(x)$ is a polynomial, the graph of $y = f(x)$ is a continuous curve. This means that it contains no gaps and is not made up of separate or disjointed parts. Since the graph is continuous and if $f(a)$ and $f(b)$ have different signs, the curve crosses the X axis an odd number of times between $x = a$ and $x = b$. Since $y = 0$ at each of these points of crossing, the abscissa of each such point is a root of $f(x) = 0$. Therefore, we have the following theorem:

■ If $f(x)$ is a polynomial and if $f(a)$ and $f(b)$ have different signs, there is an odd number of real roots of $f(x) = 0$ between $x = a$ and $x = b$.

EXAMPLE
Locate the roots of

$$x^3 - 3x^2 - 6x + 9 = 0$$

SOLUTION To locate the roots of $x^3 - 3x^2 - 6x + 9 = 0$, we consider the equation

$$y = x^3 - 3x^2 - 6x + 9$$

assign consecutive integers from -3 to 5 to x, compute each corresponding value of y, and record the results in the following table:

x	-3	-2	-1	0	1	2	3	4	5
y	-27	1	11	9	1	-7	-9	1	29

Since $f(-3) = -27$ and $f(-2) = 1$, there is an odd number of roots between $x = -3$ and $x = -2$. Similarly, there is an odd number of roots between $x = 1$ and $x = 2$ and also between $x = 3$ and $x = 4$. Furthermore, since the equation is of degree 3, it has exactly three roots. Therefore, there is exactly one root in each of the above intervals.

EXERCISE 16.3
Graphs and Roots of Polynomial Equations

Sketch the graph of the function defined in each of Probs. 1 to 12 in the indicated domain. Estimate to one decimal place the zeros of each function.

1 $y = x^3 - x^2 - 4x + 4; \; -3 \le x \le 3$

2 $y = x^3 + 3x^2 - x - 3; \; -4 \le x \le 2$

3 $y = x^3 - 3x^2 - 4x + 12; \; -2 \le x \le 4$

4 $y = x^3 - 9x^2 + 20x; \; 0 \le x \le 6$

5 $y = x^3 - 5x^2 + 5x + 3; \; -1 \le x \le 4$

6 $y = x^3 - 6x - 4; \; -3 \le x \le 3$

7 $y = x^3 - 6x^2 + 10x - 4; \; 0 \le x \le 5$

8 $y = x^3 - 3x^2 - 2x + 4; \; -2 \le x \le 4$

9 $y = x^4 - 2x^3 - 9x^2 + 10x + 20; \; -3 \le x \le 4$

10 $y = x^4 - 6x^3 + 9x^2 - 2; \; -1 \le x \le 4$

11 $y = x^4 - 7x^2 + 2x + 2; \; -3 \le x \le 3$

12 $y = x^4 + x^3 - 6x^2 - 5x + 5; \; -3 \le x \le 3$

In each of the following problems, find the pairs of consecutive integers between which the roots of the equation lie.

13 $2x^3 - x^2 - 4x + 2 = 0$

14 $3x^3 - x^2 - 15x + 5 = 0$

15 $4x^3 + 3x^2 - 14x + 5 = 0$

16 $4x^3 + 6x^2 - 1 = 0$

17 $x^3 + 3x^2 - 5x - 5 = 0$

18 $x^3 - 2x^2 - 4x + 6 = 0$

19 $x^3 - 7x^2 + 14x - 7 = 0$

20 $x^3 + 3x^2 - 10x - 23 = 0$

21 $x^4 - 2x^3 - 4x^2 + 4x + 4 = 0$

22 $x^4 - 2x^3 - 4x^2 + 6x + 3 = 0$

23 $x^4 + 2x^3 - 12x^2 - 16x + 32 = 0$

24 $x^4 - 4x^3 - 6x^2 + 12x + 9 = 0$

16.8
APPROXIMATION OF IRRATIONAL ROOTS

Basis for Approximating

We stated that the general method for solving polynomial equations of degrees 3 and 4 is long and tedious and that general methods for solving polynomials of degree greater than 4 do not exist. Therefore, we must resort to some method of approximation to obtain the irrational roots of such equations. Several methods exist, and we shall use one that depends upon the following fact: If $f(x)$ is a polynomial and $y_1 = f(x_1)$ and $y_2 = f(x_2)$, then if x_1 and x_2 are sufficiently near each other, the portion of the graph of $y = f(x)$ between (x_1, y_1) and (x_2, y_2) will lie very near the straight line, or secant, that connects these two points. Consequently, in Fig. 16.3a, if the graph crosses the X axis at $(a, 0)$ and the straight line crosses it at $(b, 0)$, the root a of $f(x) = 0$ will be very near b.

We can calculate the value of b by use of Fig. 16.3a, where the lines CD and EC are parallel to the X and Y axes, respectively, and the coordinates of $C, D, E, F,$ and B are as indicated. The triangles EFB and ECD are similar. Hence,

$$\frac{FB}{CD} = \frac{EF}{EC} \tag{1}$$

Furthermore,

$$FB = b - x_1, \quad CD = x_2 - x_1, \quad EF = 0 - y_1 = -y_1, \quad \text{and} \quad EC = y_2 - y_1$$

Substituting these values in (1), we obtain

$$\frac{b - x_1}{x_2 - x_1} = \frac{-y_1}{y_2 - y_1} \tag{2}$$

Now we solve (2) for b and get

$$\blacksquare \quad b = x_1 - \frac{y_1(x_2 - x_1)}{y_2 - y_1} \tag{16.12}$$

We obtain the same result by using Fig. 16.3b. However, in this case, using positive directions, we have

$$\frac{FB}{CD} = \frac{FE}{CE}$$

or in terms of coordinates,

$$\frac{b - x_1}{x_2 - x_1} = \frac{y_1 - 0}{y_1 - y_2}$$

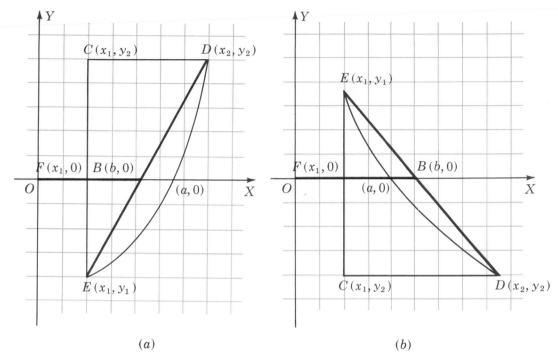

(a)   (b)

FIGURE 16.3

Nevertheless, if we solve this equation for b, we obtain (16.12).

If the graph of $y = f(x)$ is to the right of the secant, as in Fig. 16.3a, then $b < a$, but if the graph is to the left of the secant, as in Fig. 16.3b, then $b > a$. Furthermore, if $f(x_1)$ and $f(b)$ have the same sign, then $b < a$, but if $f(x_1)$ and $f(b)$ have different signs, then $b > a$.

The steps in obtaining successive approximations to the root follow:

1 Determine integral values of x_1 and x_2 so that $x_2 - x_1 = 1$ and $f(x_1)$ and $f(x_2)$ have different signs.

2 Substitute these values in (16.12), calculate b, and round the result off to one decimal place.

3 Label this value b_1. Then we say that b_1 is the *first approximation* to a. Usually b_1 will differ from a by less than 0.1. That is, usually $b_1 < a < b_1 + 0.1$ or $b_1 - 0.1 < a < b_1$. This, however, is not always true.

4 Determine to one decimal place values for x_1 and x_2 such that $x_2 - x_1 = 0.1$ and $f(x_1)$ and $f(x_2)$ differ in sign. Usually these values will be either b_1 and $b_1 + 0.1$ or $b_1 - 0.1$ and b_1. If, however, neither of these two pairs of values satisfies the conditions, other values near b_1 must be tried.

5 After x_1 and x_2 have been found and the corresponding values of y_1 and y_2 calculated, substitute them in (16.12), compute b_1, and round

387

the result off to three decimal places. Label the result b_2, and call it the second approximation to a. Usually b_2 will differ from a by less than 0.001. However, if $f(b_2)$ and $f(x_1)$ have the same signs, then $b_2 < a$ and $b_2 + 0.001$ may be a better approximation to a, so that the latter value should be tested. Likewise, if $f(b_2)$ and $f(x_1)$ have opposite signs, then $b_2 > a$, and $b_2 - 0.001$ should be tested.

6   The third approximation is obtained by the above procedure with the use of x_1 and x_2 so that $x_2 - x_1 = 0.01$. Then b_3 will usually differ from a by less than 0.00001. The calculation, however, is long and tedious. Methods that depend on calculus are generally used if approximations correct to more than three decimal places are desired.

We illustrate this procedure by calculating to three decimal places the largest positive root of

$$2x^3 - 3x^2 - 12x + 6 = 0 \tag{1}$$

We first consider the equation

$$y = f(x) = 2x^3 - 3x^2 - 12x + 6 \tag{2}$$

and assign the integers from -3 to 4, inclusive, to x and obtain the following table of corresponding values of x and y:

x	-3	-2	-1	0	1	2	3	4
y	-39	2	13	6	-7	-14	-3	38

From this table we see that there is a root between -3 and -2, between 0 and 1, and between 3 and 4. Hence, the largest positive root is between 3 and 4. Furthermore, $f(3) = -3$ and $f(4) = 38$. Hence, with

$$(x_1, y_1) = (3, -3) \quad \text{and} \quad (x_2, y_2) = (4, 38)$$

Formula (16.12) yields

$$b = 3 - \frac{-3(4 - 3)}{38 - (-3)}$$

$$= 3 + \tfrac{3}{41}$$

Therefore,

$$b_1 = 3.1$$

By synthetic division

$$
\begin{array}{rrrr|l}
2 - 3 & -12 & +6 & & (3.1 \\
+ 6.2 & + 9.92 & - 6.448 & & \\
\hline
2 + 3.2 & - 2.08 & - 0.448 & &
\end{array}
$$

and $f(3.1) = -0.448$. Hence, $f(b_1) = f(3.1)$ and $f(x_1) = f(3)$ have the same sign, so that 3.1 is less than a. Therefore, we find $f(3.2)$ by synthetic division:

$$2 - 3 \quad -12 \quad +6 \qquad (3.2$$
$$\underline{6.4 + 10.88 - 3.584}$$
$$2 + 3.4 - \quad 1.12 + 2.416$$

Thus, $f(3.2) = 2.416$.

Now, using (16.12) with $(x_1, y_1) = (3.1, -0.488)$ and $(x_2, y_2) = (3.2, 2.416)$, we obtain

$$b_1 = 3.1 - \frac{-0.448(3.2 - 3.1)}{2.416 - (-0.448)}$$

$$= 3.1 + \frac{0.0448}{2.864}$$

$$= 3.1 + 0.0156 \cdots$$

Hence,

$$b_2 = 3.116 \qquad \text{**to three decimal places**}$$

By synthetic division, we find that $f(3.116) = -0.0110$, and since $f(3.1)$ and $f(3.116)$ have the same sign, 3.116 is less than the root. Furthermore, by synthetic division, $f(3.117) = 0.0165$. Hence, we conclude that the root of (1) to three decimal places is 3.116.

To enable the reader to check his understanding of the method, we now give the results in each step of the approximation to three decimal places of the root of (1) that is between 0 and 1.

1   $f(0) = 6$, $f(1) = -7$. If $(x_1, y_1) = (0, 6)$, $(x_2, y_2) = (1, -7)$, then (16.12) yields $b_1 = 0.5$.

2   $f(0.5) = -0.5$, $f(x_1) = f(0) = 6$. Hence, $f(x_1)$ and $f(0.5)$ have different signs, so that 0.5 is greater than the root. However, $f(0.4) = 0.848$. Hence, for the second approximation, we use $(x_1, y_1) = (0.4, 0.848)$ and $(x_2, y_2) = (0.5, -0.5)$, substitute in (16.12), and get

$$b_2 = 0.463 \qquad \text{**to three decimal places**}$$

3   $f(0.463) = -0.0006$ to four decimal places. Since $f(x_1) = f(0.4) = 0.848$, 0.463 is greater than the root. Hence, we try 0.462 and find that $f(0.462) = 0.0218$. Therefore, since $f(0.463)$ is nearer zero than $f(0.462)$, we conclude that the root to three decimal places is 0.463.

EXERCISE 16.4

Irrational Roots of Polynomial Equations

Find to two decimal places the least positive root of the equation in each of Probs. 1 to 8.

1 $7x^3 + 21x^2 - 24 = 0$ **2** $x^3 - 9x^2 + 15x - 3 = 0$

3 $x^3 - 9x^2 + 18x - 6 = 0$ **4** $x^3 + 6x^2 + 6x - 6 = 0$

5 $x^3 + 9x^2 - 8 = 0$ **6** $x^3 + 9x^2 + 12x - 38 = 0$

7 $3x^3 - 9x - 2 = 0$ **8** $x^3 - 12x + 12 = 0$

In each of Probs. 9 to 16, find to two decimal places the negative root with the least absolute value.

9 $2x^3 - 6x^2 + 1 = 0$ **10** $3x^3 - 9x^2 + 2 = 0$

11 $x^3 - 3x^2 - 12x - 6 = 0$ **12** $x^3 - 6x^2 + 6x + 2 = 0$

13 $x^3 - 3x^2 - 6x + 2 = 0$ **14** $4x^3 + 12x^2 - 15 = 0$

15 $x^4 - 4x^3 + 3x^2 + 2x - 1 = 0$ **16** $x^4 + 2x^3 - x^2 + 2x + 1 = 0$

Find to two decimal places all roots of the equation in each of Probs. 17 to 24.

17 $x^3 + 3x^2 - 3x - 7 = 0$ **18** $x^3 + 3x^2 - 6x - 14 = 0$

19 $x^3 + 6x^2 - 28 = 0$ **20** $3x^3 - 9x^2 + 4 = 0$

21 $x^3 - 6x^2 + 6x + 2 = 0$ **22** $x^3 - 3x^2 - 6x + 2 = 0$

23 $3x^3 - 9x^2 + 2 = 0$ **24** $x^3 - 3x^2 - 9x - 1 = 0$

Find to three decimal places the least positive root of the equation in each of the following problems.

25 $x^3 + 9x^2 + 15x - 21 = 0$ **26** $x^3 - 3x^2 - 12x - 6 = 0$

27 $x^3 + 3x^2 - 6x - 14 = 0$ **28** $x^3 + 3x^2 - 3x - 7 = 0$

29 $4x^4 - 8x^3 + 7x^2 - 3x - 1 = 0$ **30** $x^4 - x^3 - 3x^2 - 4x - 2 = 0$

31 $x^4 - x^3 - 4x^2 - 8x - 8 = 0$ **32** $x^4 - 3x^3 - x^2 - 2x + 2 = 0$

tables

TABLE I   COMMON LOGARITHMS

N	0	1	2	3	4	5	6	7	8	9
10	0000	0043	0086	0128	0170	0212	0253	0294	0334	0374
11	0414	0453	0492	0531	0569	0607	0645	0682	0719	0755
12	0792	0828	0864	0899	0934	0969	1004	1038	1072	1106
13	1139	1173	1206	1239	1271	1303	1335	1367	1399	1430
14	1461	1492	1523	1553	1584	1614	1644	1673	1703	1732
15	1761	1790	1818	1847	1875	1903	1931	1959	1987	2014
16	2041	2068	2095	2122	2148	2175	2201	2227	2253	2279
17	2304	2330	2355	2380	2405	2430	2455	2480	2504	2529
18	2553	2577	2601	2625	2648	2672	2695	2718	2742	2765
19	2788	2810	2833	2856	2878	2900	2923	2945	2967	2989
20	3010	3032	3054	3075	3096	3118	3139	3160	3181	3201
21	3222	3243	3263	3284	3304	3324	3345	3365	3385	3404
22	3424	3444	3464	3483	3502	3522	3541	3560	3579	3598
23	3617	3636	3655	3674	3692	3711	3729	3747	3766	3784
24	3802	3820	3838	3856	3874	3892	3909	3927	3945	3962
25	3979	3997	4014	4031	4048	4065	4082	4099	4116	4133
26	4150	4166	4183	4200	4216	4232	4249	4265	4281	4298
27	4314	4330	4346	4362	4378	4393	4409	4425	4440	4456
28	4472	4487	4502	4518	4533	4548	4564	4579	4594	4609
29	4624	4639	4654	4669	4683	4698	4713	4728	4742	4757
30	4771	4786	4800	4814	4829	4843	4857	4871	4886	4900
31	4914	4928	4942	4955	4969	4983	4997	5011	5024	5038
32	5051	5065	5079	5092	5105	5119	5132	5145	5159	5172
33	5185	5198	5211	5224	5237	5250	5263	5276	5289	5302
34	5315	5328	5340	5353	5366	5378	5391	5403	5416	5428
35	5441	5453	5465	5478	5490	5502	5514	5527	5539	5551
36	5563	5575	5587	5599	5611	5623	5635	5647	5658	5670
37	5682	5694	5705	5717	5729	5740	5752	5763	5775	5786
38	5798	5809	5821	5832	5843	5855	5866	5877	5888	5899
39	5911	5922	5933	5944	5955	5966	5977	5988	5999	6010
40	6021	6031	6042	6053	6064	6075	6085	6096	6107	6117
41	6128	6138	6149	6160	6170	6180	6191	6201	6212	6222
42	6232	6243	6253	6263	6274	6284	6294	6304	6314	6325
43	6335	6345	6355	6365	6375	6385	6395	6405	6415	6425
44	6435	6444	6454	6464	6474	6484	6493	6503	6513	6522
45	6532	6542	6551	6561	6571	6580	6590	6599	6609	6618
46	6628	6637	6646	6656	6665	6675	6684	6693	6702	6712
47	6721	6730	6739	6749	6758	6767	6776	6785	6794	6803
48	6812	6821	6830	6839	6848	6857	6866	6875	6884	6893
49	6902	6911	6920	6928	6937	6946	6955	6964	6972	6981
50	6990	6998	7007	7016	7024	7033	7042	7050	7059	7067
51	7076	7084	7093	7101	7110	7118	7126	7135	7143	7152
52	7160	7168	7177	7185	7193	7202	7210	7218	7226	7235
53	7243	7251	7259	7267	7275	7284	7292	7300	7308	7316
54	7324	7332	7340	7348	7356	7364	7372	7380	7388	7396

N	0	1	2	3	4	5	6	7	8	9

TABLE I COMMON LOGARITHMS (continued)

N	0	1	2	3	4	5	6	7	8	9
55	7404	7412	7419	7427	7435	7443	7451	7459	7466	7474
56	7482	7490	7497	7505	7513	7520	7528	7536	7543	7551
57	7559	7566	7574	7582	7589	7597	7604	7612	7619	7627
58	7634	7642	7649	7657	7664	7672	7679	7686	7694	7701
59	7709	7716	7723	7731	7738	7745	7752	7760	7767	7774
60	7782	7789	7796	7803	7810	7818	7825	7832	7839	7846
61	7853	7860	7868	7875	7882	7889	7896	7903	7910	7917
62	7924	7931	7938	7945	7952	7959	7966	7973	7980	7987
63	7993	8000	8007	8014	8021	8028	8035	8041	8048	8055
64	8062	8069	8075	8082	8089	8096	8102	8109	8116	8122
65	8129	8136	8142	8149	8156	8162	8169	8176	8182	8189
66	8195	8202	8209	8215	8222	8228	8235	8241	8248	8254
67	8261	8267	8274	8280	8287	8293	8299	8306	8312	8319
68	8325	8331	8338	8344	8351	8357	8363	8370	8376	8382
69	8388	8395	8401	8407	8414	8420	8426	8432	8439	8445
70	8451	8457	8463	8470	8476	8482	8488	8494	8500	8506
71	8513	8519	8525	8531	8537	8543	8549	8555	8561	8567
72	8573	8579	8585	8591	8597	8603	8609	8615	8621	8627
73	8633	8639	8645	8651	8657	8663	8669	8675	8681	8686
74	8692	8698	8704	8710	8716	8722	8727	8733	8739	8745
75	8751	8756	8762	8768	8774	8779	8785	8791	8797	8802
76	8808	8814	8820	8825	8831	8837	8842	8848	8854	8859
77	8865	8871	8876	8882	8887	8893	8899	8904	8910	8915
78	8921	8927	8932	8938	8943	8949	8954	8960	8965	8971
79	8976	8982	8987	8993	8998	9004	9009	9015	9020	9025
80	9031	9036	9042	9047	9053	9058	9063	9069	9074	9079
81	9085	9090	9096	9101	9106	9112	9117	9122	9128	9133
82	9138	9143	9149	9154	9159	9165	9170	9175	9180	9186
83	9191	9196	9201	9206	9212	9217	9222	9227	9232	9238
84	9243	9248	9253	9258	9263	9269	9274	9279	9284	9289
85	9294	9299	9304	9309	9315	9320	9325	9330	9335	9340
86	9345	9350	9355	9360	9365	9370	9375	9380	9385	9390
87	9395	9400	9405	9410	9415	9420	9425	9430	9435	9440
88	9445	9450	9455	9460	9465	9469	9474	9479	9484	9489
89	9494	9499	9504	9509	9513	9518	9523	9528	9533	9538
90	9542	9547	9552	9557	9562	9566	9571	9576	9581	9586
91	9590	9595	9600	9605	9609	9614	9619	9624	9628	9633
92	9638	9643	9647	9652	9657	9661	9666	9671	9675	9680
93	9685	9689	9694	9699	9703	9708	9713	9717	9722	9727
94	9731	9736	9741	9745	9750	9754	9759	9763	9768	9773
95	9777	9782	9786	9791	9795	9800	9805	9809	9814	9818
96	9823	9827	9832	9836	9841	9845	9850	9854	9859	9863
97	9868	9872	9877	9881	9886	9890	9894	9899	9903	9908
98	9912	9917	9921	9926	9930	9934	9939	9943	9948	9952
99	9956	9961	9965	9969	9974	9978	9983	9987	9991	9996
	0	1	2	3	4	5	6	7	8	9

TABLE II POWERS AND ROOTS

No.	Sq.	Sq. Root	Cube	Cube Root	No.	Sq.	Sq. Root	Cube	Cube Root
1	1	1.000	1	1.000	51	2,601	7.141	132,651	3.708
2	4	1.414	8	1.260	52	2,704	7.211	140,608	3.733
3	9	1.732	27	1.442	53	2,809	7.280	148,877	3.756
4	16	2.000	64	1.587	54	2,916	7.348	157,464	3.780
5	25	2.236	125	1.710	55	3,025	7.416	166,375	3.803
6	36	2.449	216	1.817	56	3,136	7.483	175,616	3.826
7	49	2.646	343	1.913	57	3,249	7.550	185,193	3.849
8	64	2.828	512	2.000	58	3,364	7.616	195,112	3.871
9	81	3.000	729	2.080	59	3,481	7.681	205,379	3.893
10	100	3.162	1,000	2.154	60	3,600	7.746	216,000	3.915
11	121	3.317	1,331	2.224	61	3,721	7.810	226,981	3.936
12	144	3.464	1,728	2.289	62	3,844	7.874	238,328	3.958
13	169	3.606	2,197	2.351	63	3,969	7.937	250,047	3.979
14	196	3.742	2,744	2.410	64	4,096	8.000	262,144	4.000
15	225	3.873	3,375	2.466	65	4,225	8.062	274,625	4.021
16	256	4.000	4,096	2.520	66	4,356	8.124	287,496	4.041
17	289	4.123	4,913	2.571	67	4,489	8.185	300,763	4.062
18	324	4.243	5,832	2.621	68	4,624	8.246	314,432	4.082
19	361	4.359	6,859	2.668	69	4,761	8.307	328,509	4.102
20	400	4.472	8,000	2.714	70	4,900	8.367	343,000	4.121
21	441	4.583	9,261	2.759	71	5,041	8.426	357,911	4.141
22	484	4.690	10,648	2.802	72	5,184	8.485	373,248	4.160
23	529	4.796	12,167	2.844	73	5,329	8.544	389,017	4.179
24	576	4.899	13,824	2.884	74	5,476	8.602	405,224	4.198
25	625	5.000	15,625	2.924	75	5,625	8.660	421,875	4.217
26	676	5.099	17,576	2.962	76	5,776	8.718	438,976	4.236
27	729	5.196	19,683	3.000	77	5,929	8.775	456,533	4.254
28	784	5.291	21,952	3.037	78	6,084	8.832	474,552	4.273
29	841	5.385	24,389	3.072	79	6,241	8.888	493,039	4.291
30	900	5.477	27,000	3.107	80	6,400	8.944	512,000	4.309
31	961	5.568	29,791	3.141	81	6,561	9.000	531,441	4.327
32	1,024	5.657	32,768	3.175	82	6,724	9.055	551,368	4.344
33	1,089	5.745	35,937	3.208	83	6,889	9.110	571,787	4.362
34	1,156	5.831	39,304	3.240	84	7,056	9.165	592,704	4.380
35	1,225	5.916	42,875	3.271	85	7,225	9.220	614,125	4.397
36	1,296	6.000	46,656	3.302	86	7,396	9.274	636,056	4.414
37	1,369	6.083	50,653	3.332	87	7,569	9.327	658,503	4.431
38	1,444	6.164	54,872	3.362	88	7,744	9.381	681,472	4.448
39	1,521	6.245	59,319	3.391	89	7,921	9.434	704,969	4.465
40	1,600	6.325	64,000	3.420	90	8,100	9.487	729,000	4.481
41	1,681	6.403	68,921	3.448	91	8,281	9.539	753,571	4.498
42	1,764	6.481	74,088	3.476	92	8.464	9.592	778,688	4.514
43	1,849	6.557	79,507	3.503	93	8,649	9.644	804,357	4.531
44	1,936	6.633	85,184	3.530	94	8,836	9.695	830,584	4.547
45	2,025	6.708	91,125	3.557	95	9,025	9.747	857,375	4.563
46	2,116	6.782	97,336	3.583	96	9,216	9.798	884,736	4.579
47	2,209	6.856	103,823	3.609	97	9,409	9.849	912,673	4.595
48	2,304	6.928	110,592	3.634	98	9,604	9.899	941,192	4.610
49	2,401	7.000	117,649	3.659	99	9,801	9.950	970,299	4.626
50	2,500	7.071	125,000	3.684	100	10,000	10.000	1,000,000	4.642

answers

EXERCISE 1.1   Page 7

1 A, {2, 4, 8}, {3, 7}   **2** B, ∅, ∅

3 A, B, {x | x ∈ A and does not play a trumpet}

5 {x | x is in a French class *or* in an algebra class},
    {x | x is in a French class *and* in an algebra class},
    {x | x ∈ C but is not in an algebra class}

6 {x | x has black hair or likes strawberry ice cream}
    {x | x ∈ M and likes strawberry ice cream}
    {x | x ∈ M but does not like strawberry ice cream}

7 A, B, {x | x is a male student with red hair}

9 A, {1, 5, 7, 11}, {5, 7}   **10** A − C, {6}

11 {1, 2, 4, 5, 6, 7, 8, 10, 11}

EXERCISE 1.2   Page 16

1 Posit   **2** Integers   **3** Integers

5 Rati   **6** Fractions

7 Irrational numbers

9 $25 > 16,\ \ 2 < 7,\ \ 11 > 6$

10 $-1 < 4,\ \ 5 > -8,\ \ -10 < -1$ **11** $\frac{1}{5} < \frac{1}{3},\ \ \frac{2}{3} > -\frac{5}{3},\ \ -\frac{5}{2} > -\frac{7}{2}$

13 5   **14** 3   **15** -5 **17** 6

18 1   **19** 7

EXERCISE 2.1   Page 28

5 83   **6** 627   **7** 1031   **9** 13   **10** 85   **11** -10

13 81   **14** 222   **15** 12   **17** $56a - 20$ **18** $22b + 15$

19 $116 - 50x$ **21** $7ab - 4ac$ **22** $-2x^2 - 2y^2$

23 $-2x^2y + 2xy^2$ **25** $9a - 6c$ **26** $16a^2b - 7ab^2$

27 0   **29** $a + 7b$ **30** $-a + 4b + 2c$ **31** $-3x - 2y + 2z$

33 $-2ab + 2a^2b + 2ab^2 + 5a^2b^2$ **34** $-2x^2y + 2ay^2 - xy$

35 $9a^2 - 12c^2 + 10b^2$ **37** $6a^2bc - 20ab^2c - 27abc^2$

38 $4x^2yz - 16xy^2z - 21xyz^2$ **39** $4ab - 5a^2b - ab^2$

41 $-3y + 2z$ **42** $3a - 11b$ **43** $2a^2 - 7b^2$ **45** $3a + b$

46 $-3c + 2d$ **47** $x - 7y + 7z$ **49** $-r + 3s$

50 $4c - 5d + 13e$ **51** $x + 6$

EXERCISE 2.2   Page 32

1 $4a + 3c$ **2** $x^2 - y^2 + 5z^2$ **3** $9a - 4b + 3c$

5 $8xy - 3yz$ **6** $9xy - 11yz$ **7** $18a^2 - 2ab - 4c^2$

9 $-3c^2d^2 - 3cd^2$ **10** $-2x^3 + 2x^2$ **11** $3a^2 - 3b^2 - c^2$

13 $x - 8y + 3z$ **14** $7r + 4s - 2rs + 3t$

15 $10a + 2b - 7ab + 2c$ **17** $-x - xy + y - xz$

18 $9x^3 - 11x^2 + x$ **19** $-5ab^4 + 4a^3b^2$

21 $-3ab + 6a + 8c - 6ac$ **22** $14x - 2y + 5z$

23 $10a - 4b + 6c$ **25** 785   **26** -785 **27** -201

29 $a - b - 3c$ **30** $x - y - 2z$ **31** $-x^2 - 5xy + 7y^2$

33 $-a^2 + 4b^2 - d^2$ **34** $7a^2 - 2ab + b^2$ **35** $5a^2 - ab - b^2$

37 $-a - 5b + 3c$ **38** $c^4 + 16c^3d - 12c^2d^2 + 3d^3$

39 $3x^3 + 5x^2y - 5xy^2 + 3y^3$

EXERCISE 2.3   Page 40

1 $15x^9y^6$ **2** $24a^6b^8$ **3** $21c^3d^3$ **5** $-21x^5y^8z^{11}$

6 $-20a^8b^9c^{11}$ **7** $-24c^5d^4e^4$ **9** $24a^7$ $^5y^9z^7$

11 $-42c^6d^{13}e^8$ **13** $9a^4$ **14** $27x$ y^{20}

17 $125a^6b^9z^3$ **18** $-32x^{15}y^{10}$

21 $6x^6 - 9x^4$ **22** $6a^7 - 10a^6$

25 $6x^3y^3z^3 - 4x^4y^4z^3 + 10x^4y^6z^3$

26 $-8a^4b^3c^6 + 6a^6bc^5 - 4a^3b^5c^9$

27 $18x^3y^4z^7 - 30x^4y^7z^5 + 12x^5y^3z^6$

29 $x^7y^3z^2$ **30** $-15a^7b^4c^7$ **31** $12p^6q^8r^4$ **33** $22a - 6b$

34 $6x - 12y + 27z$ **35** $-2a^2 - 9ac + 2a$ **37** $40x - 120$

38 $9x^3 + 21x^2$ **39** $29a + 9$

EXERCISE 2.4   Page 42

1 $a^2 - b^2$ **2** $9x^2 - y^2$ **3** $4a^2 - 16b^2$ **5** $x^2 + 2xy + y^2$

6 $a^4 + 4a^2b^2 + 4b^4$ **7** $9y^4 + 12y^2z^3 + 4z^6$ **9** $a^2 - 2ab + b^2$

10 $4x^2 - 12xy + 9y^2$ **11** $16a^4 - 40a^2b^2 + 25b^4$

13 $6x^2 + 19xy + 15y^2$ **14** $8a^2 + 26ab + 15b^2$

15 $10c^2 + cd - 3d^2$ **17** $28p^2 - 29pq + 6q^2$

19 $63a^2 - 41ab + 6b^2$ **19** $56x^4 + x^2y^2 - y^4$

21 $x^3 + y^3$ **22** $a^3 - 8$ **23** $125a^9 - b^9$

25 $6x^3 - 5x^2y + 14xy^2 - 30y^3$ **26** $8a^3 - 2a^2b - 11ab^2 + 5b^3$

27 $2x^9 + 8x^6y^3 + 12x^3y^6 + 8y^9$ **29** $8x^4 - 10x^3y - 3x^2y^2 + 7xy^3 - 2y^4$

30 $8a^8 + 10a^6 - 6a^4 + 23a^2 - 5$

31 $3x^5y^2 + 4x^4y^3 - 29x^3y^4 + 14x^2y^5 + 8xy^6$

33 $8a^2 + 8ab - 6ac - 6b^2 + 7bc - 2c^2$

34 $12x^4 + 14x^2y^2 - 5x^2z^2 - 10y^4 + 9y^2z^2 - 2z^4$

35 $6a^2b^2 + 11a^2bc - 16ab^2c - 10a^2c^2 + 36abc^2 - 32b^2c^2$

37 $a^5 - 1$ **38** $x^5 + 32$ **39** $x^5 + 5x^2y^3 - 2y^5$

EXERCISE 2.5   Page 51

1 x^4 **2** a^5 **3** $-z^3$ **5** $4a$ **6** $3x^2$ **7** $-8ac^2$

9 $-5d^3$ **10** $-4r^2s^3t^2$ **11** $6a^7b^3c^3$ **13** $3a^4 + 4a^2$

14 $3b^4 - 2b^2$ **15** $-4x^4 + 3$ **17** $3x^3 - 2x^2y + y^2$

18 $3a^8b^4 - 2a^4b^2 - 1$ **19** $-2z^2 + 4xyz + 6x^2y^3$

21 $-4r^6s^3 + 5r^3s^2t + 7t^2$ **22** $2x^6z + 3x^4y - 4y^2z$

23 $2p^7q^4r^3 - 3p^2q^5r^4 + 5q^6r^2$ **25** $2c - 3$

26 $3x + 2y$ **27** $3a - 2b$ **29** $3x^2 - 2xy + y^2$

30 $2a^2 + ab + b^2$ **31** $3x^2 - 2xy + y^2$ **33** $4x^2 - 3x + 3$

34 $x + 3y$ **35** $x^3 - 2x^2y + 3xy^2$

EXERCISE 3.1   Page 58

1 $3x^2 + 16x + 5$ **2** $24c^2 + 10cd + d^2$ **3** $2y^2 + 7y + 3$

5 $6x^2 + 9x + 3$ **6** $6x^2 + 7x + 2$ **7** $30c^2 + 11cd + d^2$

9 $2c^2 - 7cd + 6d^2$ **10** $14r^2 - 11rt + t^2$ **11** $14x^2 - 67xy + 63y^2$

13 $72k^2 - 67km + 15m^2$ **14** $12c^2 - 52cd + 55d^2$

15 $8x^2 - 53xy + 30y^2$

17 $3x^2 - xy - 2y^2$

18 $7x^2 + 25xy - 12y^2$

19 $12a^2 - 19ab - 70b^2$

21 $14c^2 + 3cd - 27d^2$

22 $40m^2 + mp - 6p^2$

23 $15a^2 - 8ab - 55b^2$

25 $a^2 + 4ab + 4b^2$

26 $9c^2 + 6cd + d^2$

27 $4x^2 + 12x + 9$

29 $9r^2 + 6rt + t^2$

30 $81w^2 + 36wz + 4z^2$

31 $25m^2 + 20mn + 4n^2$

33 $25c^2 - 20cd + 4d^2$

34 $121x^2 - 66xy + 9y^2$

35 $81p^2 - 126pq + 49q^2$

37 $64x^2 - 48xy + 9y^2$

38 $9a^2 - 54ab + 81b^2$

39 $144x^2 - 120xy + 25y^2$ 41 $a^2 - 9$ 42 $r^2 - s^2$

43 $a^2 - 25b^2$ 45 $81w^2 - z^2$ 46 $9m^2 - 4n^2$

47 $36x^2 - 25y^2$ 49 9999   50 896   51 2499

53 391   54 375   55 1596

57 $a^2 + b^2 + c^2 + 2ab - 2ac - 2bc$ 58 $x^2 + y^2 + 1 + 2xy + 2x + 2y$

59 $x^2 + 4y^2 + 16z^2 + 4xy + 8xz + 16yz$

61 $x^2 + y^2 + z^2 + w^2 + 2xy - 2xz - 2xw - 2yz - 2yw + 2zw$

62 $a^2 + 4b^2 + c^2 + 4d^2 - 4ab + 2ac - 4ad - 4bc + 8bd - 4cd$

63 $4r^2 + s^2 + 9t^2 + u^2 - 4rs - 12rt - 4ru + 6st + 2su + 6tu$

65 $9a^2 + 24ab + 16b^2 - c^2$ 66 $9m^2 - 6mq + q^2 - 4z^2$

67 $16a^2 - 40ab + 25b^2 - 4c^2$ 69 $x^4 - x^2 - 2x - 1$

70 $16a^2 - b^2 - 2b - 1$ 71 $a^2 + 2ab + b^2 - c^2 - 2cd - d^2$

73 $4a^2 - 20ab + 25b^2 + 10ac - 25bc + 6c^2$

74 $c^2 - 2cd + d^2 - 2ec + 2ed - 15e^2$

75 $4x^2 - 4xy + y^2 + 4xz - 2yz - 24z^2$

77 $15x^4 - 2x^3 + 4x^2y - x^2 + 4xy - 4y^2$

78 $10a^2 - 11ac + 11ab - 6c^2 + 12cb - 6b^2$

79 $2x^2 - xy + xz - y^2 + 2yz - z^2$

EXERCISE 3.2   Page 63

1 $(x + 4)(x - 3)$ 2 $(b - 3)(b + 2)$ 3 $(y + 2)(y - 1)$

5 $(z + 8)(z - 7)$ 6 $(x - 9)(x + 8)$ 7 $(b + 10)(b - 9)$

9 $(a + b)(a + 7b)$ 10 $(x + 4y)(x + 3y)$ 11 $(c + 7d)(c + 2d)$

13 $(a - 11b)(a - 2b)$ 14 $(p - 5q)(p - 4q)$ 15 $(x - 6y)(x - y)$

17 $2(a + 2)(3a - 2)$ 18 $(3x - 5)(2x + 3)$ 19 $(6a - 5)(a + 1)$

21 $(8c + 7d)(c - d)$ 22 $(4r - 5s)(3r + 4s)$ 23 $(6a + 7b)(2a - 3b)$

25 $(x + 1)^2$ 26 $(a - 3)^2$ 27 $(c - 5)^2$

29 $(2x + 1)^2$ 30 $(3a + 1)^2$ 31 $(5x - 1)^2$

33 $(3a + 4b)^2$ 34 $(5a - 2b)^2$ 35 $(7c + 3d)^2$

37 $(10x + 11y)^2$ 38 $(9a - 11b)^2$ 39 $(20r + 7s)^2$

41 $(3x + 4y)(4x - 3y)$ **42** $(6a + 5b)(2a - 3b)$ **43** $(6y + z)(4y - 9z)$

45 $(6a - 5b)(3a + 4b)$ **46** $(7c - 4d)(4c + 3d)$

47 $(3x - 10y)(5x + 2y)$ **49** $(3x + 4)(4x + 9)$

50 $(6a + 1)(2a + 9)$ **51** $(a + 12b)(8a + 3b)$

53 $(5c - 6d)(c - 6d)$ **54** $(9r - 8s)(r - s)$

55 $(3a - 4b)(3a - 2b)$ **57** $(x + y + 1)^2$

58 $(a + b + c)^2$ **59** $(a - b + 1)^2$ **61** $(x - 1)(x - 2)$

62 $(y - 7)(y - 6)$ **63** $(b - 4)(b + 3)$

65 $(a + b + 2)(a + b - 1)$ **66** $(x + y + 5z)(x + y - 2z)$

67 $(a - 2b - 4c)(a - 2b + 2c)$

EXERCISE 3.3 Page 66

1 $(a + 3)(a - 3)$ **2** $(x + 4)(x - 4)$ **3** $(c + 6)(c - 6)$

5 $(2a + 3b)(2a - 3b)$ **6** $(3x + 4y)(3x - 4y)$

7 $(5c + 7d)(5c - 7d)$ **9** $(3x^2 - 8y)(3x^2 + 8y)$

10 $(4a^3 - 5b^2)(4a^3 + 5b^2)$ **11** $(7c^4 + 2c^3)(7c^4 - 2c^3)$

13 $(11h^6 + 2k^3)(11h^6 - 2k^3)$ **14** $(10r^5 + 7r^2)(10r^5 - 7r^2)$

15 $(8c^4 + 5d^5)(8c^4 - 5d^5)$

17 $(9x^9 + 4y^4)(9x^9 - 4y^4)$ **18** $(25h^5 + 4k)(25h^5 - 4k)$

19 $(6r^3 + 11s^4)(6r^3 - 11s^4)$ **21** $(a - 1)(a^2 + ab + 1)$

22 $(b + 1)(b^2 - b + 1)$ **23** $(c - 4)(c^2 + 4c + 16)$

25 $(2a + b)(4a^2 - 2ab + b^2)$ **26** $(2x - 3y)(4x^2 + 6xy + 9y^2)$

27 $(3c + 4d)(9c^2 - 24cd + 16d^2)$ **29** $(x^2 + 1)(x^4 - x^2 + 1)$

30 $(2a^2 - 1)(4a^4 + 2a^2 + 1)$ **31** $(c - 2d^3)(c^2 + 2cd^3 + 4d^6)$

33 $(2a^2 - 3b^4)(4a^4 + 6a^2b^4 + 9b^8)$

34 $(5r^5 + 2s^2)(25r^{10} - 10r^5s^2 + 4s^4)$

35 $(4h^4 - 3k^3)(16h^8 + 12h^4k^3 + 9k^6)$

37 $(4a^5 - 5b)(16a^{10} + 20a^5b + 25b^2)$

38 $(3c^4 + 7b^3)(9c^8 - 21c^4b^3 + 49b^6)$

39 $(8r^8 - 3s^2)(64r^{16} + 24r^8s^2 + 9s^4)$

41 $(x + y + 2)(x + y - 2)$ **42** $(a + b + 3)(a + b - 3)$

43 $(b + c + 2)(b - c + 2)$ **45** $(x + z - 3)(x - z - 3)$

46 $(2a - 3b + 2c)(2a - 3b - 2c)$ **47** $(5c - 2b - 3d)(5c - 2b + 3d)$

49 $(3 - a - b)(3 + a + b)$ **50** $(2 - r + s)(2 + r - s)$

51 $(x + y + 2z)(x - y - 2z)$ **53** $(a + b)(a - b)(a^2 + b^2)$

54 $(2c - 1)(2c + 1)(4c^2 + 1)$ **55** $(3x + 2y)(3x - 2y)(9x^2 + 4y^2)$

57 $(x - 1)(x^2 + x + 1)(x + 1)(x^2 - x + 1)$

58 $(a + b)(a^2 - ab + b^2)(a^6 - a^3b^3 + b^6)$

59 $(x^2 + 1)(x^4 - x^2 + 1)(x - 1)(x^2 + x + 1)(x + 1)(x^2 - x + 1)$

EXERCISE 3.4   Page 69

1 $3(a + 4)$

2 $3(2x - 3y)$

3 $5(a - 2b)$

5 $a(a + 3b)$

6 $2x(4y - 5z)$

7 $xy(x - y + z)$

9 $3(a + b)(a - b)$

10 $5(x + 2y)(x - 2y)$

11 $7(c + 3d)(c - 3d)$

13 $a(a + 2b)(a - 2b)$

14 $2x(x + 3y)(x - 3y)$

15 $3c^2(2c + d)(2c - d)$

17 $2(a - 2b)(a^2 + 2ab + 4b^2)$

18 $3(3 + x)(9 - 3x + x^2)$

19 $a(a - b)(a^2 + ab + b^2)$

21 $3(x - y)(x + 2y)$

22 $5(2a - b)(a + 2b)$

23 $6(x + 2y)^2$

25 $c(4c + d)(2c - d)$

26 $3x(3x - 5y)(x + 2y)$

27 $4d(5c + 6d)(c - d)$

29 $(a + d)(b + c)$

30 $(x + 3)(y + 2)$

31 $(c + 2d)(x + y)$

33 $(4a - b)(c + 2d)$

34 $(5r - s)(3t + 2u)$

35 $(3c + 4f)(2d - 5e)$

37 $(x - y)(x - z)$

38 $(2a - b)(a - 2c)$

39 $(3c - 4d)(c - e)$

41 $(a - b)(a + b + 1)$

42 $(x + z)(x - z - 2)$

43 $(x + 2y)(x - 2y + 1)$

45 $(a + b)(a - 2b + 2)$

46 $(x - y)(x + 3y - 1)$

47 $(c + d)(c - 3d - 3)$

49 $(a^2 + a - 1)(a^2 - a - 1)$

50 $(x^2 + x - 2)(x^2 - x - 2)$

51 $(y^2 + y + 3)(y^2 - y + 3)$

53 $(a^2 - ab + b^2)(a^2 + ab + b^2)$

54 $(x^2 + 2xy - 2y^2)(x^2 - 2xy - 2y^2)$

55 $(c^2 - cd + 2d^2)(c^2 + cd + 2d^2)$

57 $(2x^2 - xy + 3y^2)(2x^2 + xy + 3y^2)$

58 $(3m^2 + mn - n^2)(3m^2 - mn - n^2)$

59 $(a^2 + 2ab - 4b^2)(a^2 - 2ab - 4b^2)$

EXERCISE 4.1   Page 76

1 $\dfrac{-4}{x - 2}$

2 $\dfrac{-x - y}{x - y}$

3 $\dfrac{3x - y}{3y - x}$

5 $\dfrac{-x - y}{x^2 - xy}$

6 $\dfrac{-a + 2b}{a^2 - ab}$

7 $\dfrac{-a - b}{a^2 - b^2}$

9 $\dfrac{10a}{15b^2}$

10 $\dfrac{9acb}{12cb^3}$

11 $\dfrac{16x^3y^2}{20x^2y^3}$

13 $\dfrac{a^2 + 2ab + b^2}{a^2 - b^2}$

14 $\dfrac{2a^2 - 5ab - 3b^2}{a^2 - 9b^2}$

15 $\dfrac{x^2 - 9x + 18}{x^2 - 9}$

17 $\dfrac{a^2 - b^2}{a^2 + 2ab + b^2}$

18 $\dfrac{x^2 - 4y^2}{x^2 - 4xy + 4y^2}$

19 $\dfrac{x^2 - 2xy + y^2}{x^3 - y^3}$

21 $\dfrac{5}{6x^2y^2}$

22 $\dfrac{2}{3a^2b}$

23 $\dfrac{2}{3x^3yz}$

25 $\dfrac{x + 5}{x + 6}$

26 $\dfrac{a + b}{a - b}$

27 $\dfrac{x + 2}{x + 1}$

29 $\dfrac{2y}{3x}$

30 $\dfrac{2x^2}{3y^2}$

31 $\dfrac{3d}{4ce^2}$

33 $\dfrac{x}{y + 1}$

34 $\dfrac{a}{b + 3}$

35 $\dfrac{xy}{z + 2}$

37 $\dfrac{1}{a + b}$

38 $\dfrac{-x}{x + 2y}$

39 $\dfrac{3x - 2y}{x}$

41 $\dfrac{x - 2}{x + 2}$

42 $\dfrac{x - 2}{x + 2}$

43 $\dfrac{a + 2b}{2a + b}$ **45** $\dfrac{r + s}{r + 2s}$ **46** $\dfrac{2x + y}{x - 3y}$ **47** $\dfrac{m - 2n}{2m + n}$

49 $\dfrac{x + y}{y - x}$ **50** $\dfrac{c - 3d}{2c + d}$ **51** $\dfrac{y - 2z}{2y + z}$ **53** $\dfrac{x^2 - xy + y^2}{x - y}$

54 $\dfrac{x^4 + x^2 y^2 + y^4}{x^2 + y^2}$ **55** $\dfrac{1}{a + b}$ **57** $\dfrac{1}{2a + 1}$ **58** $\dfrac{1}{x + 1}$

59 $\dfrac{1}{3b + 1}$ **61** $\dfrac{-x - 3}{-x + 4}, \; -\dfrac{-x - 3}{x - 4}, \; -\dfrac{x + 3}{-x + 4}$

$+$

62 $\dfrac{-a - 1}{-a + 3}, \; -\dfrac{-a - 1}{a - 3}, \; -\dfrac{a + 1}{-a + 3}$ **63** $\dfrac{-a - b}{a - b}, \; -\dfrac{-a - b}{b - a}, \; -\dfrac{a + b}{a - b}$

EXERCISE 4.2   Page 81

1 $\frac{4}{9}$ **2** 3 **3** $\frac{5}{4}$ **5** $\frac{4}{3}$ **6** $\frac{27}{5}$ **7** $\frac{3}{2}$

9 $\dfrac{5}{3b}$ **10** $\dfrac{y^5}{6}$ **11** $\dfrac{8c^2}{5}$ **13** $\dfrac{8y^2}{5}$ **14** $\dfrac{1}{3ab}$

15 $\dfrac{2a^7 b^2}{9}$ **17** $\dfrac{3}{2x^3 y^4}$ **18** $\dfrac{2}{3b^7}$ **19** $\dfrac{8b}{5a^4 c^2}$

21 $\dfrac{x}{9yz}$ **22** $7c^2$ **23** $\dfrac{2a^2}{7b}$ **25** $\dfrac{w}{xy}$ **26** $\dfrac{3a}{2c}$

27 $\dfrac{1}{2c}$ **29** $\dfrac{2a}{5b}$ **30** $\dfrac{d}{3}$ **31** $\dfrac{2n}{3m}$ **33** $\dfrac{3c(b + c)}{2a(b + 2c)}$

34 $\dfrac{x + 3y}{xy}$ **35** $\dfrac{2a}{3b}$ **37** x^2 **38** $\dfrac{a^3}{b^3}$ **39** $c^3 d$

41 $\dfrac{1}{ab^2}$ **42** $\dfrac{a}{b^2}$ **43** $x^3 y$ **45** $\dfrac{a}{b}$ **46** $\dfrac{x(2x - y)}{y(x + 2y)}$

47 $\dfrac{2c(c + d)}{3b(c - d)}$ **49** $\dfrac{5a}{3b^2}$ **50** $\dfrac{x(x + y)}{3(x - y)}$ **51** $\dfrac{2d}{c}$

53 $\dfrac{x + y}{x - y}$ **54** $\dfrac{2a + b}{a - 2b}$ **55** $\dfrac{c + d}{c - d}$ **57** $\dfrac{x + 2}{x - 3}$

58 $\dfrac{(a + 2)(a + 1)}{(a - 1)^2}$ **59** $\dfrac{x^2 - 1}{x^2 - 9}$

EXERCISE 4.3   Page 86

1 $\frac{1}{4}$ **2** $\frac{5}{9}$ **3** $\frac{3}{10}$ **5** 2 **6** $\frac{1}{3}$ **7** $\frac{-1}{2}$

9 $\dfrac{5x^2 + 9y^2 - 4z^2}{30xyz}$ **10** $\dfrac{9c^2 + 4a^2 - 10b^2}{24abc}$ **11** $\dfrac{12c^2 + 10d^2 - 7e^2}{28cde}$

13 $\dfrac{-32}{9}$ **14** $\dfrac{x}{4}$ **15** $\dfrac{3x}{5y}$ **17** $\dfrac{c - d^2}{c^2 d^2}$ **18** $\dfrac{7}{6r^2}$

19 $\dfrac{3x^3 - y^3}{6x^2 y^3}$ **21** $\dfrac{x^2 + 3y^2}{x^2 + 3xy}$ **22** $\dfrac{9r^2 + s^2}{3rs - 2s^2}$ **23** $\dfrac{15a^2 - 8b^2}{2b(5a + 2b)}$

25 $\dfrac{3y^2}{2x(x + y)}$ **26** $\dfrac{a}{b}$ **27** $\dfrac{-w^2}{3z(z - w)}$ **29** $\dfrac{3a + 2b}{a + b}$ **30** $\dfrac{-a}{a - b}$

31 $\dfrac{x}{x-2y}$　　**33** $\dfrac{b^2}{b(a^2-b^2)}$　　**34** $\dfrac{42c^2}{(3c-2d)(c+2d)(2c+d)}$

35 $\dfrac{12t^2}{(r-2t)(r+t)(r+2t)}$　　**37** $\dfrac{9r^2}{(p+r)(2p-r)(p-2r)}$

38 $\dfrac{6}{(h+1)(h-1)(2h+1)}$　　**39** $\dfrac{6a^2}{(a-b)(a+b)(2a+b)}$

41 $\dfrac{1}{c-2d}$　　**42** $\dfrac{2}{x+2y}$　　**43** 2　　**45** $\dfrac{2r-t}{r+2t}$　　**46** $-\dfrac{w+z}{2w+z}$

47 $-\dfrac{2a+b}{a+2b}$　　**49** $\dfrac{1}{x}$　　**50** $\dfrac{c}{d(2c-3d)}$　　**51** $\dfrac{3}{b}$

53 $\dfrac{4(x-2y)}{3(x^2-y^2)}$　　**54** $\dfrac{a-b}{(3a-b)(a+b)}$　　**55** $\dfrac{c+d}{(c-d)(3c+2d)}$

57 $\dfrac{w+z}{w-z}$　　**58** $\dfrac{1}{3h+k}$　　**59** $\dfrac{3(r+7s)}{(3r-2s)(2r-5s)}$

EXERCISE 4.4　Page 91

1 $\frac{5}{2}$　**2** $\frac{4}{3}$　**3** $\frac{18}{11}$　**5** -2　**6** $\frac{7}{8}$　**7** $\frac{15}{16}$

9 $\dfrac{x}{2x-1}$　**10** $\dfrac{a^2}{a+1}$　**11** $\dfrac{a}{3a-1}$　**13** $\dfrac{3x-1}{x-2}$　**14** $\dfrac{x+2}{x}$

15 $\dfrac{x-3}{x+2}$　**17** x^2-1　**18** $a+2$　**19** $x+3$　**21** $\dfrac{r}{r-3t}$

22 $\dfrac{1}{d+e}$　**23** $\dfrac{3}{x-2y}$　**25** $\dfrac{a+b}{a-b}$　**26** $\dfrac{r}{r+t}$　**27** $\dfrac{x+2y}{x-y}$

29 $\dfrac{b+c}{b-c}$　**30** $\dfrac{c-2d}{2c+3d}$　**31** $\dfrac{a+2b}{a+b}$　**33** $\dfrac{1}{x+1}$

34 $-\dfrac{1}{w+1}$　**35** $\dfrac{x^2-y^2}{x}$　**37** $1-a$　**38** $3-a$

39 $-x$

EXERCISE 5.1　Page 97

1 {Washington, Oregon, California}　　**2** {Washington}

3 {Mexico, Hampshire, York}　　**5** {York}　　**6** {Hawaii, Alaska}

7 {General Motors, Chrysler, Ford}　　**9** {Mexico}

10 {Foot, basket, base}　　**11** {Hand, basket, volley}

13 {Mississippi, Missouri, Ohio, Arkansas, Colorado, Tennessee}　**14**　{Red, white}

15 {Cardinal, Oriole}　　**17** {13, 22, 31, 40}

18 {3, 12, 21, 30, 120, 111, 210, 102, 201, 300}

19 {3, 5, 7, 11, 13, 17, 19}

21 {42, 75}　　**22**　{171, 252, 333, 414}　　**23**　{1, 8, 27, 64}

25 {1, 5, 7}　　**26**　{3, 6, 9}　　**27**　{0, 1, 4, 9, 16, 25, 36, 49}

29 {1, 4, 6, 9, 12, 16, 18, 24, 25}   **30** {2, 4, 8, 10, 14}

31 {3, 5}

EXERCISE 5.2   Page 104

1 {2}   **2** {5}   **3** {4}   **5** {7}   **6** {9}   **7** {6}   **9** {11}

10 {13}   **11** {10}   **13** $\left\{\frac{2}{3}\right\}$ **14** $\left\{\frac{1}{2}\right\}$ **15** $\left\{\frac{3}{4}\right\}$

17 {∅}   **18** {∅}   **19** {∅}   **21** {12}   **22** {18}   **23** {36}

25 {15}   **26** {30}   **27** {45}   **29** $\left\{\dfrac{3}{a}\right\}$ **30** $\left\{\dfrac{a-1}{2}\right\}$

31 $\left\{\dfrac{a-b}{b}\right\}$ **33** {5}   **34** {7}   **35** {9}   **37** {6}   **38** {8}

39 {10}   **41** {5}   **42** {7}   **43** {11}   **45** {6}   **46** {13}

47 {15}

EXERCISE 5.3   Page 108

1 {5}   **2** {3}   **3** {7}   **5** {8}   **6** {4}   **7** {9}   **9** {7}

10 {6}   **11** {5}   **13** {12}   **14** {14}   **15** {11}   **17** {4}

18 {9}   **19** {7}   **21** {11}   **22** {13}   **23** {8}   **25** {4}   **26** {7}

27 {9}   **29** {6}   **30** {8}   **31** {3}   **33** {5}   **34** {11}   **35** {8}

37 {8}   **38** {6}   **39** {5}   **41** {7}   **42** {2}   **43** {9}

EXERCISE 5.4   Page 113

1 $\{x \mid x > 1\}$ **2** $\{x \mid x > 5\}$ **3** $\{x \mid x < 2\}$

5 $\{x \mid x < 2\}$ **6** $\{x \mid x > 1\}$ **7** $\{x \mid x < 5\}$

9 $\{x \mid x > 4\}$ **10** $\{x \mid x < -2\}$ **11** $\{x \mid x > 3\}$

13 $\{x \mid x < 3\}$ **14** $\{x \mid x > -3\}$ **15** $\{x \mid x < -2\}$

17 $\{x \mid x > 4\}$ **18** $\{x \mid x < -1\}$ **19** $\{x \mid x < \frac{7}{4}\}$

21 $\{x \mid x > -\frac{5}{4}\}$ **22** $\{x \mid x > 2\}$ **23** $\{x \mid x < \frac{3}{2}\}$

25 $\{x \mid x > 2\}$ **26** $\{x \mid x < 43\}$ **27** $\{x \mid x < \frac{4}{3}\}$

EXERCISE 5.5   Page 120

1 $\{x \mid x < -3\} \cup \{x \mid x > 2\}$ **2** $\{x \mid x > 1\} \cup \{x \mid x < -5\}$

3 $\{x \mid -6 < x < 3\}$ **5** $\{x \mid 0 < x < 1\} \cup \{x \mid x < -2\}$

6 $\{x \mid -2 < x < 2\} \cup \{x \mid x < -3\}$ **7** $\{x \mid -4 < x < -1\} \cup \{x \mid x > 5\}$

9 $\{x \mid 0 < x < 2\}$ **10** $\{x \mid -3 < x < 0\}$ **11** $\{x \mid x < 3\} \cup \{x \mid x > 4\}$

13 $\{x \mid -1 < x < 9\}$ **14** $\{x \mid -2 < x < -\frac{2}{3}\}$

15 $\{x \mid x < 5\} \cap \{x \mid x > -2\}$ **17** $\{x \mid x > -1\} \cup \{x \mid x < -7\}$

18 $\{x \mid x < \frac{-2}{3}\} \cup \{x \mid x > 2\}$ **19** $\{x \mid x < \frac{-3}{2}\} \cup \{x \mid x > 2\}$

EXERCISE 5.6   Page 127

1 23, 24, 25   **2** 5, 6, 9   **3** $36, $54   **5** $173, $386

6 140 hr   **7** Leon, 8.8 lb; Dean, 12.2 lb

9 Bill, 4 hr; Jack, 6 hr; Joe, 11 hr   **10** 35 miles per hour

11 $104   **13** 4   **14** $100   **15** 8 hr   **17** 8   **18** 16,000

19 1000   **21** 12, 24, 6   **22** 50 quarters, 65 dimes, 80 nickels

23 73°F   **25** 3 miles   **26** $6\frac{1}{2}$ hr

27 50 miles per hour, freeway; 15 miles per hour, suburbs

29 $6\frac{6}{7}$ hr   **30** $1\frac{1}{5}$ hr   **31** $4\frac{4}{9}$ hr   **33** 7 hr

34 4 hr   **35** 11:00 a.m.   **37** 6 hr   **38** 2 hr

39 6   **41** 0.4 gal   **42** 25 ml

43 30 miles per hour, 45 miles per hour   **45** 112 miles per hour

46 25 miles per hour   **47** $50

EXERCISE 6.1   Page 139

1 6, 10, 7   **2** 8, −7, −3   **3** $\frac{19}{8}$, $\frac{13}{8}$, 0   **5** 18, 42, $\frac{5}{3}$ **6** 17, 62, $-\frac{1}{3}$

7 0, −20, $\frac{16}{9}$ **9** 1, −11, 4   **10** 80, 0, $-\frac{3}{4}$ **11** 4, −7, $-\frac{15}{4}$ **13** $6x + 1$

14 $3t - 3t^2 - 1$ **15** $\dfrac{6}{s(s-3)}$ **17** $2t^2 - 1$ **18** $t^2 + t - 4$

19 $\dfrac{2t^2 - 10t + 24}{(t-2)^2}$ **21** $\{(-2, -9), (0, -5), (2, -1), (4, 3), (6, 7)\}$

22 $\{(-3, 12), (-2, 7), (0, 3), (1, 4), (2, 7)\}$

23 $\{(1, 0), (2, \frac{1}{2}), (3, \frac{2}{3}), (4, \frac{3}{4}), (5, \frac{4}{5})\}$

25 $\{(-2, 1), (-1, -2), (0, -3), (1, -2), (2, 1), (3, 6)\}$

26 $\{(-1, 0), (0, -\frac{1}{2}), (1, 0), (2, \frac{1}{2})\}$

27 $\{(0, -3), (1, -1), (2, \frac{5}{3}), (3, 6), (4, \frac{61}{5})\}$

29 $\{y \mid -11 \leq y \leq 9\}$ **30** $\{y \mid -24 \leq y \leq 1\}$ **31** $\{y \mid 2 \leq y \leq 149\}$

33 $\{(1, \sqrt{35}), (1, -\sqrt{35}), (2, 4\sqrt{2}), (2, -4\sqrt{2}), (3, 3\sqrt{3}), (3, -3\sqrt{3})$
$(4, 2\sqrt{5}), (4, -2\sqrt{5}), (5, \sqrt{11}), (5, -\sqrt{11}), (6, 0)\}$; relation

34 $\{(2, 1), (2, -1), (3, \sqrt{3}), (3, -\sqrt{3}), (4, \sqrt{5}), (4, -\sqrt{5}), (5, \sqrt{7}),$
$(5, -\sqrt{7}), (6,3), (6, -3)\}$; relation

35 $\{(0, -1), (1, 1), (2, 7), (3, 17), (4, 31)\}$; function

EXERCISE 6.2   Page 146

2 (a) The Y axis; (b) the X axis; (c) the line through the origin that bisects the second and fourth quadrants; (d) the line through the origin that bisects the first and third quadrants

3 (a) The line through (3, 0) parallel to the Y axis; (b) the line through (0, 6) parallel to the X axis; (c) the line through (−5, 0) parallel to the Y axis; (d) the line through (0, −1) parallel to the X axis

5 2   **6** -7 **7** $-\frac{5}{2}$ **9** 5   **10** -3 **11** $\frac{13}{2}$ **13** 0   **14** ± 3

15 None

17 $4, -1$ **18** $-5, 2$ **19** -2 **21** x intercept 4, y intercept -12

22 x intercept 8, y intercept 16   **23** x intercept -2, y intercept 8

EXERCISE 6.3   Page 156

9 $\{(x, x - 3) \mid x \geq 3\}$ **10** $\{(x, \frac{1}{3}(x + 1) \mid x \geq 0\}$

11 $\{(x, 3x - 9) \mid x \leq 2\}$ **13** $\{(x, \sqrt{x}) \mid x \geq 0\}$

14 $\{(x, \sqrt{x - 2}) \mid x \geq 2\}$ **15** $\{(x, \sqrt{x - 5}) \mid x \geq 5\}$

17 $y = \dfrac{2x + 3}{x - 1}, x \neq 1$ **18** $y = \sqrt{2x - 4}, x \geq \frac{5}{2}$

19 $y = (x - 3)^2 - 4, x \geq 3$ **21** $\{x \mid x > \frac{2}{3}\}$ **22** $\{x \mid x > -\frac{1}{4}\}$

23 $\{x \mid x > \frac{2}{5}\}$ **25** $\{x \mid x < 3\}$ **26** $\{x \mid x < 4\}$ **27** $\{x \mid x < -4\}$

29 $\{x \mid x < -2\} \cup \{x \mid x > 1\}$

30 $\{x \mid -3 < x < 1\}$

31 $\{x \mid x < -4\} \cup \{x \mid x > 2\}$

EXERCISE 7.1   Page 163

1 $\{(3, 2)\}$ **2** $\{(2, 2)\}$ **3** $\{(-3, 2)\}$ **5** $\{(1.5, 3)\}$

6 Inconsistent   **7** $\{(3, 1.5)\}$ **9** Inconsistent

10 $\{(1, 1.5)\}$ **11** $\{(0.5, -2)\}$ **13** $\{(1.5, -2.5)\}$

14 Inconsistent   **15** $\{(2, -0.3)\}$ **17** $\{(1.5, 1)\}$

18 $\{(2, 0.6)\}$ **19** Dependent   **21** $\{(1.5, -4)\}$

22 $\{(1.4, -1)\}$ **23** $\{(2, 2.6)\}$ **25** $\{(3.5, 0.4)\}$ **26** $(2.3, -1.3)$

27 Dependent   **29** $\{(1.4, -0.9)\}$ **30** Inconsistent

31 $\{(3, 0.6)\}$ **33** Dependent   **34** $\{(3, -0.9)\}$ **35** $\{(2.2, 0.6)\}$

EXERCISE 7.2   Page 167

1 $\{(3, -1)\}$ **2** $\{(2, -5)\}$ **3** $\{(-4, 3)\}$ **5** $\{(3, \frac{1}{2})\}$

6 $\{(5, -\frac{2}{3})\}$ **7** $\{(\frac{3}{5}, 1)\}$ **9** $\{(\frac{4}{5}, -\frac{1}{5})\}$

10 $\{(\frac{5}{4}, -\frac{1}{4})\}$ **11** $\{(1, \frac{1}{4})\}$ **13** $\{(2, -3)\}$ **14** $\{(2, 2)\}$

15 $\{(\frac{1}{2}, \frac{3}{2})\}$ **17** $\{(2, 1)\}$ **18** $\{(-\frac{1}{4}, -\frac{3}{8})\}$

19 $\{(-2, 1)\}$ **21** $\{(3, -2)\}$ **22** $\{(-5, 4)\}$ **23** $\{(6, 1)\}$

25 $\{(\frac{1}{4}, -2)\}$ **26** $\{(3, -\frac{1}{5})\}$ **27** $\{(\frac{3}{2}, 3)\}$

29 $\{(\frac{1}{3}, \frac{5}{2})\}$ **30** $\{(\frac{1}{4}, \frac{3}{5})\}$ **31** $\{(-\frac{5}{6}, \frac{3}{2})\}$

33 $\{(8, -9)\}$ **34** $\{(1, -\frac{1}{3})\}$ **35** $\{(-8, -\frac{13}{5})\}$

37 $\{(\frac{3}{7}, -\frac{4}{7})\}$ **38** $\{(-1, 2)\}$ **39** $\{(\frac{1}{4}, \frac{3}{4})\}$

41 $\{(3, 2)\}$ **42** $\{(4, 2)\}$ **43** $\{(6, -4)\}$

EXERCISE 7.3 Page 171

1 $\{(2, -1, 1)\}$ 2 $\{(1, -2, 3)\}$ 3 $\{(4, -2, -1)\}$

5 $\{(-2, 3, -1)\}$ 6 $\{(3, -1, 2)\}$ 7 $\{(1, -2, 4)\}$

9 $\{(\frac{1}{2}, 2, -3)\}$ 10 $\{(2, -\frac{2}{3}, 1)\}$ 11 $\{(4, -1, -\frac{3}{4})\}$

13 $\{(\frac{3}{2}, -2, 5)\}$ 14 $\{(-3, -\frac{4}{3}, 1)\}$ 15 $\{(1, -\frac{3}{5}, -4)\}$

17 $\{(\frac{1}{2}, \frac{3}{2}, \frac{1}{4})\}$ 18 $\{(\frac{1}{3}, \frac{2}{3}, -\frac{1}{6})\}$

19 $\{(\frac{5}{2}, -3, -\frac{3}{2})\}$ 21 $\{(\frac{1}{4}, -\frac{3}{4}, \frac{1}{2})\}$

22 $\{(\frac{1}{3}, \frac{1}{6}, -\frac{1}{2})\}$ 23 $\{(\frac{3}{2}, -\frac{2}{3}, \frac{1}{4})\}$

EXERCISE 7.4 Page 176

1 48, 24   2 18, 6   3 36, 24   5 100, 150

6 20, 15

7 40 at $35,000; 60 at $30,000   9 2 miles per hour, 4 miles per hour

10 300 miles per hour, 20 miles per hour

11 200 miles per hour, 20 miles per hour

13 27 miles by bus, 800 miles by plane

14 Horseback, $2\frac{1}{2}$ hr; car, $\frac{1}{2}$ hr; plane, 1 hr

15 Morning, 150 miles; afternoon, 130 miles

17 300 by 200 ft   18 30 by 80 ft, 60 by 80 ft

19 20,000 sq yd, 10,000 sq yd   21 15 sedans, 20 sports cars, 10 station wagons

22 100 bales, $17,500   23 A $38.50, B $25.25

25 8 hr, 6 hr   26 Older boy, 4 hr; younger boy, 5 hr

27 8 lb of the $3.50 grade, 12 lb of the $3 grade

29 Joe, $4\frac{1}{2}$ hr; Bill, $3\frac{3}{5}$ hr; Tom, 6 hr   30 24, 12, 16

31 5, 600, 4

EXERCISE 7.5 Page 184

1 $S = \{P(x, y) \mid P$ is above the graph of $y = 2x - 12$ and below the graph of $y = 3x + 6\}$

2 $S = \{P(x, y) \mid P$ is above the graph of $y = 3x + 15$ and below the graph of $y = 2x - 8\}$

3 $S = \{P(x, y) \mid P$ is above the graphs of $y = 4x + 7$ and $y = 2x - 5\}$

21 15, −9   22 3, −45   23 4, −3.4   25 4, −7

26 9, −12   27 24, −26

EXERCISE 8.1 Page 189

1 10   2 24   3 1   5 11   6 24

7 19   9 0   10 0   11 0   13 −28

14 26   15 11   21 (5, −2)   22 (14, 18)

23 Inconsistent **25** $\{(\frac{5}{2}, -\frac{7}{2})\}$ **26** $\{(-\frac{1}{3}, -1)\}$
27 $\{(\frac{2}{3}, 0)\}$ **29** $\{(-\frac{1}{6}, -\frac{1}{6})\}$ **30** $\{(\frac{9}{8}, \frac{3}{4})\}$
31 $\{(\frac{1}{7}, \frac{2}{7})\}$ **33** $\{(\frac{3}{2}, \frac{5}{2})\}$ **34** $\{(\frac{1}{9}, -\frac{11}{9})\}$
35 Inconsistent

EXERCISE 8.2   Page 194

1 21   **2** 70   **3** 32   **5** -180 **6** -20
7 76   **9** -168 **10** 126   **11** -60 **13** -17
14 6   **15** 254   **17** 30   **18** 98   **19** 108
21 29   **22** -18 **23** -54 **25** $\{-3\}$ **26** $\{1\}$
27 $\{2\}$

EXERCISE 8.3   Page 199

1 $\{(1, -1, 2)\}$ **2** $\{(-1, 2, 1)\}$ **3** $\{(2, -3, 1)\}$
5 $\{(4, -2, 1)\}$ **6** $\{(-3, 4, 2)\}$ **7** $\{(-2, 1, 4)\}$
9 $\{(\frac{1}{2}, 2, 1)\}$ **10** $\{(2, -\frac{1}{3}, -1)\}$ **11** $\{(-3, 1, \frac{3}{2})\}$
13 $\{(\frac{1}{4}, \frac{1}{2}, 1)\}$ **14** $\{(\frac{1}{3}, \frac{1}{6}, -1)\}$ **15** $\{(\frac{3}{4}, \frac{5}{8}, 2)\}$
17 $\{(\frac{1}{2}, \frac{2}{3}, \frac{1}{4})\}$ **18** $\{(\frac{3}{2}, \frac{2}{3}, \frac{1}{6})\}$
19 $\{(\frac{1}{3}, \frac{5}{6}, \frac{1}{9})\}$
21 $\{(2, -1, 3)\}$ **22** $\{(\frac{3}{2}, -\frac{1}{2}, \frac{1}{4})\}$
23 $\{(\frac{2}{3}, \frac{1}{6}, -1)\}$

EXERCISE 9.1   Page 205

1 128   **2** 2187   **3** 4096   **5** 36   **6** 49   **7** 125
9 729   **10** 512   **11** 256   **13** 324   **14** 11,664
15 $\dfrac{81}{64}$ **17** $6a^7$ **18** $6b^2$ **19** $12b^5$ **21** c^6
22 n^7 **23** s^3 **25** x^8 **26** a^6 **27** 1   **29** $a^{10}b^{15}$
30 $x^{12}y^6$ **31** $\dfrac{a^{12}}{b^6}$ **33** $8xy^2$ **34** $4a^4b$ **35** $2u^8v^2$
37 $\dfrac{2a^2b^2c^2}{5d}$ **38** $\dfrac{3x^2y}{5z}$ **39** $\dfrac{15r^5s^4u^3}{32t^2}$ **41** $\dfrac{81a^4}{b^4}$
42 $\dfrac{32}{243c^5d^{15}}$ **43** $64r^{12}s^{18}$ **45** $\dfrac{a^4}{c^6}$ **46** $\dfrac{8m^{14}t^2}{15s^5}$
47 $\dfrac{16y^{16}z^8}{x^8}$ **49** $\dfrac{512x^6y^9}{27}$ **50** $\dfrac{4a^2}{49b^4}$ **51** $\dfrac{6561}{16d^{40}}$
53 x^{4a-1} **54** y^{3s-2} **55** w^{4x-3} **57** $x^{a-3}y^{b+2}$
58 $a^{n+1}y^{2n+3}$ **59** $c^{2x+1}d^{2y-3}$ **61** $x^{2t+1}y^{u-2}$
62 $u^{3a-5}v^{b+9}$ **63** a^6b^{2m-8}

EXERCISE 9.2   Page 210

1 $\frac{1}{9}$ **2** $\frac{1}{5}$ **3** $\frac{1}{81}$ **5** $\frac{1}{3}$ **6** $\frac{1}{32}$ **7** 7

9 $\frac{1}{64}$ **10** $\frac{1}{6}$ **11** 49 **13** $\frac{1}{64}$ **14** 729 **15** 1

17 324 **18** 32 **19** 9 **21** $\frac{3}{8}$ **22** $\frac{16}{729}$ **23** 256

25 $2^{-1}a^2b^{-2}$ **26** $3a^2b^{-3}$ **27** $w^3x^2y^2z^{-4}$ **29** $6^{-1}a^{-1}b$

30 $4a^2b^2z^{-2}$ **31** $6r^{-1}s^{-1}v^4$ **33** $2/a^7$ **34** $1/3x^3$

35 $1/8y$ **37** $1/x^2$ **38** $1/c^2$ **39** m **41** b^5c^2/a^5

42 s^3v^2u/r^3t^5 **43** x^2/w^5y^3 **45** $27pqr^3/4$

46 $16b^6d/3c^4$ **47** $2x^5y^3/z^3$ **49** v^3x^2/u^2w^3

50 p^6/q^6 **51** x^9y^6 **53** $a^{10}/16b^6$ **54** $c^{24}d^{16}/81$

55 $32y^{15}/x^5$ **57** $\dfrac{3a^2+1}{a^3}$ **58** $\dfrac{d^2+c^2}{cd}$ **59** $\dfrac{26b^a}{9}$

61 $\dfrac{x^2y+xy^2}{x^2+xy+y^2}$ **62** $\dfrac{c^3d^3-1}{c^2d}$ **63** $\dfrac{1}{x^2+xy+y^2}$

65 $\dfrac{(x+2)(x-10)}{(x-2)^3}$ **66** $\dfrac{-2(a+1)^2(a+4)}{(a-2)^3}$

67 $\dfrac{2c-9}{(c-4)^4(c-3)^3}$ **69** $\dfrac{7a-7}{(3a-1)^3(a+2)^2}$

70 $\dfrac{(c+2)(2c-6)}{(3c-4)^2}$ **71** $\dfrac{(2p+5)^2(15p-29)}{(3p-2)^2}$

73 $\dfrac{3a-3}{(3a-2)^3}$ **74** $\dfrac{22x-32}{(2x-7)^4}$ **75** $\dfrac{-11x-25}{(4x-5)^3(3x+4)^2}$

EXERCISE 9.3   Page 214

1 3 **2** $\frac{1}{2}$ **3** -2 **5** 8 **6** -8 **7** 8 **9** 7

10 9 **11** 27 **13** $\frac{1}{216}$ **14** 2 **15** $\frac{1}{0.09}$

17 $8x$ **18** $3a^2b$ **19** $2r^3$ **21** $12xy$ **22** $6u^{1/3}v$

23 $15a^{4/5}$ **25** $a^{1/3}/3y$ **26** $64y^{2/7}/9x^{3/5}$ **27** $2pq$

29 $1/3z$ **30** $1/2c^{1/2}d^2$ **31** $a^{4/9}z^{2/3}/3$

33 $9y^{4/3}/x^{1/2}$ **34** $8b^{1/2}/a$ **35** $27/cd$ **37** $3b^{1/4}/a^2$

38 $2/b^{1/5}c^{1/7}$ **39** $3s^{1/4}/r^{1/8}$ **41** $2^{4/3}/a^2b^{1/5}$

42 $3y^{3/4}/x^{5/2}$ **43** $16/ab$ **45** $3y/4x^2$ **46** $64b/27a^2$

47 $7c^{1/4}/4d^{2/3}$ **49** x **50** x^s **51** a^{x+y} **53** $x-y$

54 $x-y$ **55** $x-y^3$ **57** $(y+x)^2/x^2y^2$

58 $\dfrac{ab}{b-a}$ **59** $\dfrac{1}{ab(b+a)}$ **61** $\dfrac{4x+1}{(x+1)^{1/2}}$ **62** $\dfrac{a-5}{(a-2)^{3/2}}$

63 $\dfrac{7c-11}{(c-3)^{2/3}}$ **65** $\dfrac{-4}{(d-2)^{3/2}(d+2)^{1/2}}$

66 $\dfrac{3b+2}{(b+5)^{2/3}(2b-3)^{1/2}}$ **67** $\dfrac{w-9}{(3w-4)^{4/3}(2w+5)^{3/4}}$

1 6   **2** 7   **3** 3   **5** $2\sqrt{7}$ **6** $3\sqrt{5}$ **7** $3\sqrt[3]{2}$

9 30   **10** 60   **11** 12   **13** $\frac{2}{3}$ **14** $\frac{4}{3}$ **15** $\frac{3}{2}$

17 $\sqrt[4]{8}$ **18** $2\sqrt{2}$ **19** $\sqrt[5]{4}$ **21** $3a^2 b\sqrt{3ab}$ **22** $2s^3\sqrt{2rs}$

23 $2xy^2\sqrt[3]{2xy}$ **25** $3xy^2\sqrt[4]{xy}$ **26** $2w^3z\sqrt[4]{z^3}$

27 $2b^2\sqrt[4]{a^3b}$ **29** $4x^2y^4$ **30** $6a^5b^3$ **31** $9a^3b\sqrt{b}$

33 $3c^2d^2\sqrt[3]{d}$ **34** $7ab^2$ **35** $3r^2t\sqrt[4]{12t}$ **37** $3a^2b$

38 $6x^2y^2$ **39** $2uv\sqrt{3}$ **41** $\sqrt{21b}/3a$ **42** $2\sqrt{5rt}/5rt$

43 $3\sqrt[3]{4ab^2}/2b^2$ **45** $4a^2c^2\sqrt{b}/b$ **46** $7s^2\sqrt{t}/r$

47 $\sqrt{3yz}/w^2z^2$ **49** $c\sqrt{ab}/6ab$ **50** $v\sqrt{3uvw}/3u^2$

51 $\sqrt[3]{20a}/5ac$ **53** $xyz\sqrt{2}/2$ **54** $5b^2/c^2d^2$

55 $\sqrt{3t}/5s^2t$ **57** $b\sqrt{3ab}$ **58** $y\sqrt{2xy}$ **59** $\sqrt[3]{2xy^2}$

61 $\sqrt{2}$ **62** 1   **63** $\sqrt[5]{48x^4}/2x$

EXERCISE 9.5 Page 221

1 2   **2** -5 **3** 4   **5** $-16+\sqrt{6}$ **6** $10+3\sqrt{14}$

7 $16-5\sqrt{15}$ **9** $-8\sqrt{3}$ **10** $6+9\sqrt{2}$ **11** $-10-20\sqrt{2}$

13 a^2-bc **14** $r-s^2$ **15** $x\sqrt{x}+y\sqrt{y}$ **17** $-6+2\sqrt{15}$

18 $1+2\sqrt{6}$ **19** $10-2\sqrt{35}$ **21** $\dfrac{6\sqrt{2}-11}{7}$ **22** $\dfrac{3\sqrt{5}-7}{2}$

23 $2\sqrt{30}-11$ **25** $7\sqrt{2}-5\sqrt{3}$ **26** $\dfrac{17\sqrt{2}+8\sqrt{5}}{3}$

27 $-\sqrt{5}$ **29** $\dfrac{a+2\sqrt{ab}+b}{a-b}$ **30** $\sqrt{x}-\sqrt{y}$

31 $\dfrac{w^2+(z+w)\sqrt{wz}+z^2}{w-z}$ **33** $6\sqrt{2}$ **34** $-2\sqrt{3}$

35 $5\sqrt{5}$ **37** $4\sqrt[3]{3}$ **38** $3\sqrt[3]{2}$ **39** $9\sqrt[3]{5}$

41 $5\sqrt{3}-2\sqrt[3]{3}$ **42** $3\sqrt[3]{9}-3\sqrt{3}$ **43** $2\sqrt[3]{5}-\sqrt{5}$

45 $6x^2\sqrt{3y}$ **46** $3c^2\sqrt{2d}$ **47** $7ab\sqrt{ab}$ **49** $4a^3b^4\sqrt{ab}$

50 $17b^3c^2\sqrt{bc}$ **51** $12w^3z^3\sqrt{wz}$ **53** $\dfrac{xy\sqrt{y}}{4}$ **54** $\dfrac{2xy\sqrt{6}}{3}$

55 $\dfrac{a}{2b}\sqrt{3b}$ **57** $(a+b)(\sqrt[3]{ab}+\sqrt{ab})$

58 $\left(\dfrac{b^2-1}{b^2}\right)\left(\sqrt{3b}+\sqrt[3]{3b^2}\right)$ **59** $\left(\dfrac{z-w}{3}\right)\left(\sqrt[3]{9z}-\sqrt{3z}\right)$

EXERCISE 10.1 Page 227

1 $\left\{\frac{3}{2}, -\frac{3}{2}\right\}$ **2** $\left\{\frac{5}{4}, -\frac{5}{4}\right\}$ **3** $\left\{\frac{8}{7}, -\frac{8}{7}\right\}$

5 $\{4, -4\}$ **6** $\{2, -2\}$ **7** $\{3, -3\}$ **9** $\left\{\dfrac{2\sqrt{3}}{3}, \dfrac{-2\sqrt{3}}{3}\right\}$

10 $\left\{\dfrac{4\sqrt{2}}{3}, \dfrac{-4\sqrt{2}}{3}\right\}$ 11 $\left\{\dfrac{2\sqrt{10}}{5}, \dfrac{-2\sqrt{10}}{5}\right\}$ 13 $\{2i, -2i\}$

14 $\{3i, -3i\}$ 15 $\{4i, -4i\}$ 17 $\{\sqrt{3}i, -\sqrt{3}i\}$

18 $\{\sqrt{6}i, -\sqrt{6}i\}$ 19 $\{\sqrt{10}i, -\sqrt{10}i\}$ 21 $\{3, -5\}$

22 $\{2, -4\}$ 23 $\{1, -3\}$ 25 $\{7, -4\}$ 26 $\{3, -6\}$

27 $\{8, -5\}$ 29 $\{\frac{3}{2}, -2\}$ 30 $\{\frac{2}{3}, -3\}$

31 $\{\frac{1}{4}, -2\}$ 33 $\{2, -\frac{5}{3}\}$ 34 $\{\frac{5}{2}, -3\}$

35 $\{\frac{7}{2}, -1\}$ 37 $\{\frac{2}{3}, -\frac{3}{2}\}$ 38 $\{\frac{1}{2}, -\frac{4}{3}\}$

39 $\{\frac{5}{3}, -\frac{1}{2}\}$ 41 $\{\frac{2}{3}, -\frac{1}{5}\}$ 42 $\{\frac{4}{3}, -\frac{2}{5}\}$

43 $\{\frac{3}{2}, -\frac{5}{4}\}$ 45 $\{\frac{6}{5}, -\frac{4}{3}\}$ 46 $\{\frac{7}{4}, -\frac{5}{3}\}$

47 $\{\frac{8}{5}, -\frac{7}{4}\}$ 49 $\{-\frac{7}{4}, \frac{5}{3}\}$ 50 $\{\frac{9}{4}, -\frac{5}{2}\}$

51 $\{\frac{1}{6}, -\frac{2}{7}\}$ 53 $\{-\frac{7}{4}, \frac{8}{5}\}$ 54 $\{-\frac{9}{2}, \frac{7}{2}\}$

55 $\{\frac{8}{3}, -\frac{9}{4}\}$ 57 $\{\frac{7}{6}, -\frac{9}{5}\}$ 58 $\{\frac{8}{5}, -\frac{9}{4}\}$

59 $\{\frac{5}{6}, -\frac{9}{7}\}$ 61 $\{b, 2b/3\}$ 62 $\{-2a, 5a/2\}$

63 $\{3t, 7t/2\}$ 65 $\{b, -2a\}$ 66 $\{q/p, -3q\}$

67 $\{c/b, -c/3\}$

EXERCISE 10.2   Page 234

1 $\{3, -9\}$ 2 $\{4, -2\}$ 3 $\{-7, -3\}$ 5 $\{3, -8\}$

6 $\{5, -4\}$ 7 $\{9, -6\}$ 9 $\{\frac{5}{2}, -\frac{7}{2}\}$ 10 $\{\frac{2}{3}, -\frac{5}{3}\}$

11 $\{\frac{7}{2}, -\frac{3}{2}\}$ 13 $\{-2, \frac{1}{2}\}$ 14 $\{1, \frac{2}{3}\}$ 15 $\{\frac{1}{2}, -3\}$

17 $\{\frac{1}{2}, \frac{1}{3}\}$ 18 $\{\frac{5}{4}, -\frac{3}{2}\}$ 19 $\{\frac{1}{2}, \frac{1}{5}\}$

21 $\{\frac{1}{2}, -\frac{4}{3}\}$ 22 $\{\frac{3}{2}, \frac{1}{3}\}$ 23 $\{\frac{5}{2}, -\frac{9}{4}\}$

25 $\{2 + \sqrt{6}, 2 - \sqrt{6}\}$ 26 $\{1 + \sqrt{5}, 1 - \sqrt{5}\}$

27 $\{3 + \sqrt{2}, 3 - \sqrt{2}\}$ 29 $\left\{\dfrac{-2 + \sqrt{5}}{2}, \dfrac{-2 - \sqrt{5}}{2}\right\}$

30 $\left\{\dfrac{3 + \sqrt{2}}{2}, \dfrac{3 - \sqrt{2}}{2}\right\}$ 31 $\left\{\dfrac{5 + 3\sqrt{5}}{6}, \dfrac{5 - 3\sqrt{5}}{6}\right\}$

33 $\left\{\dfrac{5 + \sqrt{5}}{4}, \dfrac{5 - \sqrt{5}}{4}\right\}$ 34 $\left\{\dfrac{-4 + \sqrt{7}}{9}, \dfrac{-4 - \sqrt{7}}{9}\right\}$

35 $\left\{\dfrac{-7 + \sqrt{5}}{4}, \dfrac{-7 - \sqrt{5}}{4}\right\}$ 37 $\{4 + 3i, 4 - 3i\}$

38 $\{5 + i, 5 - i\}$ 39 $\{3 + 4i, 3 - 4i\}$

41 $\left\{\dfrac{1 + 5i}{2}, \dfrac{1 - 5i}{2}\right\}$ 42 $\left\{\dfrac{-2 + 3i}{2}, \dfrac{-2 - 3i}{2}\right\}$

43 $\left\{\dfrac{1 + 3i}{3}, \dfrac{1 - 3i}{3}\right\}$ 45 $\left\{\dfrac{2 + \sqrt{5}i}{2}, \dfrac{2 - \sqrt{5}i}{2}\right\}$

46 $\left\{\dfrac{3 + \sqrt{6}i}{3}, \dfrac{3 - \sqrt{6}i}{3}\right\}$ 47 $\left\{\dfrac{-4 + \sqrt{5}i}{2}, \dfrac{-4 - \sqrt{5}i}{2}\right\}$

49 $\{a, -5a\}$ 50 $\{2m, -3m\}$ 51 $\{5q, -3p\}$

53 $\left\{2, \dfrac{a+b}{b}\right\}$ **54** $\left\{\dfrac{1}{rt}, t\right\}$ **55** $\left\{\dfrac{b}{2}, \dfrac{b+c}{2}\right\}$ **57** $\left\{\dfrac{a}{a-b}, -3\right\}$

58 $\left\{\dfrac{c+d}{c-d}, -1\right\}$ **59** $\left\{\dfrac{-3}{p}, 5\right\}$

EXERCISE 10.3　Page 238

1 $\{-7, -2\}$　　　**2** $\{5, -6\}$　　　**3** $\{7, -3\}$　　　**5** $\{\frac{5}{2}, -2\}$

6 $\{-\frac{4}{3}, 3\}$　　　**7** $\{\frac{7}{5}, -2\}$　　　**9** $\{\frac{7}{3}, -\frac{5}{2}\}$　　　**10** $\{\frac{5}{3}, -\frac{4}{3}\}$

11 $\{\frac{9}{4}, -\frac{5}{3}\}$　　　**13** $\{\frac{5}{6}, -\frac{6}{7}\}$　　　**14** $\{\frac{7}{5}, -\frac{9}{5}\}$

15 $\{\frac{5}{8}, -\frac{4}{7}\}$　　　**17** $\{2+\sqrt{3}, 2-\sqrt{3}\}$　　**18** $\{5+\sqrt{7}, 5-\sqrt{7}\}$

19 $\{4+\sqrt{3}, 4-\sqrt{3}\}$　　　　　　**21** $\left\{\dfrac{3+\sqrt{5}}{2}, \dfrac{3-\sqrt{5}}{2}\right\}$

22 $\left\{\dfrac{-2+\sqrt{7}}{3}, \dfrac{-2-\sqrt{7}}{3}\right\}$　　**23** $\left\{\dfrac{5+\sqrt{3}}{5}, \dfrac{5-\sqrt{3}}{5}\right\}$

25 $\left\{\dfrac{2+3\sqrt{2}}{3}, \dfrac{2-3\sqrt{2}}{3}\right\}$　　**26** $\left\{\dfrac{3+2\sqrt{5}}{2}, \dfrac{3-2\sqrt{5}}{2}\right\}$

27 $\left\{\dfrac{4+3\sqrt{7}}{3}, \dfrac{4-3\sqrt{7}}{3}\right\}$　　**29** $\left\{\dfrac{7+2\sqrt{3}}{5}, \dfrac{7-2\sqrt{3}}{5}\right\}$

30 $\left\{\dfrac{7+3\sqrt{5}}{6}, \dfrac{7-3\sqrt{5}}{6}\right\}$　　**31** $\left\{\dfrac{-5+3\sqrt{2}}{7}, \dfrac{-5-3\sqrt{2}}{7}\right\}$

33 $\{-3+i, -3-i\}$　　　　　　**34** $\{4+2i, 4-2i\}$

35 $\{2+5i, 2-5i\}$　　**37** $\left\{\dfrac{3+i}{2}, \dfrac{3-i}{2}\right\}$　　**38** $\left\{\dfrac{5+3i}{2}, \dfrac{5-3i}{2}\right\}$

39 $\left\{\dfrac{3+5i}{3}, \dfrac{3-5i}{3}\right\}$　　**41** $\left\{\dfrac{2+\sqrt{5}i}{5}, \dfrac{2-\sqrt{5}i}{5}\right\}$

42 $\left\{\dfrac{-3+\sqrt{6}i}{2}, \dfrac{-3-\sqrt{6}i}{2}\right\}$　　**43** $\left\{\dfrac{5+\sqrt{2}i}{2}, \dfrac{5-\sqrt{2}i}{2}\right\}$

45 $\left\{\dfrac{-3+\sqrt{6}i}{2}, \dfrac{-3-\sqrt{6}i}{2}\right\}$　　**46** $\left\{\dfrac{-5+\sqrt{2}i}{3}, \dfrac{-5-\sqrt{2}i}{3}\right\}$

47 $\left\{\dfrac{-6+\sqrt{2}i}{2}, \dfrac{-6-\sqrt{2}i}{2}\right\}$　　**49** $\{2a, b\}$　　**50** $\{5\ uv\}$

51 $\{2b/c, -b/2c\}$　　**53** $\{r+t, r-3t\}$　　**54** $\{c/d, 2c/3d\}$

55 $\left\{\dfrac{ab}{5}, \dfrac{3}{ab}\right\}$　　**57** $\left\{\dfrac{c+2d}{3}, 3d\right\}$　　**58** $\left\{\dfrac{3}{2m+n}, \dfrac{-2}{m+n}\right\}$

59 $\{-7/b, -b\}$

EXERCISE 10.4　Page 243

1 Irrational and unequal; $6, -15$　　　**2** Irrational and unequal; $4, -16$

3 Irrational and unequal; $-10, 15$　　**5** Rational and equal; $10, 25$

6 Rational and unequal; $3, -28$　　　**7** Rational and unequal; $-6, -16$

9 Rational and unequal; $-\frac{7}{2}, -\frac{15}{2}$ **10** Rational and equal; $1, \frac{1}{4}$

11 Rational and equal; $\frac{4}{3}, \frac{4}{9}$ **13** Imaginary; $\frac{6}{5}, \frac{2}{5}$

14 Imaginary; $-5, \frac{13}{2}$ **15** Imaginary; $\frac{4}{3}, \frac{2}{3}$

17 Rational and equal; $\frac{4}{3}, \frac{4}{9}$ **18** Rational and unequal; $-\frac{6}{5}, -\frac{8}{5}$

19 Rational and equal; $\frac{8}{3}, \frac{16}{9}$ **21** Irrational and unequal; $2, \frac{1}{9}$

22 Irrational and unequal; $3, -\frac{7}{2}$ **23** Irrational and unequal; $\frac{5}{2}, \frac{5}{4}$

25 Imaginary; $-2, \frac{6}{5}$ **26** Imaginary; $-5, 7$

27 Imaginary; $\frac{14}{5}, 2$ **29** Irrational and unequal; $\frac{14}{5}, 1$

30 Irrational and unequal; $-\frac{16}{3}, 4$ **31** Irrational and unequal; $\frac{9}{2}, \frac{9}{4}$

33 Real and unequal; $\sqrt{3}, -\frac{1}{4}$ **34** Real and unequal; $-2\sqrt{6}/3, \frac{5}{9}$

35 Real and equal $-4\sqrt{3}/3, \frac{4}{3}$ **37** $12x^2 - 5x - 3 = 0$

38 $7x^2 + 47x - 14 = 0$ **39** $6x^2 + 5x - 6 = 0$

41 $x^2 - 4x - 1 = 0$ **42** $x^2 + 6x + 2 = 0$

43 $9x^2 - 6x - 1 = 0$

EXERCISE 10.5   Page 249

1 $\{1\}$ **2** $\{5\}$ **3** $\{3\}$ **5** $\{\frac{1}{2}\}$ **6** $\{10\}$ **7** $\{17\}$

9 $\{0, -3\}$ **10** $\{7, -1\}$ **11** $\{5, -2\}$ **13** $\{-3, -2\}$

14 $\{2\}$ **15** $\{0, 4\}$ **17** $\{1, \frac{1}{2}\}$ **18** $\{3\}$ **19** $\{\frac{3}{2}\}$

21 $\{1, 5\}$ **22** $\{3, -1\}$ **23** $\{4, 8\}$ **25** $\{\frac{3}{2}, 1\}$

26 $\{5\}$ **27** $\{-1\}$ **29** $\{4\}$ **30** $\{8\}$ **31** $\{13, 1\}$

33 $\{2\}$ **34** $\{5\}$ **35** $\{3\}$ **37** $\{4\}$ **38** $\{6, 2\}$

39 $\{4, -1\}$ **41** $\{3b\}$ **42** $\{a\}$ **43** $\{2b\}$

EXERCISE 10.6   Page 253

1 $\{4, -4, 1, -1\}$ **2** $\{3, -3, 1, -1\}$ **3** $\{4, -4, 2, -2\}$

5 $\{3, -3, 2i, -2i\}$ **6** $\{3, -3, 4i, -4i\}$ **7** $\{5, -5, 2i, -2i\}$

9 $\{1, -1, \frac{1}{2}, -\frac{1}{2}\}$ **10** $\{2, -2, \frac{1}{2}, -\frac{1}{2}\}$ **11** $\{\frac{1}{2}, -\frac{1}{2}, \frac{1}{3}, -\frac{1}{3}\}$

13 $\{\sqrt{2}, -\sqrt{2}, i, -i\}$ **14** $\{\sqrt{3}, -\sqrt{3}, 2i, -2i\}$ **15** $\left\{2i, -2i, \dfrac{\sqrt{2}}{2}, -\dfrac{\sqrt{2}}{2}\right\}$

17 $\{\frac{1}{4}, 1\}$ **18** $\{\frac{1}{3}, 1\}$ **19** $\{\frac{1}{3}, -\frac{1}{3}, 1, -1\}$

21 $\{\frac{1}{2}, -1\}$ **22** $\{1, -\frac{2}{3}\}$ **23** $\{2, -1\}$ **25** $\{2, -2, 1, -1\}$

26 $\{3, -3, 2, -2\}$ **27** $\{3, -3, 1, -1\}$ **29** $\{1, -1, 3, -3\}$

30 $\{3, -3, 2, -2\}$ **31** $\{2, -2, 1, -1\}$

33 $\{1, -2, \frac{1}{2}(1 - \sqrt{13}), -\frac{1}{2}(1 + \sqrt{13})\}$ **34** $\{2, -3, 3, -4\}$

35 $\{2, -4, -1, -1\}$ **37** $\{-\frac{3}{4}, -\frac{6}{5}\}$ **38** $\{-\frac{7}{3}, -5\}$

39 $\{2, 8\}$ **41** $\{-4, \frac{11}{6}\}$

42 $\{-\frac{7}{3}, -5\}$ **43** $\{\frac{1}{2}, \frac{4}{3}\}$ **45** $\{\frac{3}{2}, 4\}$ **46** $\{\frac{10}{3}\}$

47 $\{2, -2, \frac{7}{4}, -\frac{7}{4}\}$ **49** $\{-2\}$ **50** $\{-1, 3\}$

51 $\{1, \frac{17}{12}\}$

1 $\{x \mid x > 3\} \cup \{x \mid x < -2\}$

2 $\{x \mid x > 5\} \cup \{x \mid x < 2\}$

3 $\{x \mid x > 2\} \cup \{x \mid x < -\frac{1}{2}\}$

5 $\{x \mid 1 < x < \frac{3}{2}\}$

6 $\{x \mid -7 < x < -3\}$

7 $\{x \mid -\frac{3}{2} < x < 1\}$

9 $\{x \mid -3 < x < \frac{1}{2}\}$

10 $\{x \mid -1 < x < -\frac{1}{3}\}$

11 $\{x \mid -2 < x < \frac{2}{3}\}$

13 $\{x \mid x > \frac{2}{3}\} \cup \{x \mid x < -\frac{1}{2}\}$

14 $\{x \mid x < -3\} \cup \{x \mid x > \frac{1}{2}\}$

15 $\{x \mid x < -2\} \cup \{x \mid x > \frac{3}{4}\}$

17 $\{x \mid x < -\frac{5}{2}\} \cup \{x \mid x > \frac{4}{3}\}$

18 $\{x \mid -\frac{3}{4} < x < \frac{1}{3}\}$

19 $\{x \mid x < -\frac{5}{2}\} \cup \{x \mid x > -\frac{2}{3}\}$

21 $\{x \mid x > 1\} \cup \{x \mid x < 1\}$

22 $\{x \mid x < -2\} \cup \{x \mid x > -2\}$

23 $\varnothing$

25 $\{x \mid x \text{ is any real number}\}$

26 $\varnothing$

27 $\{x \mid x \text{ is any real number}\}$

EXERCISE 10.8 **Page 263**

1 5, 6 **2** 8, −7 **3** 8, 10; −6, −8 **5** 5 10

6 $\frac{3}{2}, \frac{2}{3}$ **7** 5 10; −10, −5 **9** 50 by 50 yd

10 15 by 20 ft **11** 10 ft **13** 200 miles per hour

14 180 miles per hour **15** 35 miles per hour **17** 4, 6

18 5 hr, $7\frac{1}{2}$ hr **19** 10, 15 **21** 4 by 12 ft

22 28 by 21 ft **23** 1 in. **25** 3 by 24 ft

26 30, 40 **27** 4 **29** 6, 8 **30** 50, 60

31 45 cu ft per min

EXERCISE 11.1 **Page 275**

21 $\{(2.4, 6.2), (-9, -3.7)\}$

22 $\{(-1, 2), (-5, -6)\}$

23 $\{(3, 2), (-4, -1.5)\}$

25 $\{(4, 2), (2.9, -1.7)\}$

26 $\{(2, 2), (0.8, -1.3)\}$

27 $\{(2, 1), (3.1, 5.5)\}$

29 $\{(12, 5), (12, -5)\}$

30 $\{(1, 2), (1, -2), (4, 4), (4, -4)\}$

31 $\{(-3, 2), (-3, -2), (1.7, 3.7), (1.7, -3.7)\}$

33 $\{(1, 3), (1, -3), (-1, 3), (-1, -3)\}$ **34** $\{(4, 6), (4, -6), (-4, 6), (-4, -6)\}$

35 $\{(2.6, 3.1), (2.6, -3.1), (-2.6, 3.1), (-2.6, -3.1)\}$

EXERCISE 11.2 **Page 283**

1 $\{(9, -6), (1, 2)\}$ **2** $\{(6, 4), (3, 1)\}$ **3** $\{(0, 0), (2, 2)\}$

5 $\{(2, 1), (-1, -2)\}$ **6** $\{(2, -3), (-\frac{6}{5}, \frac{17}{5})\}$ **7** $\{(1, \frac{1}{2}), (-\frac{1}{2}, -1)\}$

9 $\{(2, 5), (1, 3)\}$ **10** $\{(2, -3), (3, 1)\}$

11 $\{(-3, 2), (5, 1)\}$ **13** $\{(3 \ 1), (-5, -2)\}$ **14** $\{(4, 1), (-6, -2)\}$

15 $\{(1, \frac{1}{3}), (-\frac{5}{9}, -\frac{19}{27})\}$ **17** $\left\{(a, a), \left(\frac{13a}{5}, \frac{-11a}{5}\right)\right\}$

18 $\{(b, a), (-b, -a)\}$

19 $\left\{\left(\dfrac{-b}{m}, 0\right), (0, b)\right\}$

21 $\{(2, 1), (-1, -2)\}$

22 $\{(0, -3), (1, -1)\}$

23 $\{(3, 2), (-\frac{8}{3}, \frac{35}{9})\}$

25 $\{(i, -1 - 10i), (-i, -1 + 10i), (10, 0), (-10, 200)\}$

26 $\{(2, 14), (-2, 10), (2i, -12 + 2i), (-2i, -12 - 2i)\}$

27 $\{(0, 1), (-2, -1), (1 + \sqrt{3}, \sqrt{3}), (1 - \sqrt{3}, -\sqrt{3})\}$

29 $\{(5, 1), (-3, -1), (14 + 4\sqrt{14}, \sqrt{14}), (14 - 4\sqrt{14}, -\sqrt{14})\}$

30 $\left\{(-2 - i, i), (-2 + i, -i), \left(\dfrac{3 - \sqrt{3}}{2}, \dfrac{\sqrt{3}}{2}\right), \left(\dfrac{3 + \sqrt{3}}{2}, \dfrac{-\sqrt{3}}{2}\right)\right\}$

31 $\left\{(0, 0), (0, 0), \left(\dfrac{-3 - i\sqrt{3}}{4}, \dfrac{i\sqrt{3}}{4}\right), \left(\dfrac{-3 + i\sqrt{3}}{4}, \dfrac{-i\sqrt{3}}{4}\right)\right\}$

33 $\left\{(2, 1), (-2, -1), \left(\sqrt{3}, \dfrac{2\sqrt{3}}{3}\right), \left(-\sqrt{3}, -\dfrac{2\sqrt{3}}{3}\right)\right\}$

34 $\left\{(3, 1), (-3, -1), \left(\dfrac{\sqrt{6}}{2}, \sqrt{6}\right), \left(-\dfrac{\sqrt{6}}{2}, -\sqrt{6}\right)\right\}$

35 $\left\{(4, -2), (-4, 2), \left(\dfrac{4i\sqrt{3}}{3}, 2i\sqrt{3}\right), \left(-\dfrac{4i\sqrt{3}}{3}, -2i\sqrt{3}\right)\right\}$

37 $\{(1, 2), (-1, -2), (-2i, i), (2i, -i)\}$

38 $\{(\frac{1}{2}, 3), (-\frac{1}{2}, -3), (\frac{3}{2}, 1), (-\frac{3}{2}, -1)\}$

39 $\left\{(1, -5), (-1, 5), \left(-\dfrac{5\sqrt{2}}{2}, \sqrt{2}\right), \left(\dfrac{5\sqrt{2}}{2}, -\sqrt{2}\right)\right\}$

EXERCISE 11.3   Page 287

1 $\{(1, -3), (-1, -3), (-1, 3), (1, 3)\}$

2 $\{(5, 2), (5, -2), (-5, 2), (-5, -2)\}$

3 $\{(2, 1), (-2, 1), (2, -1), (-2, -1)\}$

5 $\{(3, \frac{1}{2}), (3, -\frac{1}{2}), (-3, \frac{1}{2}), (-3, -\frac{1}{2})\}$

6 $\{(5, 1), (5, -1), (-5, 1), (-5, -1)\}$

7 $\{(2, 1), (2, -1), (-2, 1), (-2, -1)\}$

9 $\{(2, 2), (2, -2), (-2, 2), (-2, -2)\}$

10 $\{(3, 5), (3, -5), (-3, -5), (-3, 5)\}$

11 $\{(5, 2), (5, -2), (-5, 2), (-5, -2)\}$

13 $\{(0, \frac{1}{2}), (0, -\frac{1}{2}), (2, \frac{1}{2}), (2, -\frac{1}{2})\}$

14 $\{(3, 2), (3, -2), (2, 1), (2, -1)\}$

15 $\{(\frac{1}{2}, 1), (\frac{1}{2}, -1), (1, \frac{1}{2}), (1, -\frac{1}{2})\}$

17 $\{(2, 2), (2, -2), (3, i\sqrt{3}), (3, -i\sqrt{3})\}$

18 $\{(3, 2), (3, -2), (5, i\sqrt{5}), (5, i\sqrt{5})\}$

19 $\left\{(1, 2\sqrt{2}), (1, -2\sqrt{2}), \left(\dfrac{3}{4}, \dfrac{3\sqrt{2}}{4}\right), \left(\dfrac{3}{4}, \dfrac{-3\sqrt{2}}{4}\right)\right\}$

21 $\{(2, 3), (-2, 3), (4, 5), (-4, 5)\}$

22 $\{(\frac{1}{3}, \frac{2}{3}), (-\frac{1}{3}, \frac{2}{3}), (1, 1), (-1, 1)\}$

23 $\left\{\left(\dfrac{i\sqrt{13}}{5}, \dfrac{6}{5}\right), \left(\dfrac{-i\sqrt{13}}{5}, \dfrac{6}{5}\right), \left(\dfrac{i\sqrt{6}}{5}, \dfrac{-1}{5}\right), \left(\dfrac{-i\sqrt{6}}{5}, \dfrac{-1}{5}\right)\right\}$

25 $\{(2, -3), (-2, 1)\}$

26 $\{(2, 1), (-\frac{3}{2}, -\frac{1}{6})\}$

27 $\{(\frac{1}{2}, -2), (2, 1)\}$

29 $\{(-3, 2), (\frac{5}{6}, -\frac{16}{15})\}$

30 $\{(2, -2.5), (3, -1)\}$

31 $\{(1, 2), (-\frac{1}{6}, -\frac{3}{2})\}$

1 $\{(3, 2), (-3, -2), (1, -1), (-1, 1)\}$
2 $\{(1, 3), (-1, -3), (4, 4), (-4, -4)\}$
3 $\{(2, 1), (-2, -1), (3, -3), (-3, 3)\}$
5 $\{(3, -1), (-3, 1), (2, 2), (-2, -2)\}$
6 $\{(4, 1), (-4, -1), (2, -2), (-2, 2)\}$
7 $\{(5, -1), (-5, 1), (3, -3), (-3, 3)\}$
9 $\left\{(2, -1), (-2, 1), \left(\frac{\sqrt{2}}{2}, \sqrt{2}\right), \left(-\frac{\sqrt{2}}{2}, -\sqrt{2}\right)\right\}$
10 $\{(3, 2), (-3, -2), (\sqrt{3}, -\sqrt{3}), (-\sqrt{3}, \sqrt{3})\}$
11 $\{(5, 2), (-5, -2), (4i, -4i), (-4i, 4i)\}$
13 $\{(\sqrt{3}, 2\sqrt{3}), (-\sqrt{3}, -2\sqrt{3}), (\sqrt{2}, -\sqrt{2}), (-\sqrt{2}, \sqrt{2})\}$
14 $\{(i, 2i), (-i, -2i), (3i, -3i), (-3i, 3i)$
15 $\{(2\sqrt{3}, 3\sqrt{3}), (2\sqrt{2}, -2\sqrt{2}), (-2\sqrt{2}, 2\sqrt{2}), (-2\sqrt{3}, -3\sqrt{3})\}$
17 $\{(3, 2), (1, 4)\}$ 18 $\{(1, 3), (2, -3)\}$ 19 $\{(3, 5), (2, 6)\}$
21 $\{(1, 4), (2, 3)\}$ 22 $\{(5, 3), (2, 2)\}$ ·   23 $\{(4, -1), (2, -3)\}$
25 $\{(4, 1), (3, 2)\}$ 26 $\{(7, 3), (5, 4)\}$ 27 $\{(5, 1), (3, 4)\}$
29 $\{(3, 1), (1, 3), (2, 3), (3, 2)\}$ 30 $\{(4, 1), (1, 4), (2, 5), (5, 2)\}$
31 $\{(4, -2), (-2, 4), (3, 1), (1, 3)\}$ 33 $\{(1, 2), (2, 1), (3, 2), (2, 3)\}$
34 $\{(3, 5), (5, 3), (6, 1), (1, 6)\}$ 35 $\{(4, 5), (5, 4), (-3, 6), (6, -3)\}$
37 $\{(1, -3), (-3, 1), (5, -3), (-3, 5)\}$
38 $\{(-1, 7), (7, -1), (2, 6), (6, 2)\}$ 39 $\{(1, 5), (3, -4), (5, 1), (-4, 3)\}$

EXERCISE 11.5 Page 297

1 8, 4   2 8, 6   3 9, 3   5 16 ft, 12 ft
6 12 ft, 6 ft, 3 ft   7 Base, 16 by 16 ft; height, 4 ft
9 100, $175   10 15 by 36 ft   11 100 by 150 ft
13 5 by 12 ft   14 15 by 15 ft
15 4 by 9 in. or 3 by 12 in.   17 120 shares at $65 apiece   18 2 in., 5 in.
19 Side 8 in., radius 7 in., or side $\frac{332}{25}$, radius $\frac{91}{25}$
21 Square field, 80 by 80 rods; rectangular field, 40 by 80 rods
22 Father, 4 days; son, 6 days
23 660 yd, 70 yd

EXERCISE 12.1 Page 305

1 $\frac{1}{3}$ 2 $\frac{7}{4}$ 3 $\frac{11}{2}$ 5 $\frac{7}{2}$ 6 $\frac{264}{35}$
7 $\frac{3}{40}$ 9 $860 per month   10 48 miles per hour

11 3 doughnuts per boy **13** $\frac{1}{30}$ **14** 21.50 **15** 0.21

17 $\frac{24}{5}$ **18** $\frac{8}{5}$ **19** 3 **21** $\frac{1}{2}$ **22** $-\frac{2}{5}$

23 3, -2 **25** ± 12 **26** ± 14 **27** ± 18 **29** $12\frac{4}{5}$

30 $21\frac{1}{3}$ **31** 36 **33** 16 **34** 6 **35** 3

37 $x = 21$, $y = 6$ **38** $x = 12$, $y = 3$ **39** $x = 10$, $y = 12$

41 8712 ft **42** 8 **43** 140 miles

EXERCISE 12.2   Page 309

1 (a) $p = kq$ (b) $a = k/b$ (c) $x = kyz$ (d) $u = kw^2/v$

2 6 **3** 30 **5** 2 **6** 120 **7** 112 **9** 5 sq in.

10 7 g **11** \$121.50 **13** 2 liters

14 The second post requires $\frac{4}{5}$ as much as the first.

15 205 cu ft **17** 3.5 in.

18 The safe load of the first is $\frac{2}{3}$ that of the second.

19 1.8 lb **21** The second jet has four times the power of the first.

22 10 in. **23** 15 lb **25** 670

26 300 rpm **27** $\frac{1}{200}$ sec

EXERCISE 13.1   Page 316

1 62.56 **2** 723.9 **3** 3.156 **5** 708,900 **6** 90,530,000 **7** 611,900

9 9078 **10** 318.8 **11** 339,700 **13** 47.34 **14** 516.3 **15** 9062

17 $5.917(10^3)$ **18** $5.629(10)$ **19** $6.95752(10^5)$ **21** $5.76(10^{-2})$

22 $4.987(10^{-4})$ **23** $9.35(10^{-6})$ **25** $8.133(10^{-3})$ **26** $1.280(10^3)$

27 $2.9700(10^4)$ **29** $7.51(10^3)$ **30** $9.71(10^4)$ **31** $1.83(10^7)$ **33** $8.7(10)$

34 $7.7(10^{-1})$ **35** $1.3(10^3)$ **37** $5.07(10^4)$ **38** $5.62(10^2)$ **39** $3.6(10^2)$

41 $2.5(10)$ **42** $1.4(10^{-1})$ **43** $2.25(10^{-2})$ **45** $2.60(10)$ **46** $9.47(10^2)$

47 $9.820(10)$ **49** $7.81(10^2)$ **50** $1.02(10)$ **51** 3.7

EXERCISE 13.2   Page 322

1 $\log_3 27 = 3$ **2** $\log_2 16 = 4$ **3** $\log_3 243 = 5$ **5** $\log_{1/3} 9 = -2$

6 $\log_{1/4} 64 = -3$ **7** $\log_{1/2} 32 = -5$ **9** $\log_4 \frac{1}{16} = -2$ **10** $\log_3 \frac{1}{81} = -4$

11 $\log_5 \frac{1}{625} = -4$ **13** $\log_{36} 6 = \frac{1}{2}$ **14** $\log_{125} 5 = \frac{1}{3}$ **15** $\log_4 8 = \frac{3}{2}$

17 $3^2 = 9$ **18** $5^5 = 3125$ **19** $6^3 = 216$ **21** $10^{-1} = \frac{1}{10}$ **22** $4^{-2} = \frac{1}{16}$

23 $5^{-3} = \frac{1}{125}$ **25** $(\frac{1}{3})^{-2} = 9$ **26** $(\frac{1}{8})^{-3} = 512$ **27** $(\frac{1}{2})^{-8} = 256$ **29** 6

30 4 **31** 5 **33** $\frac{3}{2}$ **34** $\frac{2}{3}$ **35** $\frac{3}{4}$ **37** 125 **38** 32 **39** $\frac{1}{81}$

41 16 **42** 216 **43** $\frac{4}{3}$ **45** 5 **46** 6 **47** 2 **49** $\frac{3}{5}$ **50** $\frac{16}{25}$

51 $\frac{125}{216}$ **53** 18 **54** 20 **55** 28 **57** 2 **58** 4 **59** 8 **61** 30

62 48 **63** 7 **65** 13 **66** 12 **67** 2

1 1.4409	**2** 0.5011	**3** 2.7024	**5** $8.8609 - 10$
6 $9.9289 - 10$	**7** $9.9969 - 10$	**9** 29.0	**10** 396

11 4.31 **13** 6.44 **14** 74.7 **15** 883 **17** 0.767

18 0.0592	**19** 0.0493	**21** 282	**22** 46.8
23 7.85	**25** 2.8405	**26** 1.8542	**27** 3.4810
29 $9.3995 - 10$	**30** $8.4123 - 10$	**31** $7.8514 - 10$	
33 1.5553	**34** 0.7735	**35** 2.6747	**37** $9.7861 - 10$
38 $8.8839 - 10$	**39** $7.3707 - 10$	**41** 46.53	
42 2.612	**43** 346.6	**45** 627.3	**46** 7518
47 97.28	**49** 0.6464	**50** 0.04778	**51** 0.002317

EXERCISE 13.4 Page 335

1 102 **2** 425 **3** 76.1 **5** 0.00356 **6** 0.00375 **7** 0.00150 **9** 0.00456

10 3.49 **11** 9.79 **13** 0.0548 **14** 0.833 **15** 0.0890 **17** 2.53

18 0.886 **19** 0.830 **21** 4.86 **22** 5.37 **23** 0.000000167 **25** 28.6

26 14.3 **27** 0.458 **29** 2.78 **30** 0.733 **31** 13.5 **33** 4.433 **34** 3.457

35 0.8714 **37** 6.897 **38** 1.325 **39** 0.4819 **41** 0.5699 **42** 241,300

43 2,107,000 **45** 0.01838 **46** 1.380 **47** 8.258

49 $\log c + \log x - \log(x + 4)$

50 $2 \log c + \log x - \log(x - 5)$

51 $\log c + 4 \log x + \frac{1}{2} \log(x + 1)$

53 $\log c + 3 \log(x + y) - \frac{1}{2} \log(x - y)$

54 $\log c + 2 \log x + \log y - 3 \log(x + y)$

55 $\log c - \frac{3}{4} \log x - 2 \log(x + y)$

EXERCISE 13.5 Page 340

1 3.59 **2** 2.10 **3** 7.21 **5** 1.81 **6** 1.39 **7** 3.07

9 2.99 **10** 1.60 **11** 6.30 **13** 9.75 **14** 3.78

15 2.44 **17** 5.10 **18** 2.32 **19** 3.30 **21** 5.300

22 1.410 **23** 3.103 **25** 0.2642 **26** 1.793 **27** 9.962

29 0 **30** -1 **31** 3 **33** 1 **34** 2 **35** 4 **37** 1

38 -2 **39** 8 **41** 3 **42** ± 3 **43** 2, 1

45 $x = 1.363, \ y = 0.7325$ **46** $x = 2.37, \ y = 0.7928$

47 $x = 1.971, \ y = 1.004$ **49** $x = 3.017, \ y = 1.057$

50 $x = 2.079, \ y = 1.897$ **51** $x = 2.890, \ y = 1.014$

EXERCISE 14.1 Page 346

1 5, 8, 11, 14, 17, 20 **2** 8, 6, 4, 2, 0 **3** 5, 7, 9, 11, 13, 15, 17

5 $l = 13, \ s = 48$ **6** $l = 5, \ s = -5$ **7** $l = 8, \ s = 50$ **9** $s = 35, \ d = 2$

10 $s = 36, d = -2$ **11** $a = 1, d = 3$ **13** $s = 10, n = 4$ **14** $l = 17, d = 3$

15 $a = -3, s = 4$ **17** $a = 21, l = -9$ **18** $n = 6, l = -7$

19 $a = -12, n = 7$ **21** 3721 **22** 240 ft, 1024 ft **23** $6840

25 $66,750 **26** $3.12 **27** 65 **30** $\frac{1}{3}$, 1 **31** $x = 9, y = 27$

EXERCISE 14.2   Page 351

1 3, 6, 12, 24, 48, 96   **2** 2, 6, 18, 54, 162

3 1, -3, 9, -27, 81 **5** 5, 15, 45, 135, 405, 1215 **6** 4, -8, 16, -32, 64

7 -2, 4, -8, 16, -32, 64, -128 and -2, -4, -8, -16, -32, -64, -128

9 $-\frac{4}{3}$, 4, -12, 36, -108 **10** 3, 9, 27, 81, 243, 729

11 ± 4, -8, ± 16, -32, ± 64, -128, ± 256 **13** 64 **14** -729

15 -64 **17** 1 **18** $\frac{1}{4}$ **19** 256

21 63 **22** 363 **23** 1094 **25** 364 **26** 86

27 1094 **29** $s = 242, n = 5$ **30** $s = 400, n = 4$

31 $a = \frac{1}{9}$, $s = \frac{40}{9}$ **33** $r = \frac{1}{2}$, $s = 511$

34 $r = 2$, $s = \frac{127}{8}$ **35** $a = 343, n = 4$

37 $l = 162, n = 5$ **38** $l = 25, n = 6$ **39** $n = 4, r = -2$

41 $a = 625, l = 1$ **42** $a = \frac{1}{8}, l = 8$

43 $a = 256, n = 3$

49 Yes, one **50** 4094 **51** $24,300 **53** $2048n$

54 $\frac{729}{4096}$ **55** 2, $-\frac{2}{15}$ **57** AP is 2, 5, 8; $r = 3$

58 $1379.66

EXERCISE 14.3   Page 358

1 8 **2** $\frac{8}{3}$ **3** 4 **5** 25 **6** $\frac{36}{5}$ **7** 64

9 13.5 **10** 16 **11** $\frac{256}{3}$, $\frac{256}{5}$ **13** $\frac{8}{9}$ **14** $\frac{5}{9}$

15 $\frac{20}{33}$ **17** $\frac{25}{33}$ **18** $\frac{70}{33}$ **19** $\frac{139}{66}$

21 $\frac{1}{2}$ **22** $\frac{1}{4}$ **23** 1.5 **25** 85 ft

26 24 ft **27** 128 sq in. **29** 49 in. **30** $605

31 $470,000 **33** $\dfrac{1}{x - 1}$ **34** $\dfrac{1}{2x}$ **35** $\dfrac{2}{3x - 6}$

37 8 **38** 7.5 **39** 1, 4 **41** 0, -3, -6

42 48, 55, 62, 69 **43** -10, -3, 4, 11

45 ± 4 **46** ± 9 **47** 2, 1 **49** 1, $-\frac{1}{2}$, $\frac{1}{4}$; -1, $-\frac{1}{2}$, $-\frac{1}{4}$

50 $\frac{5}{6}$, 1, $\frac{6}{5}$; $-\frac{5}{6}$, 1, $-\frac{6}{5}$ **51** 2, 4, 8, 16

EXERCISE 15.1   Page 364

1 $x^4 + 4x^3y + 6x^2y^2 + 4xy^3 + y^4$ **2** $c^5 + 5c^4d + 10c^3d^2 + 10c^2d^3 + 5cd^4 + d^5$

3 $a^7 + 7a^6y + 21a^5y^2 + 35a^4y^3 + 35a^3y^4 + 21a^2y^5 + 7ay^6 + y^7$

5 $m^6 - 6m^5t + 15m^4t^2 - 20m^3t^3 + 15m^2t^4 - 6mt^5 + t^6$

6 $s^5 - 5s^4a + 10s^3a^2 - 10s^2a^3 + 5sa^4 - a^5$

7 $a^7 - 7a^6y + 21a^5y^2 - 35a^4y^3 + 35a^3y^4 - 21a^2y^5 + 7ay^6 - y^7$

9 $a^6 + 12a^5x + 60a^4x^2 + 160a^3x^3 + 240a^2x^4 + 192ax^5 + 64x^6$

10 $x^3 + 9x^2y + 27xy^2 + 27y^3$

11 $16,384x^7 - 28,672x^6a + 21,504x^5a^2 - 8960x^4a^3 + 2240x^3a^4 - 336x^2a^5 + 28xa^6 - a^7$

13 $3125x^5 + 3125x^4y + 1250x^3y^2 + 250x^2y^3 + 25xy^4 + y^5$

14 $729a^6 + 1458a^5b + 1215a^4b^2 + 540a^3b^3 + 135a^2b^4 + 18ab^5 + b^6$

15 $s^5 - 25s^4t + 250s^3t^2 - 1250s^2t^3 + 3125st^4 - 3125t^5$

17 $16a^4 + 96a^3b + 216a^2b^2 + 216ab^3 + 81b^4$

18 $243x^5 + 1620x^4y + 4320x^3y^2 + 5760x^2y^3 + 3840xy^4 + 1024y^5$

19 $15,625t^6 - 56,250t^5b + 84,375t^4b^2 - 67,500t^3b^3 + 30,375t^2b^4 - 7290tb^5 + 729b^6$

21 $x^8 + 8x^6y + 24x^4y^2 + 32x^2y^3 + 16y^4$

22 $s^{15} + 10s^{12}h + 40s^9h^2 + 80s^6h^3 + 80s^3h^4 + 32h^5$

23 $t^{24} - 12t^{20}s + 60t^{16}s^2 - 160t^{12}s^3 + 240t^8s^8 - 192t^4s^4 + 64s^6$

25 $16a^4 + 96a^3y^2 + 216a^2y^4 + 216ay^6 + 81y^8$

26 $243x^5 + 2025x^4y^2 + 6750x^3y^4 + 11,250x^2y^6 + 9375xy^8 + 3125y^{10}$

27 $8x^3 - 84x^2y^3 + 294xy^6 - 343y^9$

29 $8x^6 + 36x^4y^4 + 54x^2y^8 + 27y^{12}$

30 $243x^{10} + 810x^8y^3 + 1080x^6y^6 + 720x^4y^9 + 240x^2y^{12} + 32y^{15}$

31 $15,625x^{18} - 56,250x^{15}y^2 + 84,375x^{12}y^4 - 67,500x^9y^6 + 30,375x^6y^8 - 7290x^3y^{10}$
$$+ 729y^{12}$$

33 $\dfrac{x^4}{16} + x^3y^{1/3} + 6x^2y^{2/3} + 16xy + 16y^{4/3}$

34 $\dfrac{a^3}{27} + a^2y^{2/3} + 9ay^{4/3} + 27y^2$

35 $\dfrac{16x^4}{625} - \dfrac{16x^3y^{1/2}}{25} + 6x^2y - 25xy^{3/2} + \frac{625}{16}y^2$

37 $x^{11} + 33x^{10}y + 495x^9y^2 + 4455x^8y^3$

38 $a^{17} + 34a^{16}b + 544a^{15}b^2 + 5440a^{14}b^3$

39 $x^{19} - 57x^{18}a + 1539x^{17}a^2 - 26,163x^{16}a^3$

41 $x^{42} + 21x^{40}y + 210x^{38}y^2 + 1330x^{36}y^3$

42 $x^{54} + 18x^{51}y^2 + 153x^{48}y^4 + 816x^{45}y^6$

43 $a^{32} - 16a^{29} + 120a^{26} - 560a^{23}$

45   1.1041   **46**   1.1255   **47**   1.1249   **49** $40,095a^8y^4$

50 $14,784x^5y^6$ **51** $-189x^2b^5$ **53** $672x^{12}y^3$

54 $-56x^9y^5$ **55** $1716(3^7)a^7b^{18}$ **57** $-20x^{3/2}y^3$

58 $1120a^4b^{4/3}$ **59** $8064c^5b^{5/2}$ **61** $-672x^2y^5$

62 $135x^4y^2$ **63** $448x^2a$

EXERCISE 16.1   Page 371

1   3   **2** -24 **3** -6 **5**   1   **6**   249   **7**   2   **17** $\{-1, 3, -5\}$

18 $\{5, -2, -7\}$ **19** $\{-\frac{1}{2}, \frac{1}{3}, -\frac{2}{3}\}$ **21** $\{2, -4, 3\}$ **22** $\{2, -1, -5\}$

23 $\{\frac{3}{2}, -1, \frac{2}{3}, -2\}$ **25** $x^2 - 4x - 2, 1$ **26** $x^2 + x + 2, -2$

27 $2x^2 - 2x + 3, -4$ **29** $x^4 + 2x^2 + x, -1$ **30** $-2x^3 + 2x^2 - 3x + 5, -12$

31 $x^4 - x^3 + 7x^2 + 11x + 6, 2$ **33** $-x^4 + 4x^3 - 2x^2 - 2x - 3, -2$

34 $x^5 + x^4 + x^3 + x^2 + 4x + 4, 1$ **35** $x^4 - 3x^3 - 3x^2 - 2x - 6, -4$

EXERCISE 16.2   Page 380

1 $5; -3, 2; 2, 3$ **2** $6; 1, 3; -2, 1; 4, 2$

3 $7; -5, 4; -1, 2; 3, 1$ **5** $5; -\frac{1}{3}, 2; \frac{1}{2}, 3$

6 $6; \frac{5}{2}, 3; -2, 2; -\frac{1}{5}, 1$ **7** $10; \frac{7}{2}, 1; -\frac{8}{3}, 4; -\frac{5}{4}, 5$

9 -1 and 5 **10** -4 and 1 **11** -2 and 4

13 -2 and 2 **14** -1 and 4 **15** -4 and 5

17 -6 and 4 **18** -4 and 4 **19** -5 and $5;$

21 $\{-1, 2, 3\}$ **22** $\{-3, 1, 4\}$ **23** $\{-4, -2, 3\}$ **25** $\{-3, \frac{1}{2}, 2\}$

26 $\{-1, \frac{2}{3}, 3\}$ **27** $\{-2, \frac{3}{2}, 1, 4\}$ **29** $\{-3, -\frac{5}{2}, \frac{1}{2}\}$ **30** $\{-\frac{3}{2}, \frac{1}{3}, 3\}$

31 $\{-3, -\frac{1}{3}, \frac{3}{2}, 2\}$ **33** $\{-\frac{3}{2}, \frac{1}{4}, \frac{3}{4}\}$ **34** $\{-\frac{3}{2}, -\frac{1}{2}, \frac{2}{3}\}$

35 $\{-\frac{2}{3}, -\frac{1}{2}, -\frac{1}{3}, \frac{1}{2}\}$ **37** $\{1 - \sqrt{2}, 2, 1 + \sqrt{2}\}$

38 $\{1 - \sqrt{3}, 1 + \sqrt{3}, -2\}$ **39** $\{-3 - \sqrt{2}, -3 + \sqrt{2}, 1\}$

EXERCISE 16.3   Page 385

1 $1, 2, -2$ **2** $1, -1, -3$ **3** $3, 2, -2$ **5** $0.4, 2.4, 3$ **6** $-2, -0.7, 2.7$

7 $0.6, 2, 3.4$ **9** $-2.2, -1.2, 2.2, 3.2$ **10** $-0.4, 0.6, 2.4, 3.4$

11 $-2.7, -0.4, 0.7, 2.4$ **13** $-2, -1; 0, 1; 1, 2$

14 $-3, -2; 0, 1; 2, 3$ **15** $-3, -2; 0, 1; 1, 2$

17 $-4, -3; -1, 0; 1, 2$ **18** $-2, -1; 1, 2; 2, 3$

19 $0, 1; 2, 3; 3, 4$ **21** $-2, -1; -1, 0; 1, 2; 2, 3$

22 $-2, -1; -1, 0; 1, 2; 2, 3$ **23** $-4, -3; -3, -2; 1, 2; 2, 3$

EXERCISE 16.4   Page 390

1 0.93 **2** 0.23 **3** 0.42 **5** 0.90 **6** 1.42 **7** 1.83 **9** -0.38

10 -0.44 **11** -0.61 **13** -1.58 **14** -1.70 **15** -0.62

17 $-3.26, -1.34, 1.60$ **18** $-3.58, -1.71, 2.29$ **19** $-4.77, -3.12, 1.88$

21 $-0.26, 1.66, 4.60$ **22** $-1.58, 0.29, 4.29$ **23** $-0.44, 0.52, 2.92$

25 0.884 **26** 5.419 **27** 2.290 **29** 1.207 **30** 2.732 **31** 3.236

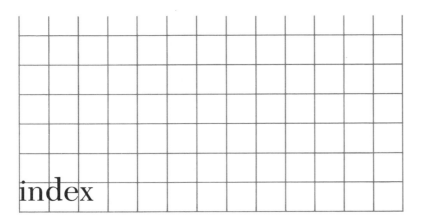

index

DEFINITIONS, AXIOMS, AND THEOREMS

(2.1) $a = a$

(2.2) If $a = b$, then $b = a$

(2.3) If $a = b$ and $b = c$, then $a = c$

(2.4) If $a = b$, then $a + c = b + c$

(2.5) If $a = b$, then $ac = bc$

(2.6) If $a = b$, then a can be replaced by b in any statement including mathematical expressions, without affecting the truth or falsity of the statement

(2.7) $b < a$ if and only if $a > b$

(2.8) One and only one of the statements $a = b$, $a > b$, and $a < b$ is true

(2.9) If $a > b$ and $b > c$, then $a > c$

(2.10) If $a > b$, then $a + c > b + c$

(2.11) If $a < b$, then $a + c < b + c$

(2.12) If a and b are numbers in the set of real numbers, there exists a unique real number c such that $a + b = c$

(2.13) $a + b = b + a$

(2.14) $a + (b + c) = (a + b) + c$

(2.15) $(a + b)c = ac + bc$

(2.16) $c(a + b) = ca + cb$

(2.17) $0 + a = a + 0 = a$

(2.18) $a + (-a) = a - a = -a + a = 0$

(2.19) $a \cdot 1 = 1 \cdot a = a$

(2.20) If $a + b = a + c$, then $b = c$

(2.21) If $a + b = d$ and $a + c = d$, then $b = c$

(2.22) $- -a = a$

(2.23) $-(a + b) = -a - b$

(2.24) $a > b$ if and only if $a - b > 0$

(2.25) $a < b$ if and only if $a - b < 0$

(2.26) $a \cdot 0 = 0 \cdot a = 0$

(2.27) If a and b are real numbers, there exists a real number c such that $ab = c$

(2.28) $a \cdot b = b \cdot a$

(2.29) $(ab)c = a(bc)$

(2.30) $a^m a^n = a^{m+n}$

(2.31) $(x^m)^n = x^{mn}$

(2.32) $(ab)^n = a^n b^n$

(2.33) If $a > c$ and $b > 0$, then $ab > cb$

(2.34) $\dfrac{a}{b} = x$ if and only if $bx = a$, with $b \neq 0$

(2.35) $a \cdot \dfrac{1}{a} = \dfrac{1}{a} \cdot a = 1$ $a \neq 0$

(2.36) If $a \neq 0$ and $ab = ac$, then $b = c$

(2.37) If $ab = 0$ and $a \neq 0$, then $b = 0$

(2.38) $\dfrac{0}{b} = 0$ $b \neq 0$

(2.39) $\dfrac{a}{a} = 1$ $a \neq 0$

(2.40) $\dfrac{a}{1} = a$

(2.41) $\dfrac{a}{b} = a\left(\dfrac{1}{b}\right) = \dfrac{1}{b}(a)$ $b \neq 0$

(2.42) $\dfrac{ab}{cd} = \dfrac{a}{c} \cdot \dfrac{b}{d}$ $c \neq 0,\ d \neq 0$

(2.43) $\dfrac{a^m}{a^n} = a^{m-n}$ $a \neq 0$

(2.44) $a^0 = 1$ $a \neq 0$

(2.45) $\dfrac{a + b - c}{d} = \dfrac{a}{d} + \dfrac{b}{d} - \dfrac{c}{d}$ $d \neq 0$

(3.1) $(ax + by)(cx + dy) = acx^2 + (ad + bc)xy + bdy^2$

(3.2) $(x + y)^2 = x^2 + 2xy + y^2$

(3.3) $(x - y)^2 = x^2 - 2xy + y^2$

(3.4) $(a + b)(a - b) = a^2 - b^2$

(3.5) $acx^2 + (ad + bc)xy + bdy^2 = (ax + by)(cx + dy)$

(3.6) $a^2 - b^2 = (a + b)(a - b)$

(3.7) $x^3 + y^3 = (x + y)(x^2 - xy + y^2)$

(3.8) $x^3 - y^3 = (x - y)(x^2 + xy + y^2)$

(4.1) $\dfrac{m}{n} = \dfrac{p}{q}$ if and only if $mq = np$
$n \neq 0,\ q \neq 0$

(4.2) $\dfrac{m}{n} = \dfrac{mf}{nf} = \dfrac{m/d}{n/d}$ $f \neq 0,\ d \neq 0,$
$n \neq 0$